SEPARATION AND PURIFICATION TECHNIQUES IN BIOTECHNOLOGY

SEPARATION AND PURIFICATION TECHNIQUES IN BIOTECHNOLOGY

by

Frederick J. Dechow

Reed & Carnrick Pharmaceuticals
Piscataway, New Jersey

Library of Congress Catalog Card Number: 88-34502
ISBN: 0-8155-1197-3
Printed in the United States

Published in the United States of America by
Noyes Publications
Mill Road, Park Ridge, New Jersey 07656

10 9 8 7 6 5 4 3 2 1

Library of Congress Cataloging-in-Publication Data

Dechow, Frederick J.
Separation and purification techniques in biotechnology.

Includes bibliographies and index.
1. Biomolecules--Separation. 2. Biomolecules--Purification. 3. Biotechnology--Technique. I. Title.
TP248.25.S47D43 1989 660.2'842 88-34502
ISBN 0-8155-1197-3

Preface

Sorptive separation techniques are found in almost every product separation or purification scheme treating fermentation or biochemical feedstreams. These sorptive techniques include adsorption, ion exchange and liquid chromatography on solid supports. The major objective of this book is to place these different methods in perspective, relative to each other, so that selection of the appropriate technique or combination of techniques may be readily made. While the emphasis has been placed on laboratory evaluation techniques, the scale-up of these techniques and their industrial applications; it is hoped that sufficient theory has been provided so that the interested reader may use the references to pursue the selectivity and kinetic considerations for sorptive procedures.

The first chapter provides a brief sketch of the nature of the biochemical feedstream and all the processes which might be involved in isolating the desired products from that feedstream.

Chapter 2 covers adsorptive separation, which is the oldest of the sorptive techniques. While many people now regard adsorption as strictly for the removal of unwanted impurities or color bodies, this chapter shows that there are many applications where adsorption is useful in isolating biochemical products. The theory of column processes developed in this chapter is also applicable to the column operations for the sorptive processes described in the other chapters.

Ion exchange procedures are described in Chapter 3. This chapter builds upon what was presented in Chapter 2 by demonstrating the effects of the additional sorptive specificity associated with the exchange of ions. Operating parameters and equipment developed for water treatment and metal recovery applications have also been included since fermentation broths have characteristics which may benefit from the use of resin-in-pulp, fluidized bed and the other procedures presented.

Chapter 4, Column Chromatography Processes, covers the use of sorptive materials to create an environment that allows the separate recovery of two or more solutes. The biospecific recognition of a solute for a ligand attached to the column material is covered in the last chapter on Affinity Chromatography.

It is my hope that this book will serve as a useful guide to the solution of the practical problems associated with separating and purifying fermentation and biotechnology products.

I would like to acknowledge the helpful suggestions of Henry C. Vogel, the patience of George Narita and the support and understanding of my wife, Joan Dechow. I also wish to recognize the assistance of Audrey Wildeck for secretarial support and of Bart Alazio for preparing the illustrations.

April 1989 Frederick J. Dechow

NOTICE

Contents

1

Introduction

The recovery of products from fermentation or biochemical processes has been cited (1) as the last hurdle to be overcome in bringing biotechnology from the laboratory to commercial status. This text will attempt to describe how adsorptive materials, ion exchange resins, column chromatography and affinity chromatography can be utilized in these recovery and purification operations. This chapter will examine the nature of the fermentation broth and will serve to put these recovery operations in perspective with other purification techniques not covered by this text. It is essential to understand the relative advantages of each and their interrelationships since most purifications will require combinations of different techniques.

1.1 FERMENTATION BROTH

The fermentation broth is the combination of insoluble, gelatinous biomass, the nutrient fluid, and the soluble metabolites resulting from the fermentation operation. When the fermentation is carried out without any form of inert support for the biomass, the limit of fluidity for stirring or aeration is approximately 3 to 7% wv dry weight of biomass. Physically, biomass is a compressible, gelatinous solid with surface layers of polysaccharide material which make it cohere and adhere. Downstream processing, therefore, has to deal with a viscous, highly non-Newtonian slurry as its feedstock.

For example, a bacterial fermentation for single cell protein will produce a broth of 3% wv in which the slurry is about 60% (by volume) wet biomass and 40% interparticle fluid. When biomass supports are used, somewhat higher operating biomass concentrations are possible and, in waste disposal fermentations, the concentration has been raised from 2 to 5 g/liter to 10 to 40 g/liter.

Compared with the feed streams to recovery processes in conventional chemical processing, fermentation broths are very dilute aqueous systems (see Table 1.1). Therefore, it will be particularly important to avoid energy intensive thermal operations and to select processes which give large concentration increases in the first stage or stages.

Table 1.1: Typical Product Concentrations Leaving Fermenters

Product	Grams per Liter
Antibiotics (e.g., Penicillin G)	10-30
Enzyme protein (serum protease)	2-5
Ethanol	70-120
Lipids	10-30
Organic acids (citric, lactic)	40-100
Riboflavin	10-15
Vitamin B12	0.02

The fluid volume in microbiological processes must be reduced by at least an order of magnitude between the broth and the final fluid stages of the recovery processes and, in some cases, by very much more. For vitamin B-12 the reduction ratio is over 1000:1. Consequently, the plant design and the range of acceptable unit operations changes significantly as the fluid progresses from the broth handling stages to the final isolation stage.

Many fermentation broths are unstable. Once any broth leaves the controlled, aseptic environment of the fermenter, it is exposed to a drastic change of conditions. An actively growing biomass from an aerated culture will be suddenly deprived of oxygen and will experience a rapidly falling concentration of nutrients. This frequently produces rapid changes in physical properties, leading to destruction of desired product. Lipids may be consumed as an alternative energy source for continuing metabolic activity. Enzymes may be destroyed by proteases released from the deteriorating cells. The broth also becomes susceptible to contamination from foreign organisms which can have the same effects.

Similar problems can occur if the recovery operations of a batch fermentation are delayed. The problems can be reduced by chilling to around 5°C. This is commonly done for enzymes and other relatively small output processes. However, this is to be avoided, if possible, with larger fluid volumes since chilling from a typical fermentation temperature of 35°C to 5°C requires refrigeration energy of about 40 kWh/m^3 of broth and considerable capital expenditure. The time for appreciable product loss to occur can be as little as 20 minutes at fermentation temperatures.

The necessary and practical recovery operations employed and the order in which they are used in downstream processing can be reduced to deceptively simple looking recovery sequences. Figure 1.1 (2) shows a schematic of the recovery sequence and the techniques associated with each process.

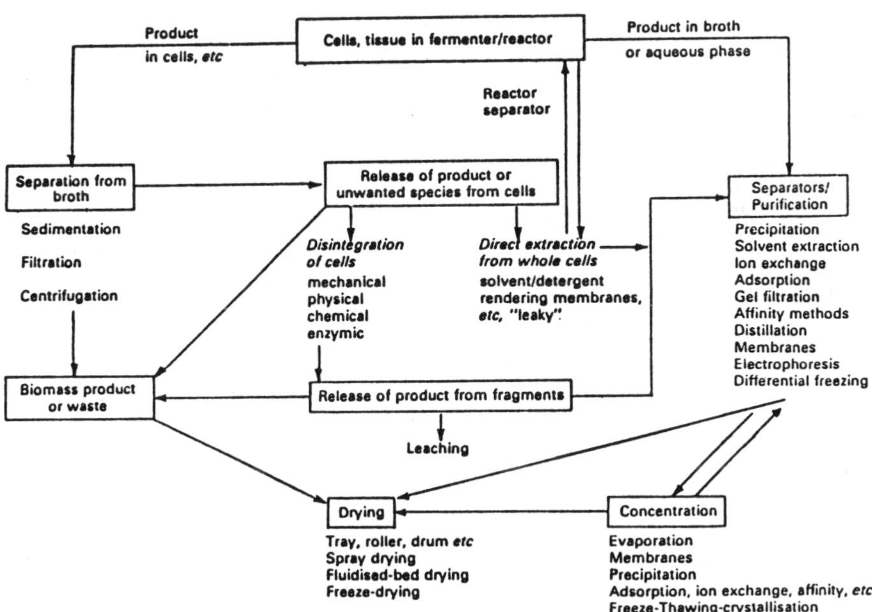

Figure 1.1. Schematic of the processes which may be involved in the separation and fractionation of fermentation products (Reference 2).

If the desired product is extracellular, it is only necessary to filter the biomass from the broth and isolate the product from the fluid. If a product is intracellular, a cell disruption step must first be employed. If the product is also water soluble, this disruption should be performed while the biomass is still in a slurry form. The chemical stability and solubility of the product will dictate the most suitable recovery techniques. Most microbiological products have limited chemical stability. This puts severe restrictions on the temperatures, the reactants, and pH levels which can be used.

An important consideration in determining the appropriateness of a recovery technique is the actual purity requirement for the product. Many products, such as enzymes and vaccines, will not need to be isolated as pure compounds. An appropriate product in these cases would be a complex mixture having the desired properties. However, removal of specific materials, such as pyrogens in injectable medicinals, would be necessary.

The major recovery difficulties arise when it is necessary to separate specific compounds from other chemically and physically similar materials. The isolation of enhanced purity enzyme protein from other protein requires highly specific physico-chemical effects to be used. Today it is the exception rather than the norm for the separated compounds to be that similar. Normally there is a range of alternative recovery techniques and a selection can be made on the basis of cost, familiarity, and reliability criteria.

1.2 RECOVERY UNIT OPERATIONS

The various methods for treating the fermentation broth can be divided into mechanical or chemical unit operations. Table 1.2 lists the processes normally included in each category.

Filtration and centrifugation are unit operations in which the suspended solids are separated from the fluid phase. Drying is the removal of moisture or solvent from solid particles, while evaporation is the removal of moisture or solvent from a solution. In crystallization, the conditions of a solution are adjusted to change the solubility of one of

Table 1.2: Separation and Recovery Techniques

Mechanical	Chemical
Filtration	Adsorption
Centrifugation	Ion exchange
Evaporation	Column chromatography
Crystallization	Affinity chromatography
Drying	Solvent extraction
Reverse osmosis	Electrophoresis
Ultrafiltration	Electrodialysis

the dissolved compounds so that it leaves the solution as a solid.

Microfiltration, ultrafiltration, and reverse osmosis are membrane processes in which separation is based on differences in ability to flow through a thin barrier that separates two fluids. Microfiltration is a hydraulically driven process using a membrane with a pore size in the 100 to 3000 Å range. For ultrafiltration, the pore size is from 10 to 125 Å, while for reverse osmosis, the pore size is from 3 to 10 Å.

Adsorption, ion exchange, column chromatography, and affinity chromatography can be grouped as recovery techniques in which the removed compound or solute establishes an equilibrium between sites on a solid phase material and the solution. In adsorption, the removed species is bonded to the solid phase material by polarity or weak chemical bonds. Ion exchange recovers material by the interchange of ions between the liquid and solid phases. Column chromatography may use adsorptive, ion exchange or molecular sieve materials to separate solutes which are first loaded onto a column of the separation material and then eluted in such a manner that the individual solutes are collected in separate fractions. In affinity chromatography, the removed species is bound with a high level of selectivity to ligands covalently attached to a solid matrix.

In solvent extraction, the removed compound establishes an equilibrium distribution between immiscible solvents, usually water and an organic liquid.

Electrophoresis and electrodialysis are separation techniques that separate charged molecules or ions using an electric field. Electrophoresis separates charged components by accentuating small differences in ionic mobility in an electric field using a moving carrier fluid. Electrodialysis concentrates components on the basis of electromigration through ionic membranes.

1.2.1 Mechanical Operations

1.2.1.1 Filtration: Filtration is typically the first step in the isolation of any product from the fermentation broth. This process separates the biomass cells, the cell debris, and any precipitates from the broth fluid. The mathematical representation of the incremental time dt to filter an additional incremental volume dV after a volume V has been filtered is given by:

$$dt/dV = a + bV \qquad (1.1)$$

The right side of the equation contains two components, a and b. The a term is $\eta r_s/AP$ and the b term, which depends on V, is equal to $\eta r_c/A^2PW$, where η is the liquid viscosity, r_s is the specific cake resistance of the filter material, A is the filter area, P is the constant applied pressure difference, r_c is the specific cake resistance, and W is the cake dry weight per unit volume of filtrate. Normally the resistance of the filter material term includes the resistance contribution of any filtration aid, pipes, and valves.

According to this equation, the resistance to filtration is due initially to only the constant term a. In theory, as filtration proceeds and the biomass cake becomes thicker, the resistance would be expected to increase linearly according to the b term with this dependence on V. Although many practical considerations must be taken into account when applying this equation, its simplicity, the complexity of a more exact description, and the uniqueness of many industrial applications result in Equation 1.1 being the most useful filtration representation.

The specific practical limitations that must be considered are the blockage or blinding of the filter, the compressibility of the biomass cake, and the variable pore structure of the cake. Blinding of the filter may be prevented by starting the filtration at a low hydraulic pressure by partially by-passing the pump. This will avoid driving the first solids into the filter support. The biomass cake's compressibility will usually be proportional to the applied hydraulic pressure up to a certain pressure. Beyond that pressure, the cake will collapse to a new compressed form so that throughput is reduced with the incremental pressure increase.

Table 1.3 shows filtration design and operation for different fermentation broths. In two of the cases noted in Table 1.3 a precoat was used. The filter will often be precoated with a filtration aid such as diatomaceous earth to reduce blinding of the filter and to increase filtration rates. The filtration aid might be added to the broth but then the quantity of filtration aid required is more than double the precoat amount (3).

Table 1.3: Representative Design and Operating Results for Fermentation Broth (Vacuum 0.68–0.85 Bar)

	Bacillus licheniformis	*Streptomyces erythreus*	*Penicillium chrysogenum*
Filter type	Vacuum precoat	Vacuum drum	Vacuum drum or precoat
Design filtration rate (ℓ/hr-m^2)	160–320	400	1,400–1,800
Solids in slurry (wt %)	8	25	2–8
Cycle time (min/rev)	0.5	3	3
Filter medium	Precoat	Nylon	Polypropylene
Cake discharge mechanism	Precoat	String	String or precoat

The equipment in this operation can be as simple as an enlarged laboratory vacuum filter to more elaborate rotary vacuum filters. These latter filters essentially consist of a hollow segmented drum covered with a filter cloth. The drum rotates in a bath of the broth to be filtered while a vacuum inside the drum sucks liquid through the filter cloth, forming a coating of solids on the outside of the drum. Provisions are usually made to wash the filter cake during filtration followed by removing the solid from the cloth. Instead of vacuum, pressure can be used to drive the fluid through a filter cake. Plate and frame pressure filters consist of wire or perforated metal frames which act as the mechanical support for the filter medium which can be fine wire mesh, woven cloth or cellulosic pads. With deep frame patterns, these filters can be used with diatomaceous pre-coats.

Example 1.1

Figure 1.2 shows the influence of pH on the filtration rate of *S. griseus* broth, 1.5 centipoise, using 100 cm^2 cotton cloth filter, diatomaceous earth filter aid and a constant pressure difference of 2 bar (4). What size of filter is needed to filter 1000 L of fermentation broth at pH 3.8 in 15 minutes?

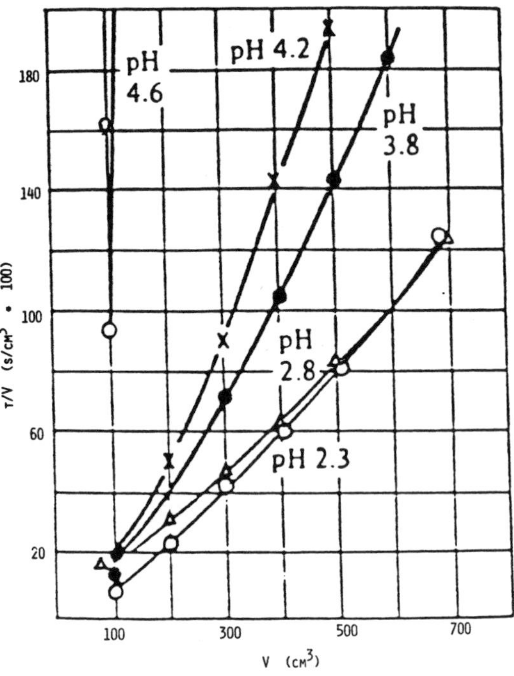

Figure 1.2. The filtration time (t) divided by the volume of filtrate collected is plotted against the volume of filtrate collected for different pH values (Reference 4).

For the sake of simplicity, it is assumed that the filter cake is incompressible so that Equation 1.1 can be integrated to give:

$$\frac{t}{V} = \frac{\eta r_s}{P}\left(\frac{1}{A}\right) + \frac{\eta r_c V}{2PW}\left(\frac{1}{A^2}\right) \qquad (1.2)$$

The slope and intercept for the pH 3.8 curve can be used to obtain the values for r_s and r_c/W from the information given. Using the data points (300, 0.7) and (600, 1.82), r_s = -55.73 and r_c/W = 99.55.

Equation 1.2 is a quadratic equation in (1/A), so that:

$$\left(\frac{1}{A}\right) = \frac{-\left(\frac{\eta r_s}{P}\right) \pm \sqrt{\left(\frac{\eta r_s}{P}\right)^2 + \frac{2\eta r_c t}{PW}}}{\frac{\eta r_c V}{PW}} \qquad (1.3)$$

With V = 1000 L and t = 900 sec, this equations gives 1/A = 5.5×10^{-6}, or A = 18.18 m^2.

The slight curvature of the experimental lines in Figure 1.2 indicates that the cake has some degree of compressibility. This is to be expected for cakes formed from fermentation cells. As the volumes, and therefore required times, for the same filter area increase, the curvature will become more pronounced. While decreasing the pH would speed up the filtration, one must be careful so that the product is not adversely affected.

Several reviews and texts (3, 5-10) exist which should be consulted for additional information on this integral part of any fermentation broth recovery system. A list of industrial filtration equipment suppliers is given in Table 1.4.

1.2.1.2 Centrifugation: Centrifugation, although widely used for cell recovery, is not nearly as ubiquitous as filtration. This process enhances the gravitational settling of the suspended solids in the fermentation broth. The mathematical representation for the centrifugation process is given by:

$$\Phi = \frac{18\eta}{d^2(\rho_S - \rho_L)\, g} \cdot \frac{sg}{\omega^2 rV} \qquad (1.4)$$

In this equation Φ is the time required for complete particle removal, d is the particle diameter, ρ_s is the particle density, ρ_L is the fluid density, η is the fluid viscosity, s is

Table 1.4: Industrial Filtration Equipment Suppliers

Company	Location
Allied Filtration Co.	Kingsley, PA
Avery Filter Co.	Westwood, NJ
Bird Machine Company	South Walpole, MA
Carl C. Brimmekamp & Co.	Stamford, CT
Denver Equipment Company	Colorado Springs, CO
Dorr-Oliver, Inc.	Stamford, CT
EIMCO (Div. of Envirotech)	Salt Lake City, UT
B C Hoesch Industries	Wharton, NJ
Inlay, Inc.	Califon, NJ
Komline-Sanderson, Inc.	Peapack, NJ
Lenser America, Inc.	Lakewood, NJ
Peterson Filters, Inc.	Salt Lake City, UT
R & R Filtration Systems	Marietta, GA
Serfilco, Ltd.	Glenview, IL
Sparkler Filters, Inc.	Conroe, TX
D.R. Sperry & Co.	North Aurora, IL

the thickness of the liquid layer in the centrifuge, ω is the angular velocity, r is the rotation radius, V is the total volume of liquid in the centrifuge and g is the gravitational constant.

The right-hand side of the equation is shown divided into two terms. The first corresponds to the terminal velocity of the particle with diameter d under the force of gravity. This term is solely concerned with properties of the fluid and suspended particles. The second component is concerned only with fixed characteristics of the centrifuge operating at a fixed speed, ω. This term is normally written as sigma, Σ, the equivalent area of sedimentation centrifuge. If it can be assumed that the liquid layer is thin and that there are no interactions between solid particles, then Σ provides a basis for comparing different sizes and types of centrifuges (11).

The equation points out that, with respect to the feed material, centrifugation is enhanced by large particle diameter, a large difference in density between particles and the fluid, and low viscosity. Likewise, separation is favored by centrifugation equipment which is operated at a high angular speed, has a large centrifuge radius, and can hold a large volume of liquid while at the same time has a thin liquid sedimentation layer.

Example 1.2

For a fermentation broth that is a dilute suspension of cells in an essentially aqueous solution, the diameter of the cells is 5 microns. They have a density of 1.001 g/cc. The viscosity of the broth is 2 centipoise. If 50 ml of the broth is placed in a centrifuge tube whose bottom in 10 cm from the rotation axis, what is the time required for complete separation of the cells from the liquid supernatant at 3000 rpm?

If one assumes that the fermentation broth contains 5% (by volume) suspended cells and that the cross-sectional area of the centrifuge is 5 cm^2, the liquid layer through which the cells must travel is a maximum of 9.5 cm. Therefore, using Equation 1.4, the time for complete removal of the suspended cells is 304 sec.

The difficulties associated with centrifugation become apparent when the typical fermentation broth characteristics are considered in the light of these centrifugation enhancement factors. The biomass particles, typically, are of very small size and of low density. The fluid can be highly viscous and, depending on the amount of dissolved solids, of density comparable to the biomass. There are physical limitations on the equipment as well. Attempts to increase capacity by increasing the centrifuge radius soon reach the point where mechanical stress and safe operation place a limit. The increase in bowl capacity and the continuous flowing through of fluids restrict the safe angular velocity. When solids are discharged during operation, the inherent imbalance this causes will further restrict the maximum safe angular velocity.

Within these limitations, several centrifuge equipment designs have been developed or adapted for the fermentation industry. The centrifuges utilized for fermentation product recovery fall into two basic types, the perforated bowl or basket type and the solid bowl high speed types. The basket centrifuge (Figure 1.3a) normally is used with a filter bag of nylon, terylene, or cotton. The feed is added continually. When the bowl is filled with the biomass cake, fresh washing liquid can be added to displace the residual broth fluid retained in the cake. This has many of the characteristics of filtration but has considerably faster throughput.

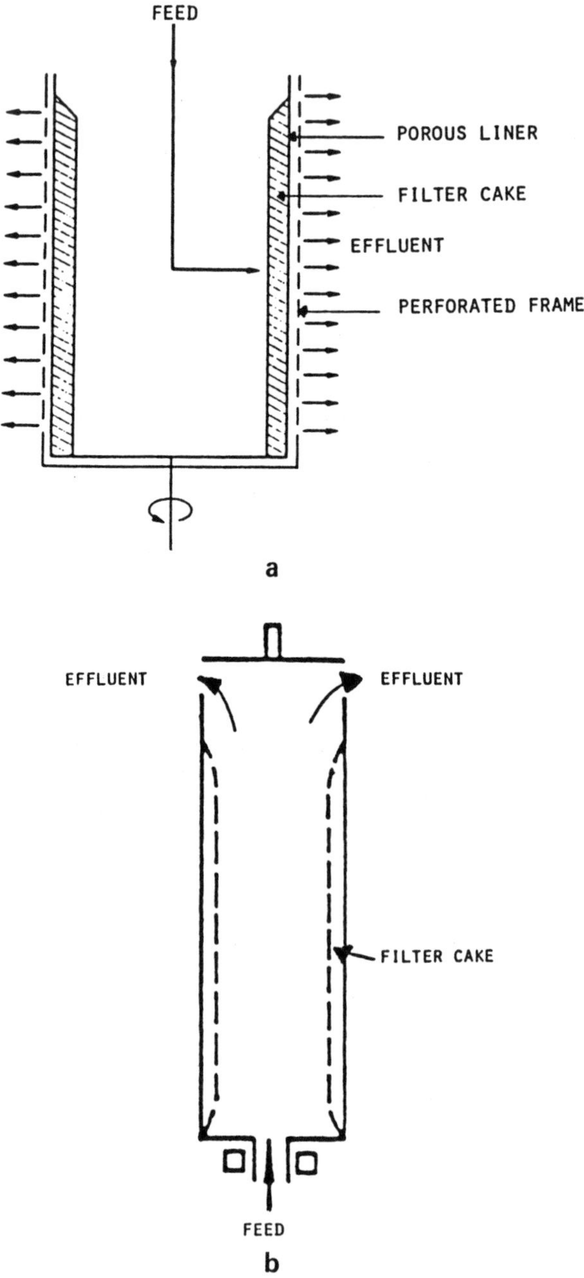

Figure 1.3. Centrifuges. (a) Perforated bowl or basket type centrifuge. (b) Tubular bowl centrifuge. (c) Multi chamber solid bowl centrifuge. (d) Nozzle disk centrifuge. (e) De-sludger or intermittent discharge centrifuge. (f) Scroll centrifuges.

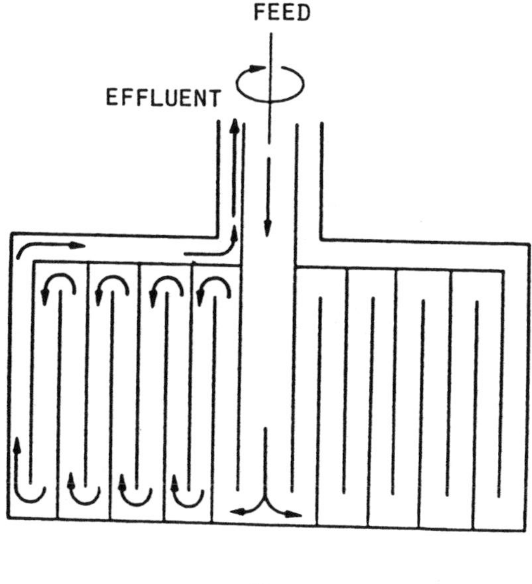

c

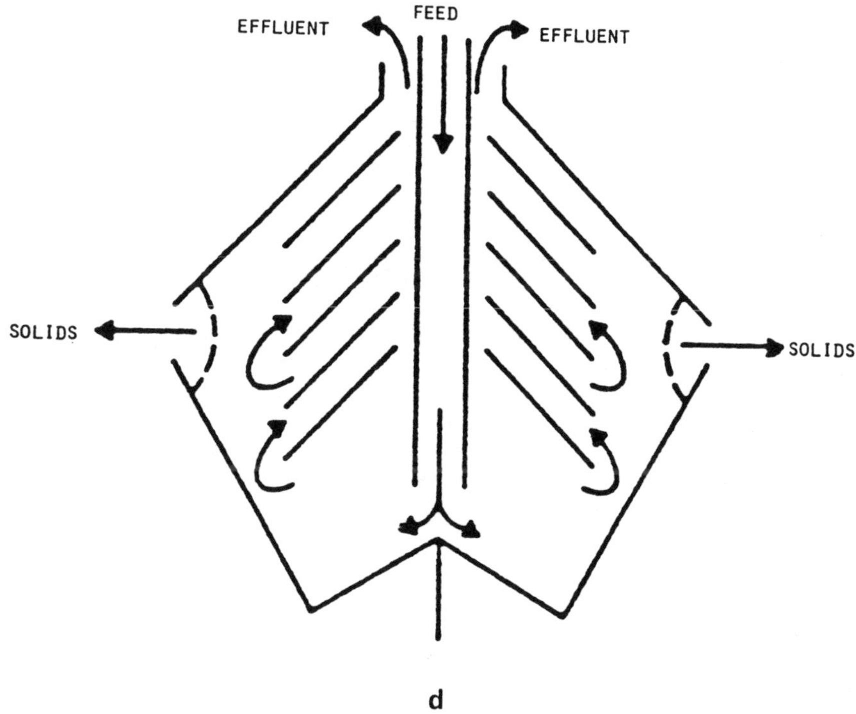

d

Figure 1.3. (continued)

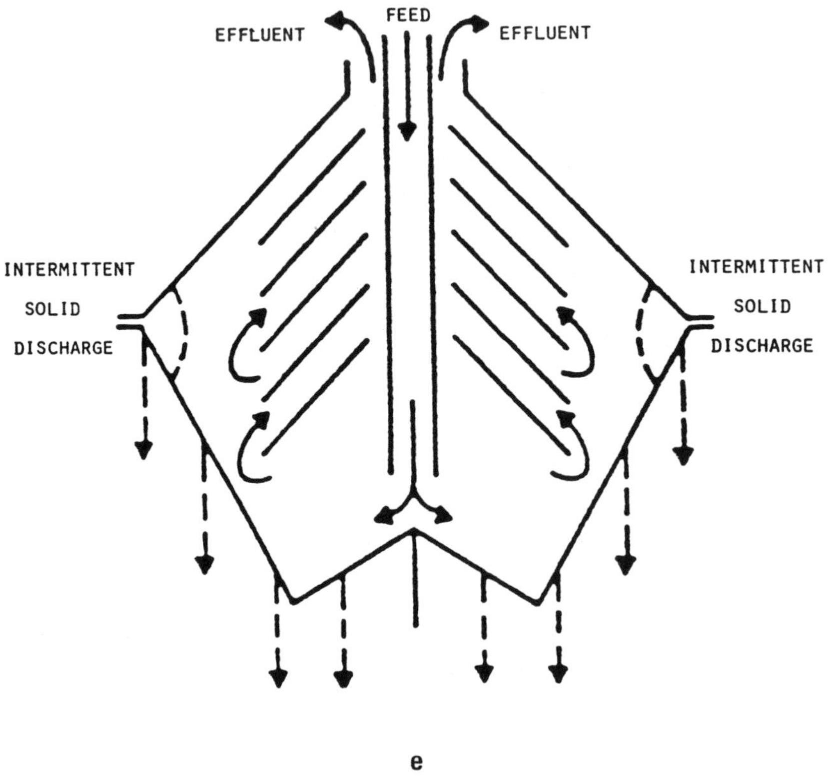

e

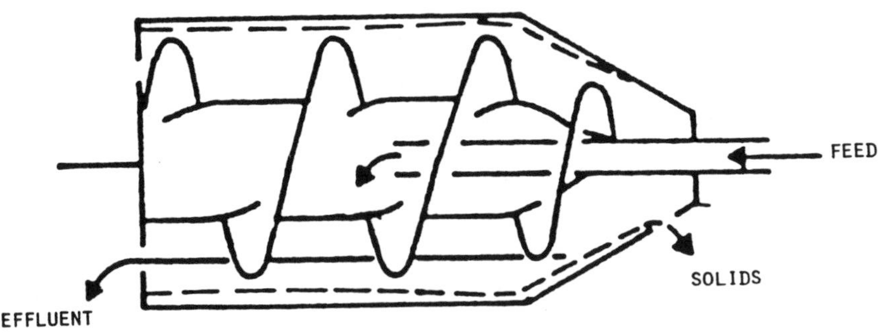

f

Figure 1.3. (continued)

Solid bowl centrifuges are available in five basic types. The first type is called tubular bowl (Figure 1.3b) and operates like a batch laboratory centrifuge. A continuous feed of material passes through the machine. As soon as the bowl is full of the biomass cake, the centrifuge is stopped, stripped down, the cake removed, the machine cleaned and the sequence repeated. Typical batch size is about 4 kg wet weight. A variation on this type is the multichamber centrifuge (Figure 1.3c) which allows three or more time the capacity of the tubular bowl. However, the disassembly and cleaning of this centrifuge is more difficult.

The third and fourth types operate continuously since the solids are automatically removed from the bowl. The one is called a nozzle disk centrifuge (Figure 1.3d). This centrifuge discharges a continuous stream of concentrated solid-slurry. The other disk centrifuge is termed a de-sludger (Figure 1.3e). This centrifuge has an intermittent discharge of solids with lower water content than the nozzle centrifuge.

The final type is called a scroll centrifuge (Figure 1.3f). This centrifuge uses a screw within a rotating bowl to allow continuous removal of the solid and the liquid portions. This type of centrifuge is recommended when the solids are so coarse that they would blind the discharge systems of the disk type of continuous centrifuges.

Table 1.5 shows the characteristics of the different types of centrifuges. There is a specific particle size range, maximum centrifugal force and fluid capacity for each type of centrifuge. Reviews and texts on centrifugation (11-15) can be used for additional information. A list of industrial centrifugation equipment suppliers is given in Table 1.6.

Table 1.5: Characteristics of Different Types of Centrifuges

Type	Particle Size (μm)	Centrifugal Force (ω^2r/g)	Capacity (ℓ/min)
Basket	10-8,000	1,800	300*
Tubular bowl	0.01-100	16,000	80*
Multichamber	0.01-100	6,000	250*
Disk, nozzle	0.1-100	9,000	4,000
De-sludger	0.1-100	6,000	1,000
Scroll	2-3,000	3,000	1,150

*With interruptions for removal of solids.

Table 1.6: Industrial Centrifuge Manufacturers

Company	Location
Alfa Laval, Inc.	Fort Lee, NJ
Amtek, Inc.	El Cajon, CA
Baker Perkins, Inc.	Saginaw, MI
Barrett Centrifugals	Worcester, MA
Bird Machine Company	South Walpole, MA
Centrico, Inc.	Northvale, NJ
Clinton Centrifuge, Inc.	Hatboro, PA
Commercial Filters	Lebanon, IN
Dorr-Oliver, Inc.	Stamford, CT
Heinkel Filtering Systems	South Norwalk, CT
Krauss Filtering Systems, Inc.	Charlotte, NC
IEC Division, Damon Corp.	Needham, MA
Pennwalt Corp., Stokes Division	Warminster, PA
Quality Solids Separation Co.	Houston, TX
Robatel, Inc.	Pittsfield, MA
Sanborn Associates, Inc.	Wrentham, MA
Tema Systems, Inc.	Cincinnati, OH
Western States Machine Company	Hamilton, OH

1.2.1.3 Evaporation: Evaporation is the concentration of a feed stream by the removal of the solvent through vaporization of the solvent. This process is distinguished from distillation in that the vapor components are not separately collected in evaporation.

Many fermentation fluids are dilute aqueous streams which require the evaporation of large amounts of water. This requirement is complicated by many fermentation products being damaged or destroyed if exposed to too great a temperature over an extended period of time. Highly efficient evaporators with short residence times are required for evaporation of these heat sensitive products.

The efficiency of an evaporator is determined by its ability to transfer heat, to separate liquid from vapor and its energy consumption per unit of solvent evaporated.

Heat transfer rates can be treated with conventional heat transfer calculations (16) if the evaporator has an outside heating element where the solution or slurry is pumped through the tubes. Under steady state conditions of flow rate, temperature, pressure and composition, the energy transferred across a heat exchanger or evaporator surface is given by:

$$Q = UA\Delta T. \tag{1.5}$$

In this equation, Q is the overall rate of heat transfer, U is the overall heat transfer coefficient for the system, A is the evaporator surface area and ΔT is the difference between the steam temperature and the liquid temperature leaving the vapor head for natural circulation equipment. For forced circulation equipment, ΔT is replaced with the log mean temperature difference to correct for the fact that the temperature difference is not constant along the evaporator surface. The overall heat transfer coefficient, U, is obtained from pilot plant data or calculated from the individual heat transfer coefficients, h_i

$$1/U = \Sigma\, 1/h_i\,. \tag{1.6}$$

These theoretical rates are greatly reduced by the presence of air in other configurations such as multiple effect evaporators. These venting problems make it difficult to translate small scale evaporation equipment data to full scale equipment.

The vapor-liquid separation portion of the evaporator is designed to prevent entrainment or liquid droplets from being carried through the evaporator system. Entrainment is a function of the vapor velocity, temperature difference, mechanical design and the physical properties (viscosity and surface tension) of the fluid being concentrated.

The vapor head or flash chamber is that area of the evaporator that provides a region for the liquid phase and the vapor phase to separate. In natural circulation evaporators, the boiling process causes liquid to move across the heating surface. The two phase mixture of liquid and vapor is less dense than the solution which pushes it forward and upward to the flash chamber. More viscous liquids and those with a high solids content require a centrifugal pump to circulate them through a loop around the heating unit.

The energy consumed per unit of solvent evaporated is best obtained by forming an energy balance and a mass balance around each stage or effect of an evaporator (Figure 1.4). For a single stage evaporator, the material balance is simply the splitting of the feed entering the evaporator into the portion discharged as vapors (the distillate) and that portion remaining (the concentrate). Entrainment is normally ignored in material balance calculations. The energy balance uses the temperature of the evaporator as the reference temperature. The steam entering the heat exchanger supplies the heat to increase the temperature of the feed stream, to form crystals (if any) and to vaporize the evaporated solvent. Radiation may require an additional 1 to 5% heat.

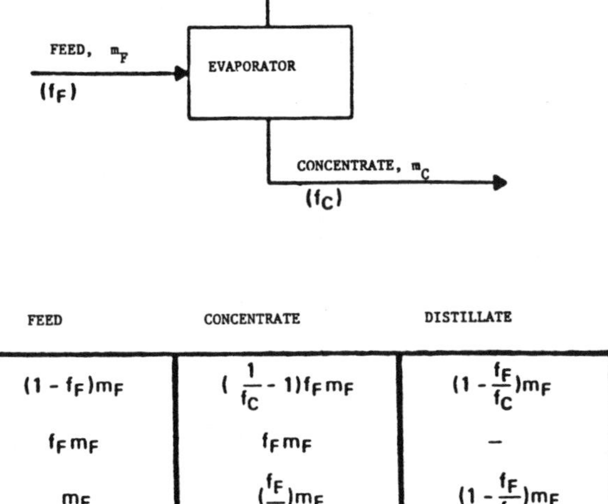

	FEED	CONCENTRATE	DISTILLATE
WATER	$(1 - f_F)m_F$	$(\frac{1}{f_C} - 1)f_F m_F$	$(1 - \frac{f_F}{f_C})m_F$
SOLIDS	$f_F m_F$	$f_F m_F$	–
TOTAL	m_F	$(\frac{f_F}{f_C})m_F$	$(1 - \frac{f_F}{f_C})m_F$

BOUNDARY CONDITIONS: $0.0 < m_F$

$0.0 < f_C \leq 1.0$

Figure 1.4. Energy and material balance of a single evaporator stage (Reference 18).

An evaporator can be something as simple as a jacketed vessel, but is usually more complex such as horizontal tube evaporators, thin film evaporators, falling-firm evaporators, long- or short-tube vertical evaporators, propeller calandrias, forced circulation evaporators, plate evaporators, flash evaporators and multiple effect evaporators.

Long-tube vertical evaporators are the most widely used evaporators today. This evaporator (Figure 1.5a) has a vertical heat exchanger topped by a vapor head. Just above the tubes is a deflector to act as a primary liquid-vapor separator. These evaporators are the most economical large scale evaporators provided the product neither is heat sensitive nor forms a precipitate during the evaporation.

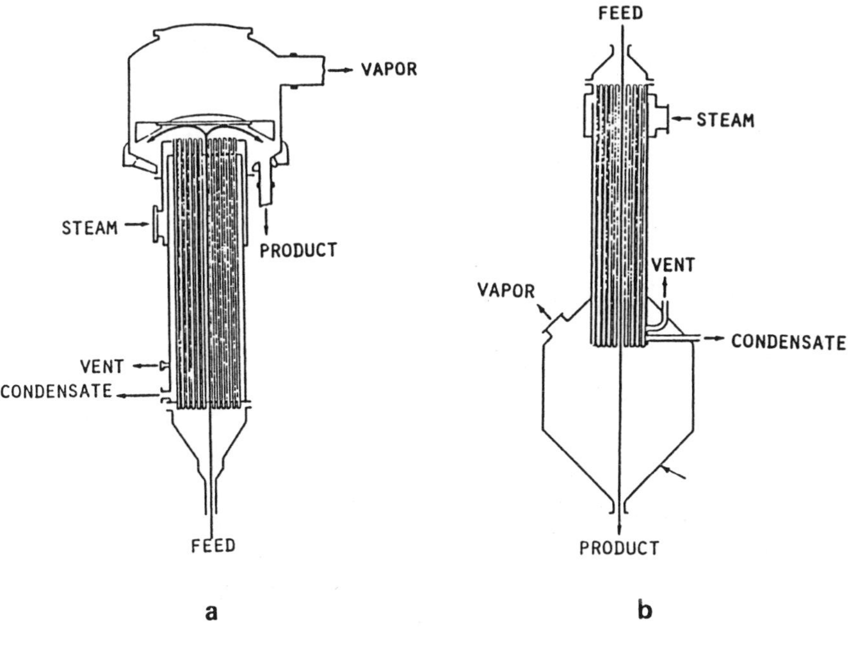

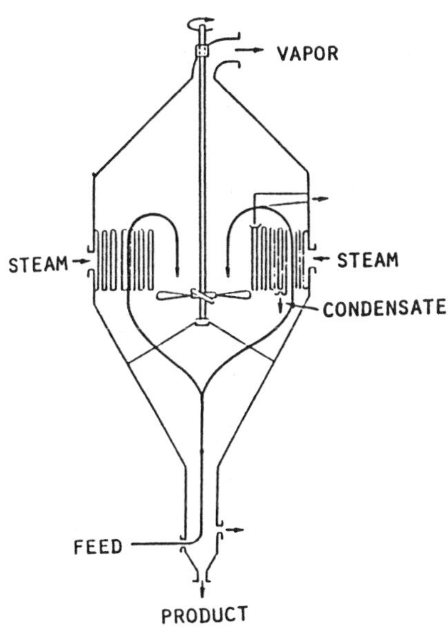

Figure 1.5. Evaporators. (a) Long-tube vertical evaporator. (b) Falling-film evaporator. (c) Propeller calandria evaporator. (d) Plate evaporator (e) Thin film evaporator.

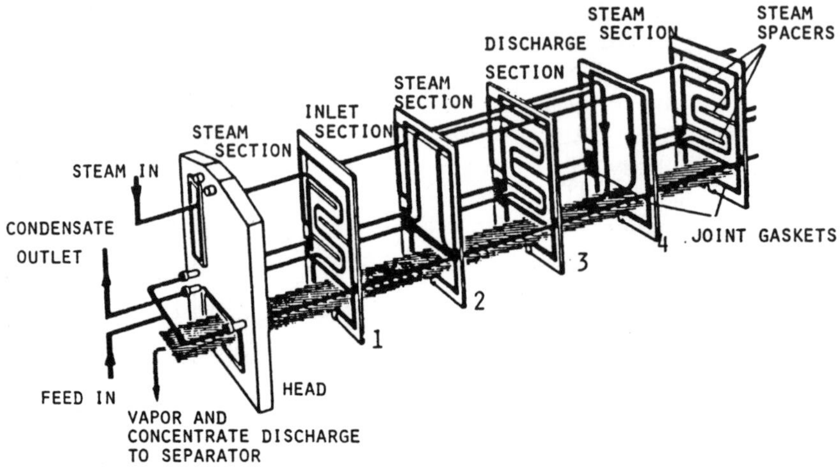

d

BELT DRIVE
CARTRIDGE-TYPE DOUBLE MECHANICAL SEAL
MOTOR
INTERNAL SEPARATOR (RADIAL TYPE)
ROTOR
HEATING JACKET
HEATING JACKET
BOTTOM CONE
MAIN BEARING
VAPOR NOZZLE
FEED NOZZLE
HEATING JACKET
A: HEATING NOZZLE
B: HEATING NOZZLE
LOWER ROTOR GUIDE BUSHING
DISCHARGE NOZZLE

e

Figure 1.5. (continued)

The falling-film evaporator (Figure 1.5b) is similar to long tube vertical evaporators except in the location of the heat exchanger. To operate properly a significant amount of liquid must be circulated through the tubes. This design is very useful for fermentation fluids that are heat sensitive because of its very short liquid retention time.

A propeller calandria evaporator (Figure 1.5c) is basically a short-tube vertical evaporator with a propeller in the center of the column. The propeller keeps any solids suspended in the liquid phase even if boiling stops. These evaporators are useful for evaporating small volumes of liquid from solutions either with or without suspended solids.

Plate evaporators (Figure 1.5d) use flat or corrugated plates as the heat transfer surface. Plate units may be used for systems that have potential scale problems. In some designs, the flat surfaces serve alternately as the fermentation fluid side and the steam side. Scale deposited when in contact with the feed solution can be removed when in contact with the steam condensate.

In thin film evaporators (Figure 1.5e), the treated fluid is moved at high speed over a heated surface by a mechanical wiper arm. The particular advantage of this type of equipment is its ability to handle high viscosity fluids and it requires only short retention times.

Multiple effect evaporators are based on the principle that the heat given up by condensation in one stage can be used to provide the reboiler heat for a different stage. These large, complex devices are only justified when large amounts of dilute aqueous fluids must be evaporated.

Example 1.3

Figure 1.6 (21) shows a comparison of heat transfer coefficients for aqueous solutions and slightly more viscous solutions. What are the relative surface areas required for long tube vs. short tube evaporation at 80°C boiling temperature for the aqueous solution when the ΔT is 10°C? How does this change as the viscosity of the solution is increased?

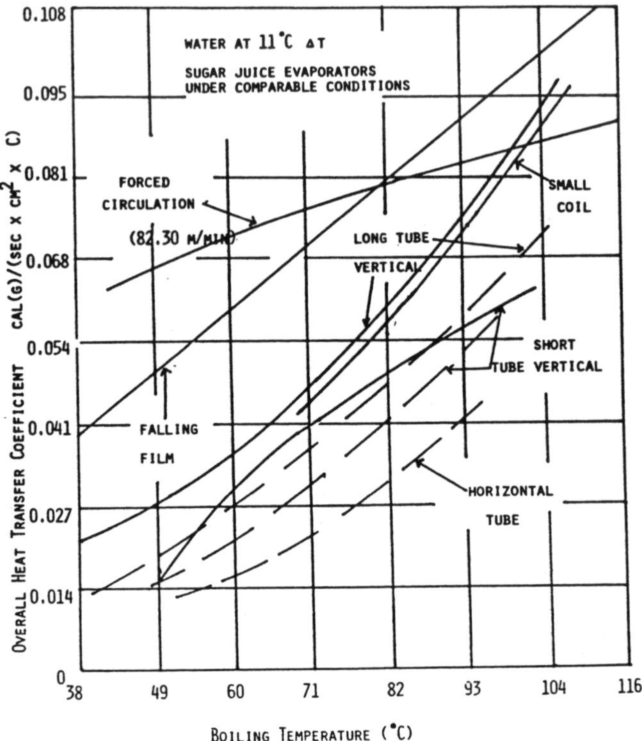

Figure 1.6. Comparison of heat transfer coefficients (Reference 21).

The graph shows that the heat transfer coefficient is 0.05877 and 0.04950 cal/(sec-cm^2-°C) for long tube and short tube vertical evaporators, respectively, in aqueous solutions. The long tube requires 18.7% less surface area. When the viscosity of the solution is increased, the heat transfer coefficient for both types of evaporators is reduced. However, the advantage in surface area for the long tube vertical evaporator is reduced to 17.0%.

Review articles (18-21) about evaporators can be consulted for additional information. Manufacturers are listed in Table 1.7.

1.2.1.4 Crystallization: Crystallization separates a material from a supersaturated solution by creating crystal nuclei and growing these nuclei to the desired size. This separation technique is applicable to those fermentation

Table 1.7: Industrial Equipment Manufacturers

Company	Location
Alfa-Laval	Fort Lee, NJ
APV Company, Inc.	Tonawanda, NY
Dedert Corp.	Olympia Fields, IL
Distillation Engineering Co.	Livingston, NJ
Goslin-Birmingham	Birmingham, AL
HPD, Inc.	Naperville, IL
Industrial Filter and Pump	Cicero, IL
The Kontro Co., Inc.	Orange, MA
Luwa Corporation	Charlotte, NC
Paul Mueller Co.	Springfield, MO
Niro Atomizer, Inc.	Columbia, MD
Pfaudler Co.	Rochester, NY
Swenson Process Equipment	Harvey, IL
Unitech, Div. of Graver Co.	Union, NJ
Henry Vogt Machine Company	Louisville, KY

products that have a low solubility in the solvent utilized. The separation is usually accomplished at low temperatures so that there is the advantage of minimal product loss for thermally sensitive materials.

The solution's supersaturation, which precedes the actual crystallization, is usually accomplished by the removal of solvent or by lowering the solution temperature. The driving force for the formation and growth of crystals is related to the extent of supersaturation.

Phase equilibrium data are essential in the determination of the most efficient method of creating the supersaturated solution and the optimum degree of supersaturation. The three most common methods of forming a supersaturated solution are shown in Figure 1.7. Cooling may be accomplished indirectly with build-in cooling surfaces if there is no possibility of incrustation formation. When there is the possibility of incrustation formation, vacuum cooling may be used. This approach has other requirements on the solution, such as a minimum amount of dissolved gases and a solution boiling temperature near that of the solvent. Evaporation to reach supersaturation differs from vacuum cooling in the larger amount of solvent removed by the evaporator and in the steady supply of heat to the evaporator. Should either of these techniques prove inadequate, the crystalline product may be salted out by the addition of chemicals.

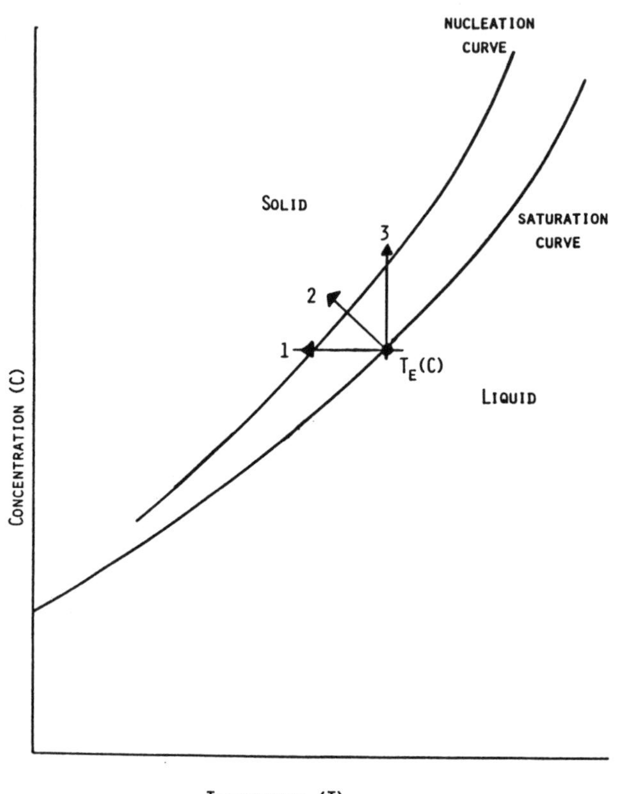

Figure 1.7. Supersaturation to obtain crystal nucleation may be obtained by: 1. cooling; 2. adiabatic evaporation; or 3. isothermal evaporation.

The amount and particle size distribution of crystals produced from a solution with a given crystallizer is a function of the nucleation and crystal growth process. These processes depend on a great number of factors, including the hydrodynamics of the system, the presence of trace impurities, temperature gradients and concentration gradients in the equipment. Many mathematical models (22-24) have been developed to describe the functioning of a crystallizer for a supersaturated solution. A typical approach uses a combination of equations describing the rate of formation and the rate of growth of crystals which are then removed from the system and the steady state supersaturation with the introduction of fresh solution (25). A general review is given by Nyvlt (26).

Crystallizer equipment is usually grouped according to the method of suspending the growing crystals. There are four groups of crystallizers: circulating magma (Figure 1.8a), circulating liquor (Figure 1.9), scraped surface (Figure 1.10) and tank crystallizer (Figure 1.11). It should be noted that there have been lists of as many as sixty-eight different types of crystallizers using other classification techniques (26).

In the circulating magma type of crystallizer, the feed enters the circulating pipe below the product discharge at a point sufficiently below the free-liquid surface to prevent flashing. The magma is the suspension of small crystals in solvent. The crystal suspension and the feed are both circulated through the heat exchanger where the temperature is increased. The heated magma re-enters the crystallizer near the liquid surface, raising the temperature locally to cause vaporization of the solvent. The cooling which occurs after vaporization causes the solution to become supersaturated to the extend necessary for crystal nucleation and growth. The product is continuously withdrawn.

A variation on this flow sequence is shown in Figure 1.8b. Here the feed enters the base of the crystallizer; baffles and an agitator are used to cause the larger crystals to be separated and removed continuously from the crystallizer. The magma type of crystallizer produces quite large (500 to 1800 microns) crystals.

The circulating liquor type of crystallizer retains the crystals in the suspension vessel and only the solution is circulated. The upflow solution keeps the crystals in suspension and allows for the uniform growth of crystals. The three distinct parts of these crystallizers are the crystal suspension chamber where the crystal growth occurs, the supersaturation chamber where the solution is adjusted to the appropriate concentrations and temperature, and the circulation pump system. These are usually the most expensive type of crystallizer.

Scraped-surface crystallizers, such as the Swenson-Walker crystallizer shown in Figure 1.10, are very important in crystallizing a fermentation product from a high viscosity solution. In this system, the supersaturation occurs by the exchange of heat between the slurry and the coolant through a jacket or double wall. An agitator is fitted with scrapers to scrape the heat transfer surface to prevent the build-up of solids which would reduce the efficiency of heat transfer through the heat exchange surface. The main limitation on capacity for these units is the amount of heat transfer surface area.

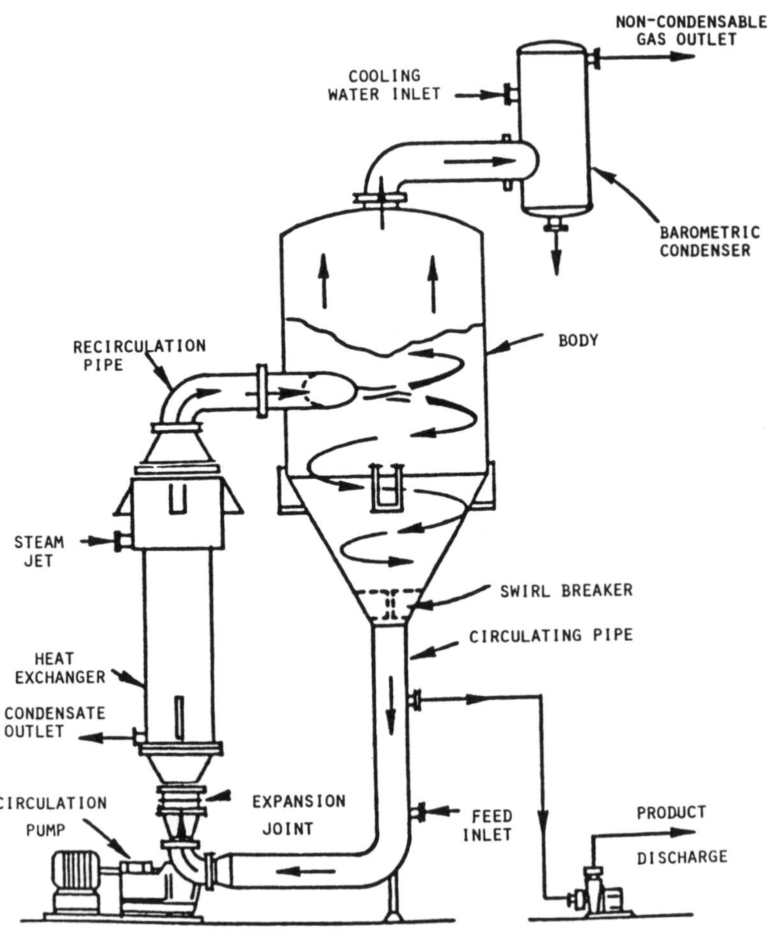

a

Figure 1.8. (a) Magma forced-circulation (evaporative) crystallizer. (b) Draft-tube-baffle crystallizer with circulating magma.

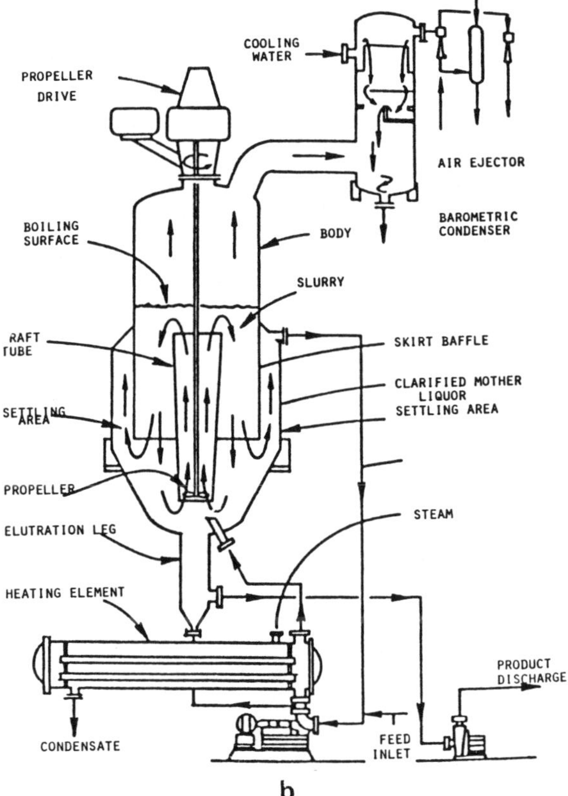

b

Figure 1.8. (continued)

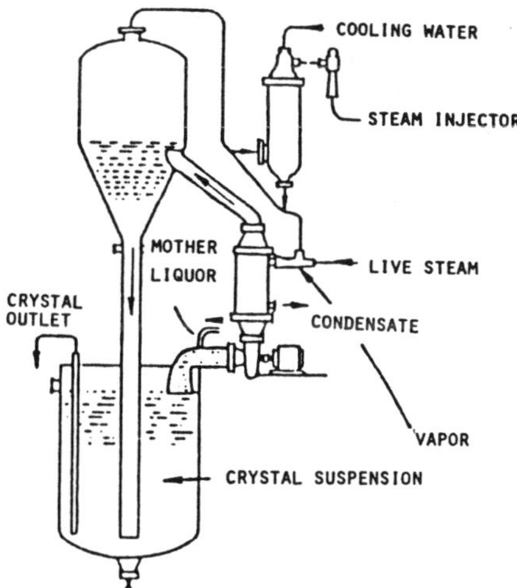

Figure 1.9. Vacuum-cooling circulating liquor crystallizer.

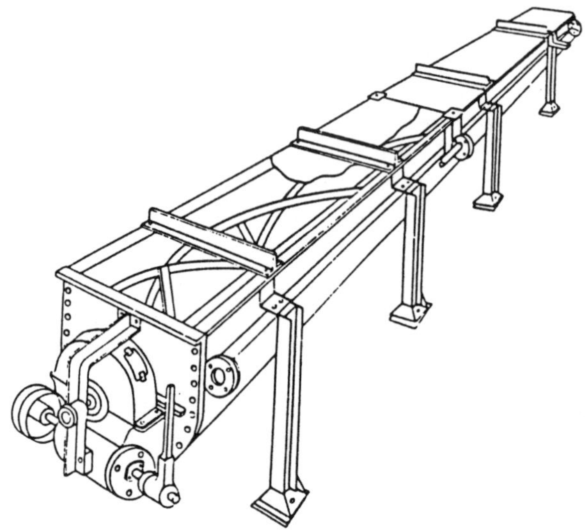

Figure 1.10. Swenson-Walker scraped surface crystallizer.

Tank crystallizers are used for smaller quantities of solutions with solutes of normal solubilities. Although it is possible to form the supersaturated solution in the tank crystallizer without agitation, it is much more difficult to control the crystal size distribution and the amount of crystals produced. The agitated heat transfer coefficients are 0.0027 to 0.027 cal/(s-cm^2-°C).

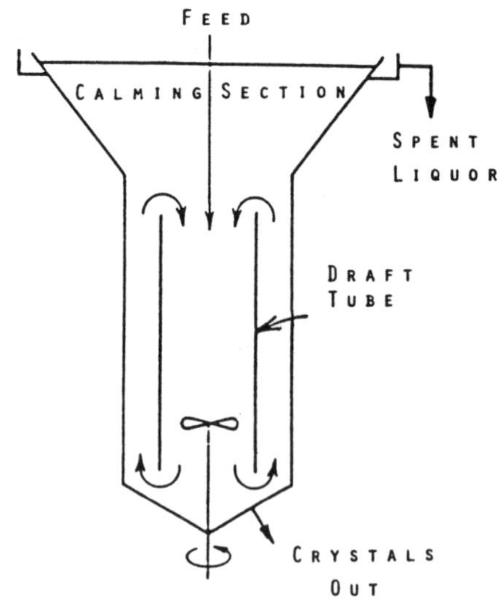

Figure 1.11. Tank crystallizer.

Books and review articles on the subject should be consulted for additional information. Table 1.8 shows manufacturers of crystallization equipment who may be helpful in providing assistance with crystallization of fermentation products.

Table 1.8: Industrial Crystallizer Equipment Manufacturers

Company	Location
APV CREPACO, Inc.	Tonawanda, NY
Aqua-Chem, Inc.	Milwaukee, WI
Blaw-Knox F & C E Co.	Buffalo, NY
Dedert Corp.	Olympia Falls, NY
Goslin-Birmingham	Birmingham, AL
HPD Incorporated	Naperville, IL
Niro Atomizer, Inc.	Columbia, MD
Pfaudler Corp.	Princeton, NJ
Swenson Division (Whiting Corp.)	Harvey, IL
Unitech, Division of Graver	Union, NJ

1.2.1.5 Drying: Drying is usually the last step in the recovery of either the cell biomass or other fermentation products. Drying is a difficult part of the recovery process since high temperatures can cause inactivation of heat labile materials while a certain amount of heat is needed to prevent enzymatic autolysis or breakdown. The drying process can be carried out in a non-adiabatic manner or in an adiabatic manner. In non-adiabatic dryers, the heat of vaporization is supplied through a heated wall to the wet solids. Adiabatic driers pass hot gases through or across the wet solids to vaporize the moisture from the solids.

For non-adiabatic drying, the rate of solvent removal is given by the equation:

$$N_V = \frac{U_o A}{\lambda_V} (T_M - T_S) \tag{1.7}$$

In this equation, U_O is the overall heat transfer coefficient, A is the utilized heat transfer surface area, λ_V is the latent heat of vaporization, T_M is the heating medium temperature and T_S is the temperature of the heating surface in contact with the solid being dried. This equation is only valid for an instant in time since this is a non-steady state process. The usefulness of this equation is that it allows comparison of U_O

for different types of dryers and, therefore, their effectiveness for a specific product.

A similar equation represents the rate of solvent removal for adiabatic drying:

$$N_V = f\, K_o\, (Y_W - Y_G) \qquad (1.8)$$

In this equation, f is the relative drying rate obtained from psychometric charts, K_o is a coefficient which depends upon the geometric configuration of the dryer and the airflow conditions (similar to U_o in Equation 1.7), Y_W is the humidity of the air next to the wet solids and Y_G is the humidity of the bulk air. Psychometric charges are plots of humidity versus enthalpy, with indications of wet bulb temperature and dry bulb temperature. While typical charts involve standard air-water vapor mixtures, separate charts must be constructed for any other solvent-gas system being studied.

The most common types of dryers are drum dryers, vacuum dryers, fluidized bed dryers, spray dryers and flash dryers. Drum dryers and vacuum dryers are non-adiabatic types of dryers, while fluidized bed, spray and flash dryers are adiabatic dryers.

Drum dryers (Figure 1.12a) consist of a hollow revolving drum which is heated internally by steam and revolves slowly in a trough containing the fluid to be dried. As the drum rotates, a thin film of fluid adheres to the drum, is dried by the heat of the drum, and finally is removed from the drum by a scraper blade.

Vacuum dryers may be of either the tray type (Figure 1.12b), the conical type (Figure 1.12c) or the rotary drum type (Figure 1.12d). Vacuum is used to reduce the vaporization temperature of the solvent. This method is only useful if the low temperature does not cause denaturation of the proteins present.

In fluidized bed dryers (Figure 1.13a), a stream of heated air is passed through material resting on a screen of wire gauze or a sintered plate. The flow of air is adjusted to fluidize the particles. A filter bag is used to prevent escape of the dried solids.

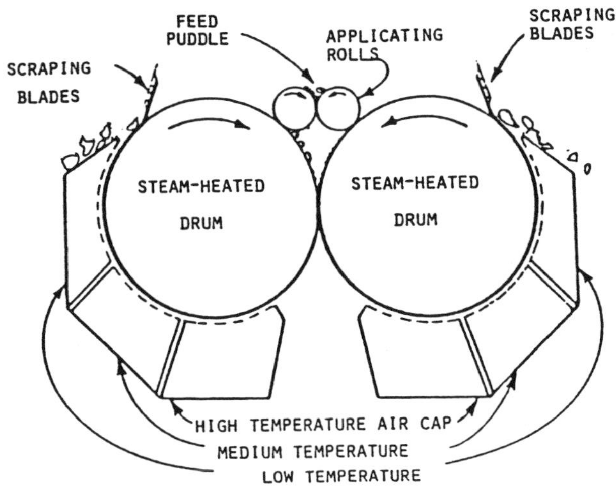

a

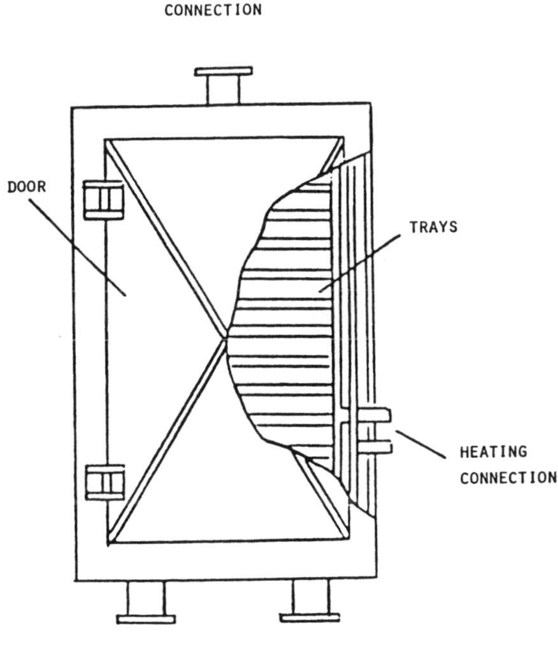

b

Figure 1.12. Dryers. (a) Drum dryer. (b) Tray-type vacuum dryers. (c) Conical vacuum dryers. (d) Rotary drum vacuum dryer.

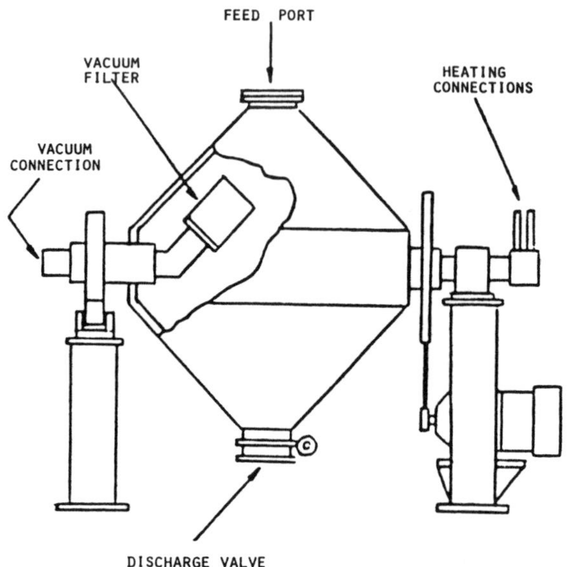

c

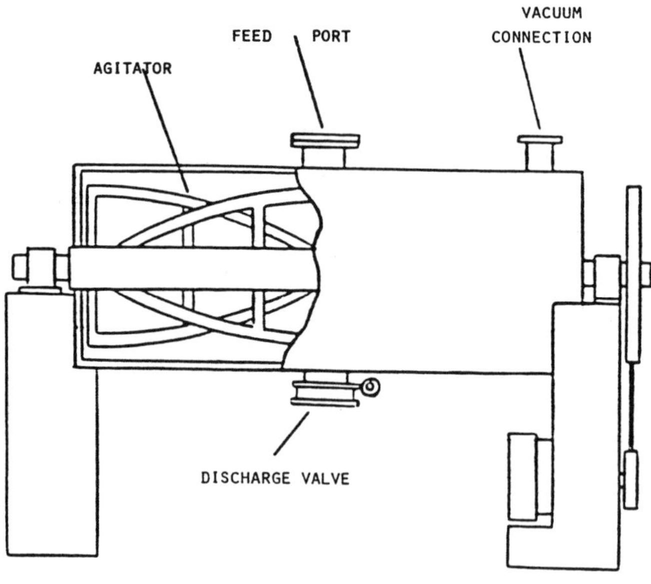

d

Figure 1.12. (continued)

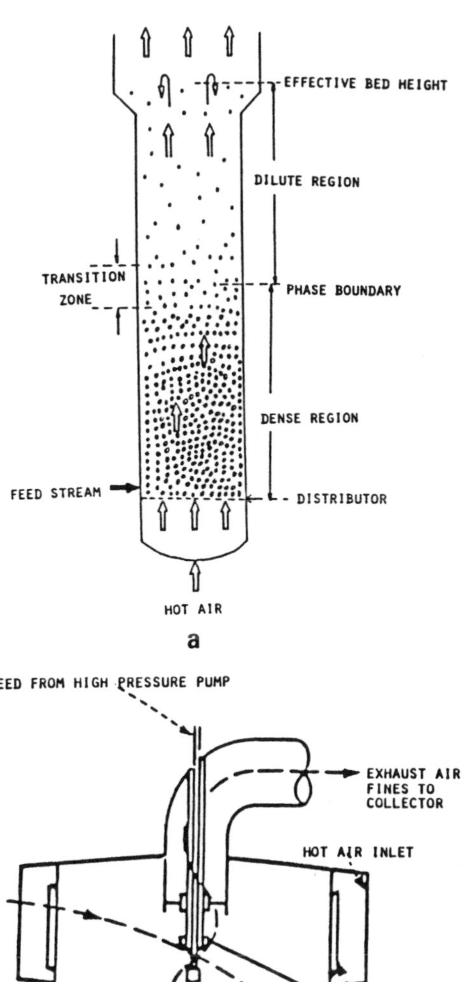

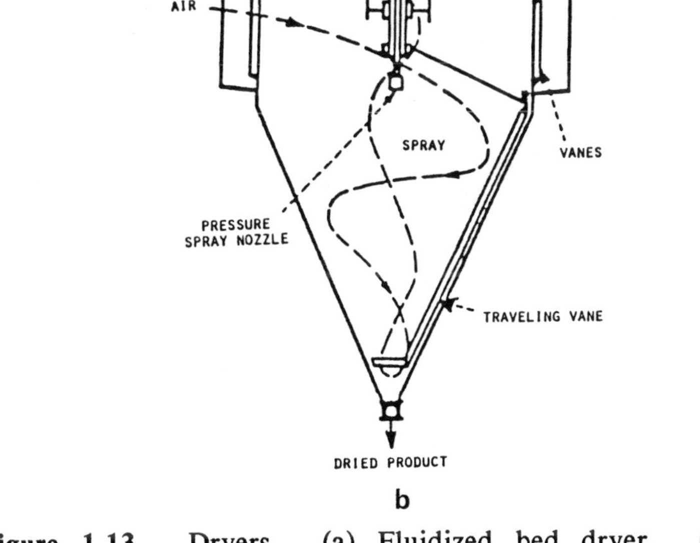

Figure 1.13. Dryers. (a) Fluidized bed dryer. (b) Spray dryer. (c) Flash dryer.

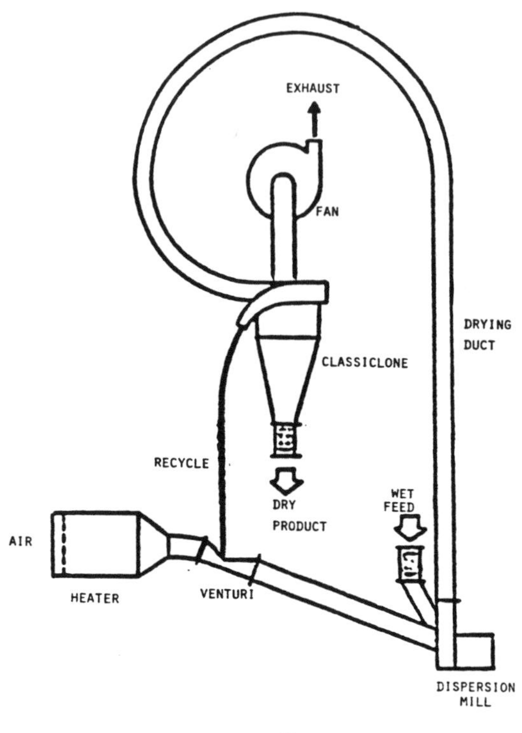

c

Figure 1.13. (continued)

Spray dryers (Figure 1.13b) have the fluid material to be dried added to a stream of hot gas. The inlet and outlet temperatures are controlled to dry the material instantaneously without any product decomposition or denaturization.

Flash drying equipment (Figure 1.13c) requires that the particulate to be dried be introduced into a hot gas stream in a duct. The particulate is both dried and conveyed by the hot gas during the very short transit time in the duct. Separation of the dried material and the spent gas occurs at the discharge of the duct.

Manufacturers of drying equipment are given in Table 1.9. Reviews and texts (27-33) should be consulted for additional information.

1.2.2 Membrane Processes

Ultrafiltration and reverse osmosis, along with

Table 1.9: Industrial Drying Equipment Manufacturers

Company	Location
Aeromatic	Towaco, NJ
A1 Jet Equipment Company	Plumsteadville, PA
APV Anhydro, Inc.	Attleboro Falls, MA
C.E. Raymond, Combustion Engineering, Inc.	Chicago, IL
Dedert Corp.	Olympia Falls, IL
Dorr-Oliver, Inc.	Stamford, CT
Glatt Air Techniques, Inc.	Ramsey, NJ
Komline-Sanderson Engineering Corporation	Peapack, NJ
Krauss Maffei Corp.	Charlotte, NC
Luwa Corp., Process Div.	Charlotte, NC
Mikropul Corp., Micron Products Division	Summit, NJ
Niro Atomizer	Columbia, MD
Procedyne Corp.	New Brunswick, NJ
Renneburg Div., Heyl and Patterson	Pittsburgh, PA
Swenson Process Equipment Inc.	Harvey, IL
Wyssmont Corporation	Fort Lee, NJ

microfiltration, are membrane separation processes. The membrane is a barrier between two fluids. Its barrier properties are its permeability or rate of transfer for a component through the membrane and its permselectivity or relative permeability flux for two components under the same operating conditions.

In traditional filtration and microfiltration, the feed solution flows directly onto the filter (membrane) and particles larger than the pores accumulate on the membrane surface (Figure 1.14a). The limitations of filtration are realized as this boundary layer builds and gradually cuts off the flow of solution through the membrane.

Ultrafiltration and reverse osmosis both separate a solute from a solution by forcing the solvent to flow through a membrane using a hydraulic pressure gradient (Figure 1.14b). The permselectivity of the membrane depends strongly on molecular size: small molecules pass through the membrane while large molecules are retained. When the dimensions of the solute are within an order of magnitude of the solvent dimensions, the process is called reverse osmosis. When the dimensions of the solute are from ten times the solvent dimension to less than 0.5 microns, the process is called ultrafiltration.

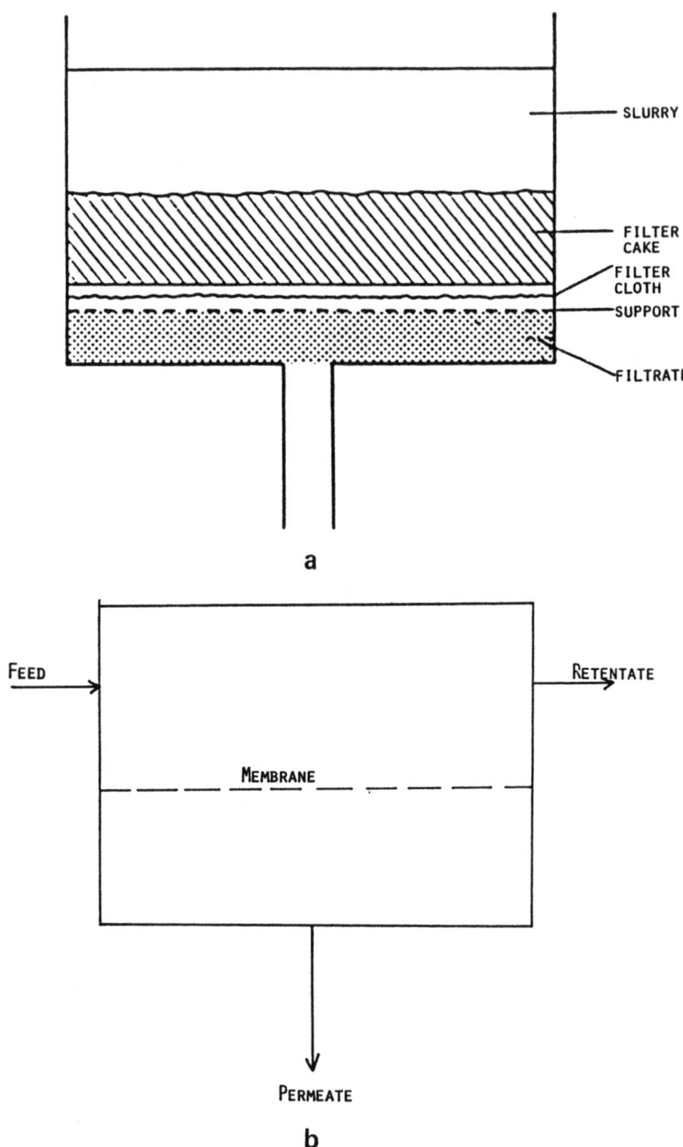

Figure 1.14. (a) Schematic of a simple filtration system. (b) Idealized membrane filtration system.

In ordinary filtration, the applied hydrostatic pressure ranges from a fraction of an atmosphere to an atmosphere. In ultrafiltration one to ten atmospheres and in reverse osmosis 10 to 100 atmospheres of hydrostatic pressure are applied as the driving force.

For ultrafiltration, the flux of the solute, J_s, is a function of the pressure driving force and the concentration driving force. This is represented mathematically as:

$$J_s = J_v C_f (1 - \sigma) - D_s^m \frac{dC_s^m}{dx} \qquad (1.9)$$

In this equation, J_v, the solvent flux, is equal to:

$$J_v = \frac{\varepsilon r^2}{8 \eta} \frac{\Delta P}{\tau \lambda} \qquad (1.10)$$

where ϵ is the membrane porosity, r is the effective pore radius, η is the solution viscosity, ΔP is the pressure difference across the membrane, τ is the tortuosity factor for the membrane and λ is the membrane thickness. C_f is the concentration of the solute in the feed solution, σ is the fraction of the pure solvent flux which passes through pores smaller than those retaining the solute molecules, D_s^m is the diffusion coefficient and dC_s^m/dx is the concentration gradient of solute across the membrane.

For most ultrafiltration systems, the pressure term of equation 1.9 contributes much more than the diffusion term. Therefore, the solvent flux will be directly proportional to the applied pressure.

The percent rejection rate, R, of a given ultrafiltration membrane for a solute is given by:

$$R = \left(1 - \frac{J_s}{J_v C_f}\right) \times 100 \qquad (1.11)$$

If one can ignore the diffusion driving force, the rejection rate is seen to be independent of the applied force.

In reverse osmosis, molecular diffusion through the membrane is the rate limiting condition. A solution in which the solute has a molecular weight of 500 or less will have a significant osmotic pressure, even as high as 100 atmospheres, which the hydrostatic pressure must first overcome before transport will occur through the membrane.

Example 1.4

Figure 1.15 (34) shows the flux through an ultrafiltration membrane in a stirred cell as a function of transmembrane pressure. Why does the stirrer rpm affect the flux?

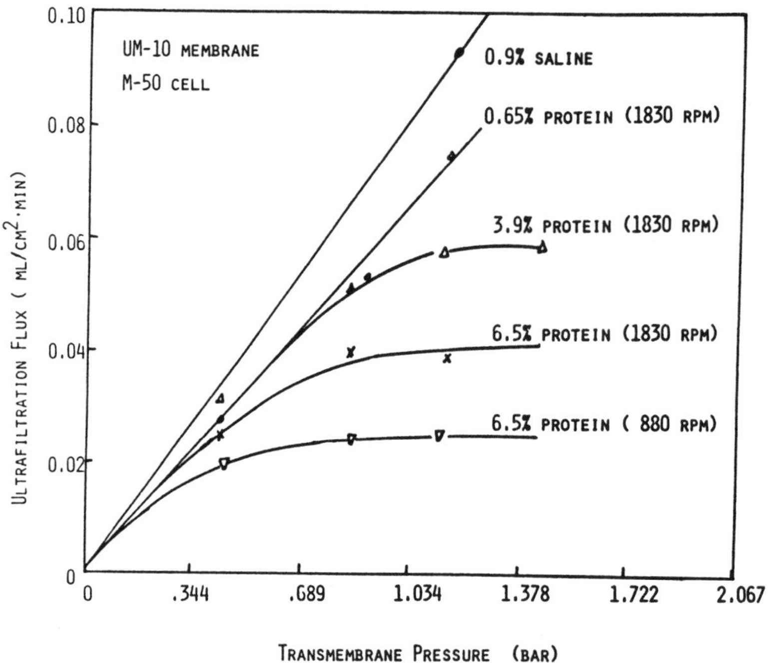

Figure 1.15. Flux pressure relationships for bovine serum albumin solutions in a stirred batch cell (Reference 34).

When protein solutions are passed by an ultrafiltration membrane, a protein gel layer builds up on the membrane. The faster stirrer speed reduces the thickness of this layer and allows greater flux.

For reverse osmosis, the flux of the solute, J_S, and of the solvent, J_V, are conveniently represented by:

$$J_s = \frac{K_1}{\lambda}(c_f - c_p) \qquad (1.12)$$

and

$$J_v = \frac{K_2}{\lambda} (\Delta P - \Delta \pi) \qquad (1.13)$$

In these equations, K_1 and K_2 are the transport coefficients of the membrane for the solute and the solvent, respectively; λ, C_f and ΔP are as defined for the ultrafiltration system; C_P is the solute concentration in the product stream; and $\Delta\pi$ is the osmotic pressure difference between the two solutions.

Using Equations 1.12 and 1.13 in Equation 1.11, the rejection rate is given by:

$$R = \frac{K_1/K_2 (\Delta P - \Delta \pi)}{1 + K_1/K_2 (\Delta P - \Delta \pi)} \times 100 \qquad (1.14)$$

Thus it can be seen that for reverse osmosis, both the rejection and the solvent flux increase with increases in the hydrostatic pressure.

Membrane separation units have one of four different configurations: tubular, planar spiral or hollow fiber (35).

Tubular modules (Figure 1.16) consist of bundles of rigid-walled, porous tubes which are 1 to 3 cm in diameter. The inside of the tube walls is lined with the control membrane. The feed solution is introduced under pressure to the inside of the tubes. The permeate is collected from the outside surfaces of the tube while the remaining solution, the retentate, is removed from the exit port of the tube.

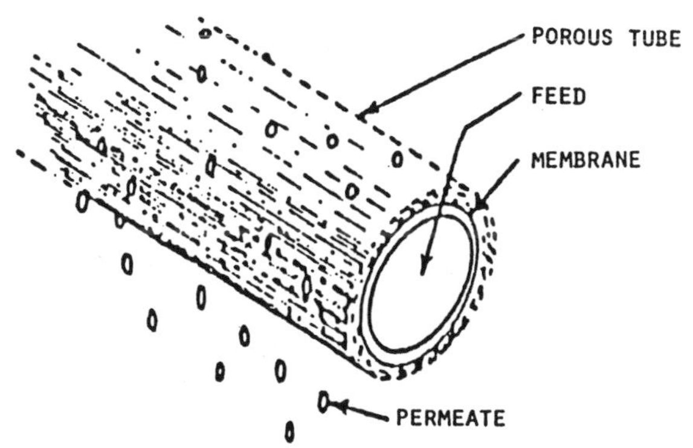

Figure 1.16. Tubular membrane configuration.

Plate and frame or planar modules (Figure 1.17) consist of a stack of grooved sheets which are covered on both sides by control membranes. Each membrane covered sheet alternates with a spacer sheet to form the module stack. The membrane edges are sealed to prevent the mixing of the permeate and the feed solution. The feed solution, under pressure, flows tangentially along each of the porous sheets as it is directed in a serpentine manner through the stack. The permeate collects in the spacer regions and flows to the permeate outlet.

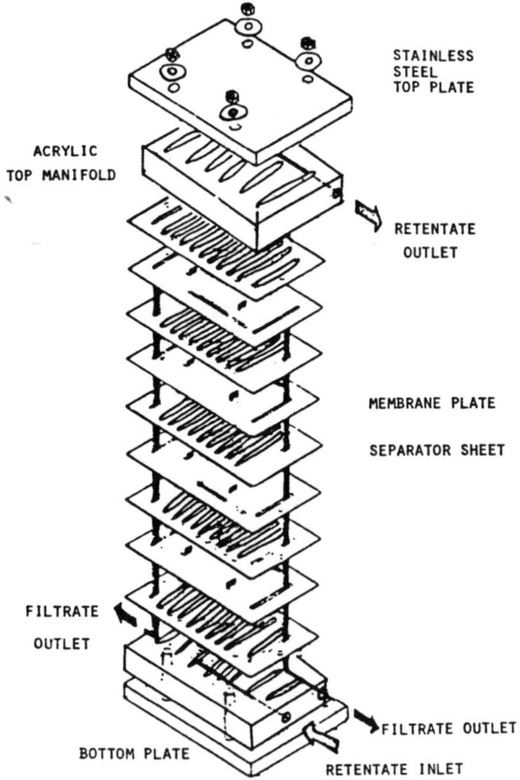

Figure 1.17. Plate and frame or planar membrane modules.

Spiral wound modules (Figure 1.18) consist of an envelope of membrane covered sheets and separator sheets which are wound concentrically around a hollow core and then inserted into a canister. The pressurized feed stream is introduced at one end of the canister and flows tangentially along the membrane to the exit port of the canister. The permeate collects in the separator area and flows to the hollow center where it is removed.

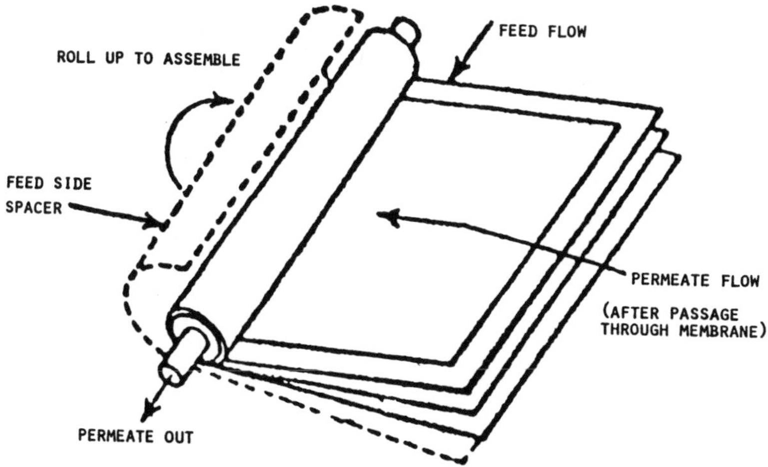

Figure 1.18. Spiral wound membrane system.

Hollow fiber membrane systems (Figure 1.19) consist of bundles of fibers with an outside diameter of 15 to 250 microns. Whereas the feed solution flows inside the tubes in the tubular modules, here it flows on the outside surface of the fibers. The permeate flows along the inside of the fibers to the end of the unit where the product is collected.

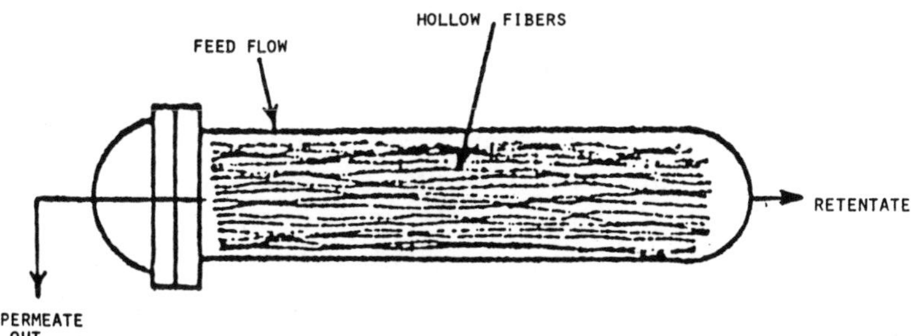

Figure 1.19. Hollow fiber membrane module.

Ultrafiltration is most often used for concentration, but it can also be used to fractionate mixtures of large and small solutes. Fairly high concentrations of suspended solids can be processed by ultrafiltration if there is sufficient agitation to prevent cake formation on the membrane. Reverse osmosis has its greatest utility in concentrating such low molecular solutes as salts and sugars.

The following review articles should be consulted for additional information on membrane processes (36-41). Membrane equipment suppliers are listed in Table 1.10.

Table 1.10: Industrial Membrane Equipment Manufacturers

Company	Location
Amicon, Div. of W.R. Grace	Danvers, MA
Culligan, Div. of Beatrice	Northbrook, IL
Dow Chemical	Midland, MI
Du Pont	Wilmington, DE
Enka America, Inc.	Ashville, NC
Gelman Sciences	Ann Arbor, MI
HPD Inc.	Naperville, IL
Koch Membrane Systems, Inc.	Wilmington, MA
Memtek Corp.	Billerica, MA
Millipore	Bedford, MA
Monsanto	St. Louis, MO
Osmonics	Minnetonka, MN
Pall	East Hills, NY
Romicon	Woburn, MA
Separex Corp.	Anaheim, CA
UOP	Des Plaines, IL
Vaponics, Inc.	Plymouth, MA

1.2.3 Solvent Extraction

Solvent extraction is a method of separation based on the transfer of a solute from one solvent into another solvent when the two solvents are brought into contact. It is essential that the two solvents be immiscible. A solute which is soluble in both phases will distribute between the two phases in a definite proportion. The desired separation is achieved by adjusting the chemical parameters of the system: pH, solvent selection or ion pair formation.

The process consists of two steps: (1) intimately mixing the two solvents until the solute has been distributed between the liquids and (2) separating the two phases. The distribution of the solute between the two phases is given by the distribution coefficient, K_D:

$$2.3\ RT \log K_D = \bar{V}_s \left[(\delta_s - \delta_1)^2 - (\delta_2 - \delta_s)^2 \right] \quad (1.15)$$

where $\bar{V}_S$ is the molar volume of the distributing solute, δ_S is its solubility parameter and δ_1 and δ_2 are the solubility parameters of the two immiscible solvents.

Most fermentation products are uncharged, although those having acidic or basic functional groups can undergo proton-transfer reactions that result in charged species, such as $RCOO^-$ and RNH_3^+. The solvent extraction of such compounds can be carried out by means of pH control.

There is the possibility that the solute extracted into the organic phase may become involved in a chemical interaction. Such interactions lower the activity of the solute and drive the overall equilibrium in the direction of greater extractability, assuming the reaction product is more soluble in the organic phase. As an example, the dimerization of carboxylic acids in non-oxygen-containing organic solvents results in higher distribution coefficients. Ligand complexes, which are important for ion-pair formation in metal salt extraction, can increase the distribution coefficient by several orders of magnitude. Quaternary ammonium salts dissolved in the organic phase have been used to form ion-pairs with strongly acidic antibiotics so that the resulting complex is hydrophobic. The efficiency of the extraction depends on the lipophilicity of the quaternary ammonium salts as shown in Table 1.11 (42).

Table 1.11: Comparison of Quaternary Ammonium Salts as Ion Pair Complexing Agents for the Olivanic Acids

Quaternary Ammonium Salt	Extraction Efficiency (%)
Benzyldimethyl-n-hexadecyl-ammonium chloride	65
Trioctylmethylammonium chloride	73
Tetra-n-butylammonium chloride	Trace
Dimethyldioctylammonium chloride	41
Dimethyldidecylammonium chloride	73

The fraction of solute extracted in a single stage is given by the following equation:

$$\Theta = \frac{C_oV_o}{C_oV_o + C_wV_w} = \frac{K_DV}{1 + K_DV} \tag{1.16}$$

where C_O and C_W represent the organic and aqueous phase concentrations, respectively; V_O and V_W are the volumes of the organic and aqueous phase, respectively; and V is the phase volume ratio, V_O/V_W. From this equation it is apparent that one can increase the extent of extraction, for a given K_D value, by increasing the phase ratio. Additional extraction stages may be carried out on the same aqueous solution using successive additions of organic solvent such that V remains the same. After n such extractions, the fraction of solute remaining in the aqueous phase is given by:

$$\Theta_n = (1 - \Theta)^{n-1} \tag{1.17}$$

The relationships discussed so far are for the extraction of a single solute from an aqueous phase to concentrate it in an organic phase. It is also possible to use solvent extraction to separate two solutes, A and B, which are present in the aqueous phase.

If the initial concentration ratio of these solutes is C_A/C_B, then the concentration ratio is reduced to $C_A\Theta_A/C_B\Theta_B$ after extraction with the organic phase, where Θ_A and Θ_B are the respective fractions extracted. The separation factor for the two solutes is given by the ratio Θ_A/Θ_B, which is the change in the initial concentration due to the extraction. An alternate measure would be the ratio of the fraction of each solute remaining in the aqueous phase after extraction, $(1 - \Theta_A)/(1 - \Theta_B)$.

Example 1.5

In a solvent extraction, what is the number of transfer stages required to obtain 99% purity when the phase ratio volume, V_o/V_w, is 0.05 for the following separation factors: (a) 1.2; (b) 1.5; (c) 2.0; (d) 2.5?

Equation 1.17 can be rearranged to give the number of transfer stages (n) required:

$$n = 1 + \frac{\log(\Theta_n)}{\log\left(1 - \frac{K_DV}{1 + K_DV}\right)} \tag{1.18}$$

For K_D = 1.2, n = 80; for K_D = 1.5, n = 65; for K_D = 2.0, n = 49 and for K_D = 2.5, n = 40. If the organic phase ratio were allowed to increase so that V = 0.10, n could be reduced to 42 for K_D = 1.2. This cannot be carried too far or the advantage of solvent extraction in increasing the solute's concentration is lost.

Solvent extraction may be applied at any stage of a purification process but is usually most useful at the beginning of an isolation procedure. High yields can be obtained if the solute is stable and recovery from the solvent is not difficult. In large scale manufacturing operations, the use of more than two successive extractions with different solvent pairs is unusual because of the large solvent volume and expensive equipment required for each extraction.

The extraction processes can be grouped into batch, continuous or countercurrent distribution processes.

When the extraction conditions can be adjusted so that the fraction extracted is greater than 0.99 ($VK_D \geqq 100$), then a single stage or batch extraction is all that is needed to place the bulk of the solute in the organic extract. When VK_D is reduced to 10, it is necessary to carry out the extraction twice to obtain 99% of the solute in the organic phase. As Figure 1.20 shows, this equipment can be as simple as an enlarged separatory funnel.

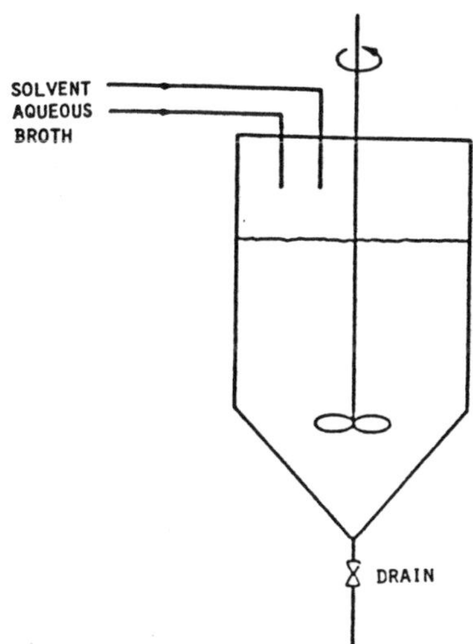

Figure 1.20. Single stage extraction unit.

When the VK_D values of the solutes are quite small, multiple-batch extractions cannot be conveniently or economically carried out since the large amount of organic solvent added offers little improvement over the original solute concentration in the aqueous phase. Continuous extraction using volatile solvents can be carried out in equipment in which the solvent distilled from an extract collection vessel is condensed, contacted with the aqueous phase and returned to the extract collection vessel in a continuous loop. Figure 1.21 shows an example of such a process scheme.

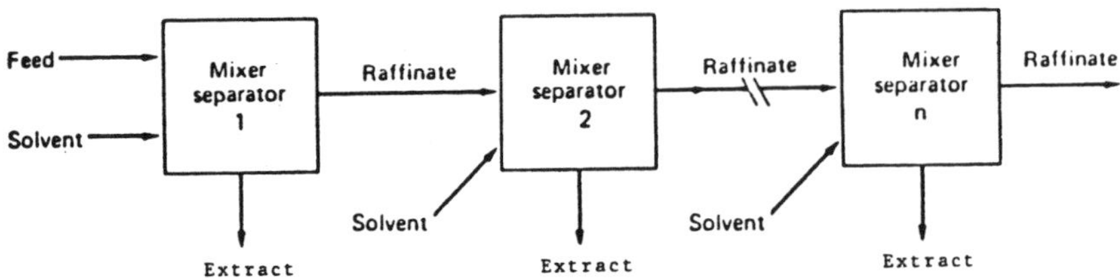

Figure 1.21. Continuous co-current extraction scheme for use with a volatile solvent.

Countercurrent distribution extraction is a special multiple-contact extraction used to separate two solutes whose K_D values are very similar (Figure 1.22). The extracted aqueous phase passes from vessels 1 to n while the product enriched solvent phase is flowing from vessels n to 1.

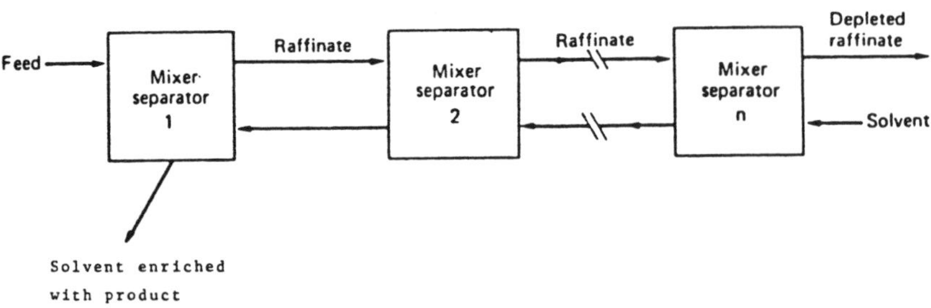

Figure 1.22. Counter-current extraction scheme.

Special centrifuges (Figure 1.23) which contain mixing and settling sections may be used to carry out several stages simultaneously. This Podbielniak centrifugal extractor has played a major role in the recovery of antibiotics.

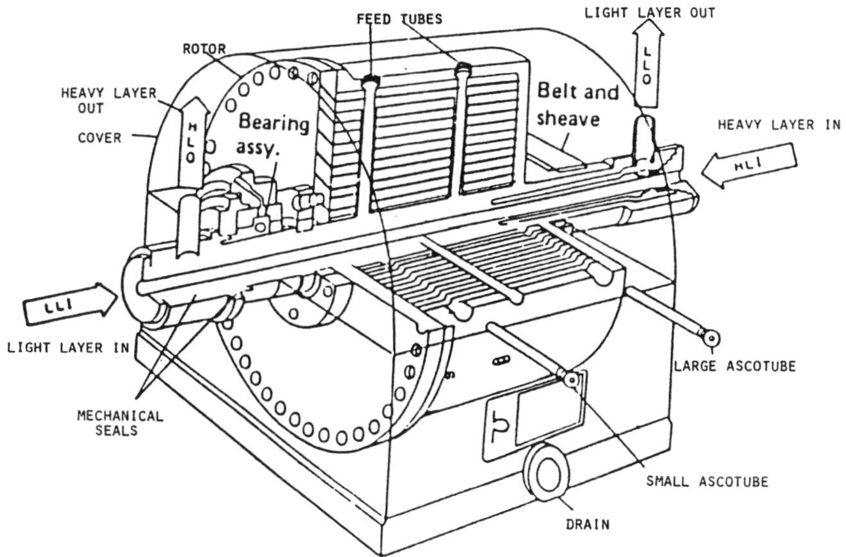

Figure 1.23. Podbielniak centrifugal extractor.

This extractor consists of a horizontal cylindrical drum revolving at up to 5000 rpm. The heavier liquid phase is introduced into the vessel on the shaft along the rotating axis while the lighter liquid phase, also introduced at the shaft, is guided internally to the periphery of the drum. As the drum rotates, the liquid phases flow countercurrently through the channels in the interior of the drum. The light liquid phase moves toward the center where it is removed. The heavy liquid phase moves to the periphery and then is guided back to the shaft for removal. Flow rates greater than 100,000 dm^3/h are possible with this type of equipment. The advantage of this type of extractor is the low hold-up volume of liquid in the equipment compared to the throughput.

Equipment manufacturers are listed in Table 1.12. For additional information, the following texts and reviews should be consulted: References 43-48. Table 1.13 shows the many enzymes which may be extracted from broths of disrupted cells using solvent extraction (49).

1.2.4 Electrophoresis and Electrodialysis

Electrophoresis is the separation technique in which electrically charged particles are transported in a direct-

Table 1.12: Industrial Extraction Equipment Manufacturers

Company	Location
Alfa-Laval/DeLaval Co.	Fort Lee, NJ
Baker Perkins, Inc.	Saginaw, MI
Escher B.V.	The Hague, Netherlands
Kuhni	Switzerland
Liquid Dynamics Co.	Hempstead, NC
Luwa A.G.	Zurich, Switzerland
Westfalia Separators	West Germany

Table 1.13: Extractive Separation of Enzymes from Disrupted Cells Using PEG 1540/Potassium Phosphate Solutions

Enzyme	Cell Concentration (%)	Partition Coefficient	Yield (%)	Purification Factor
From *Saccharomyces cerevisiae*				
α-Glucosidase	30	2.5	95	3.2
Glucose-6-phosphate dehydrogenase	30	4.1	91	1.8
Alcohol dehydrogenase	30	8.2	96	2.5
Hexokinase	30		92	1.6
From *Escherichia coli*				
Fumarase	25	3.2	93	3.4
Aspartase	25	5.7	96	6.6
Penicillin acylase	20	2.5	90	8.2
From *Brevibacterium ammoniagenes*				
Fumarase	20	3.3	83	7.5
From *Candida Boidinii*				
Formate dehydrogenase	33	4.9	90	2.0
From *Leuconostoc* species				
Glucose-6-phosphate dehydrogenase	35	6.2	94	1.3

current electric field. The mediums used in biochemical applications are usually aqueous solutions, suspensions or gels. It is the differential migration of solutes in these mediums due to the electric force that separates the solutes. The highest degree of resolution is obtained when an element of discontinuity is introduced into the system, such as a pH gradient, cellulose acetate membranes or the sieving effect of high density gels. When the discontinuity elements are charged membranes, the separation technique is known as electrodialysis.

It is the ionization of functional groups of the protein or fermentation product that provides the charged species which moves in the electric field. For large molecules, there may be several functional groups which cause the molecule to have a net positive charge at acid pH and a net negative charge at basic pH. The pH where the molecule has zero electrophoretic mobility is called the isoelectric point.

An important concept in the theory of electrophoresis is that of the electric double layer (Figure 1.24). The electric double layer is the structure formed by the fixed charges of the macromolecule (or colloid) with the relatively mobile counterions of the surrounding fluid. The total charge of the double layer is zero. However, the spatial distribution of charges is not random and gives rise to the electric potential.

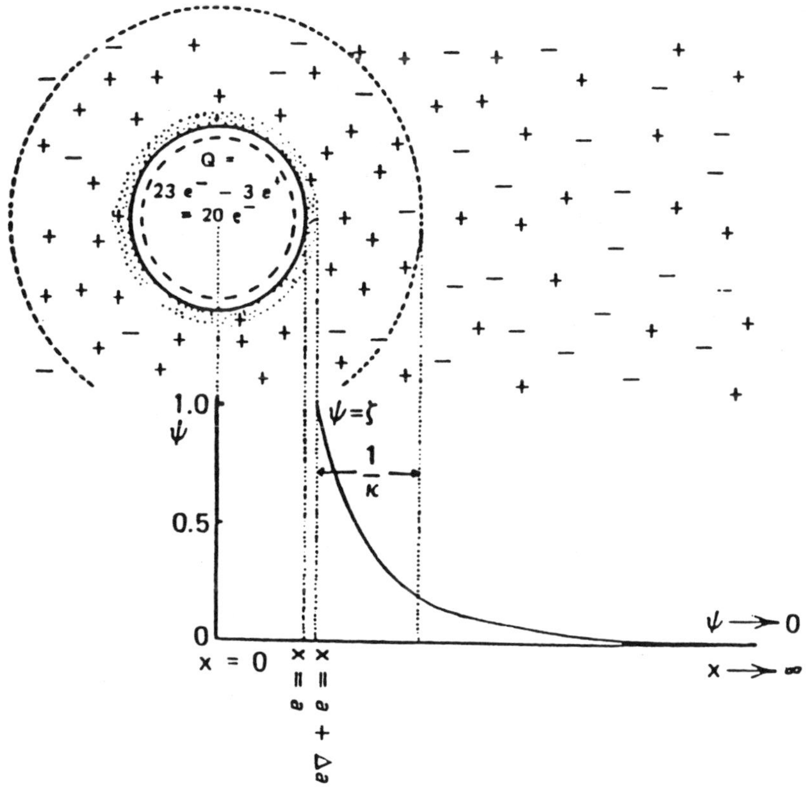

Figure 1.24. Schematic of electrical double layer.

The thickness of the double layer is normally given by the inverse of the Debye-Huckel constant, k:

$$\frac{1}{k} = \left(\frac{\varepsilon kT}{8 \pi e^2 n_o z^2} \right)^{\frac{1}{2}} \qquad (1.19)$$

where e is the electronic charge, k is the Boltzmann constant, n_o is the bulk concentration of each ionic species, z is the valence of the electrolyte and ϵ is the dielectric constant.

The mobile counterions are bound sufficiently tightly to the macromolecule so that they move with it during electrophoresis, defining a surface of shear larger than that of the macromolecule itself. The potential at this shear boundary determines the electrophoretic mobility of the macromolecule and is called its zeta potential.

The zeta potential cannot be measured directly. It is calculated on the basis of theories for the electric double layer which take into account the electrophoretic attraction, the electrophoretic retardation, the Stokes friction and the relaxation effect.

While many experimental adaptations of electrophoresis have been used for analytical purposes, zone electrophoresis and isoelectric focusing are the two variations which are also useful for preparative separations of protein mixtures.

Zone electrophoresis is characterized by the complete separation of charged solutes into separate zones. Many of the earlier separations of this type were carried out using filter paper in a Durum cell (Figure 1.25). Draped over glass rods are strips of filter paper with their ends in contact with wicks in the electrode buffer. The support medium most commonly used in this type of cell now is cellulose acetate membranes.

High density gels with specific pore sizes may be used as the resolving medium to separate proteins with similar dimensions. This technique is also quite simple in design and operation and gives excellent resolution because of the uniform porosity of the medium.

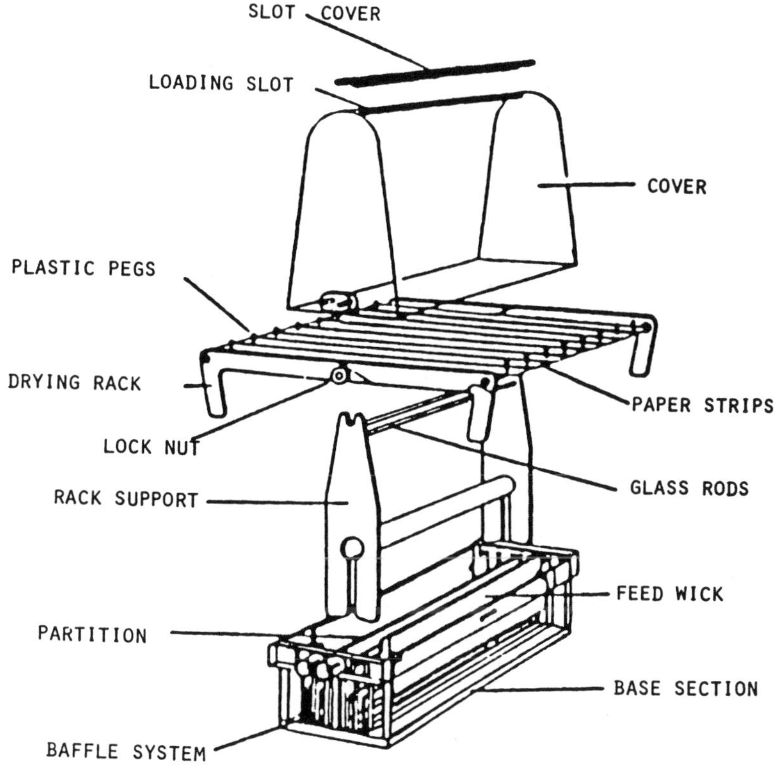

Figure 1.25. Durrum cell for electrophoresis separations.

In isoelectric focusing a pH gradient is superimposed on a column filled with an electrically neutral medium which exhibits a density gradient along the column. The density gradient, to be effective, must be greater than the gradient caused by the protein-buffer boundaries. The usual such medium is sucrose at concentrations as high as 50%.

After the sucrose density gradient has been established, the pH gradient is formed by the prolonged electrolysis of a suitable mixture of amphoteric substances which have a wide range of isoelectric points. The most acidic of the ampholytes will be near the anode end of the column, the most basic near the cathode end and the remaining ampholytes covering the range between, depending on their individual isoelectric points.

The protein mixture to be separated can be applied to

the column for electrophoretic fractionation. After such fractionation, the column is emptied into separate fractions. The protein fractions can be separated from the sucrose and the ampholytes by dialysis or ultrafiltration.

Mixtures of a large number of aminocarboxylic acids covering the most appropriate pH range (pH 3 to 10) are commercially available in molecular weights (300 to 600) sufficiently low for easy subsequent separation from the isolated proteins (52).

References for additional details on the application of this technique are the collection of papers in the books edited by Bier (51) and the text by Shaw (53).

Electrodialysis is the diffusive transfer of an ionic solute across a membrane due to an applied electric field. A necessary feature for the utilization of this technique is the solute's potential for ionization. In electrodialysis the membranes are usually positioned parallel to one another in a planar configuration. This is the typical plate and frame stack. This arrangement allows the most attractive geometry for the efficient application of the electrical driving force.

As Figure 1.26 (54) shows, anion and cation membranes are arranged alternately between two electrodes. Into alternating chambers are placed an aqueous solution containing M cations and X anions. The imposition of an electric current causes the M cations to move toward the cathode and the X anions toward the anode. However, each ionic type cannot progress more than one chamber toward the electrodes without finding its path blocked by a membrane which is impermeable to that ionic type. Therefore, the feed stream will be depleted of the ionic solute and the adjacent stream will be enriched in the ionic solute.

This technique requires that the membranes separating the chambers have a sufficient ion exchange capacity as well as pores of a sufficiently small size (~30 Å) to repel oppositely charged ions. The alternating series of dilute and concentrating cells are manifolded together. The degree of removal of the ionic solute from the feed stream is determined by the ionization potential of the solute, the rate of hydraulic flow and the applied current density.

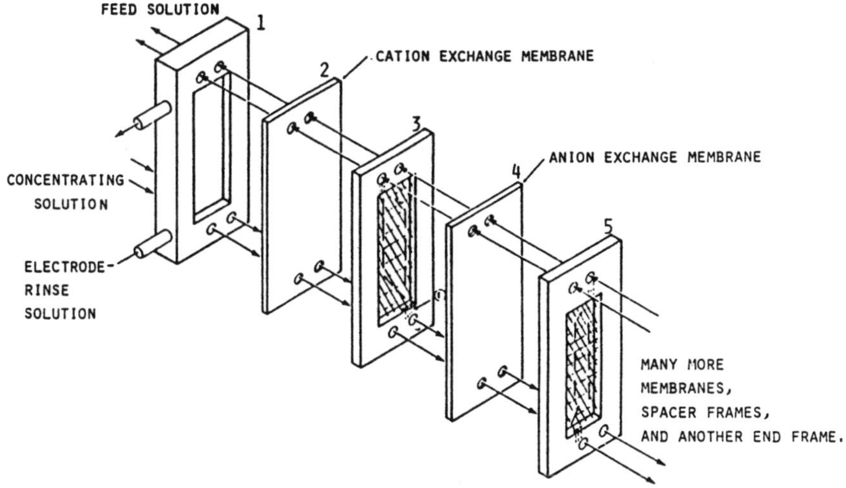

Figure 1.26. Electrodialysis modules (Reference 54).

The limiting factor for most products from fermentation applications will be very low ionization potential and concentration of the solutes. Even for an ionic compound such as sodium chloride, the solute concentration must be greater than 300 ppm. At concentrations less than this, the voltage drop which occurs across the dilute compartments will require excessive power input for economical operation.

Two parameters of importance in characterizing the ion exchange membranes for an electrodialysis system are its electrical resistance and its permselectivity. The membrane resistance per unit area, measured in sea water, should be less than 20 ohm/cm^2. The perselectivity, P_S, is the measure of the fractional increase in flux for the mobile counterion due to the presence of a perfectly selective membrane and in relation to the flux of the ion in the absence of the membrane:

$$P_s = \frac{\phi_m - \phi_s}{1 - \phi_s} \qquad (1.20)$$

where ϕ_m is the flux within the membrane and ϕ_s is the flux in free solution. In practice, permselectivity is measured by comparing the concentration gradient generated by placing the membrane between NaCl electrolyte at two different concentrations and comparing the actual potential with the theoretical Nernst potential.

Most of the electrodialysis equipment in the United States have been manufactured by Ionics (Watertown, MA). Other manufacturers are Aqua-Chem (Waukesha, WI) and Toyo Soda (Tokyo, Japan).

1.3 RECOVERY PROCESSES

A recovery process is the specific combination of the recovery unit operations needed to attain the desired degree of purity and yield for a specific solute.

The sequence of unit operations in the recovery process is typically: (1) the removal of insoluble components, (2) primary isolation, (3) purification, and (4) final product isolation (55).

The removal of insoluble components is usually carried out by filtration, centrifugation or simply decanting. This step does not affect the concentration (0.1 to 5 g/L) or the purity (0.1 to 2.0%) of the biochemical solute in the fermentation broth.

The solute concentration is increased substantially (5.0 to 10.0 g/L), with a corresponding increase in purity (1.0 to 10%), during the primary isolation. The solute is separated from substances with widely differing polarities or size by solvent extraction, adsorption, ion exchange, precipitation and ultrafiltration.

Purification operations are techniques for the removal of impurities closely related to the desired solute. Adsorption, chromatography or fractional precipitation unit operations are used for purification. The solute concentration is increased to 50 to 200 g/L while the purity is increased to 50 to 80%.

The final product isolation yields the desired solute in the form needed for final formulation, blending or shipping. Techniques include centrifugation, crystallization or drying. The purity becomes 90 to 100%.

Strategies for the development of large scale purification systems for proteins have recently been proposed by Wheelwright (56). Figure 1.27 (55) shows a recovery process flowsheet for antibiotic production. The complex process requires fifteen different separation techniques to produce both a crude and a highly purified antibiotic product.

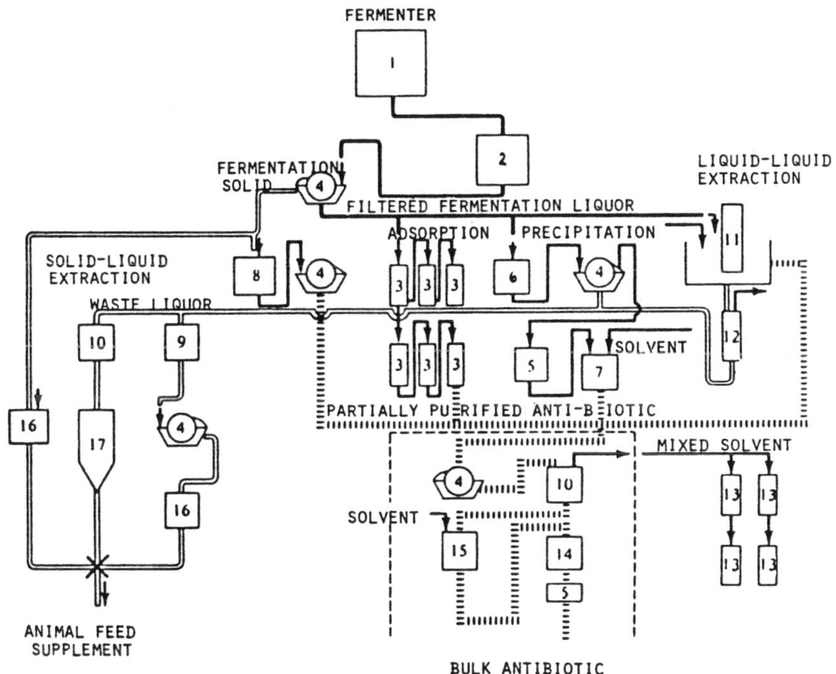

Figure 1.27. Process flowsheet for recovery of an antibiotic (Reference 55).

Flow sheets for the recovery of citric acid, sodium clavulanate, micrococcal nuclease and three human proteins synthesized in recombinant *E. Coli* are shown in Figures 1.28, 1.29, 1.30, and 1.31 to illustrate the range of techniques which are used in the recovery of biochemical products.

Citric acid (Figure 1.28) (57) is recovered by filtration of the biomass followed by a precipitation of calcium citrate by the addition of calcium hydroxide. The precipitate is treated with sulfuric acid to dissolve the calcium citrate and attain citric acid. The calcium sulfate formed is filtered off. The citric acid solution is decolorized, deionized and concentrated using activated carbon and ion exchange resins. Crystallization, centrifugation and drying operations are used for the final recovery of pure citric acid crystals.

HARVESTED BROTH
↓
FILTER OFF *A. NIGER* MYCELIUM USING ROTARY VACUUM FILTER
↓
ADD $CA(OH)_2$ TO FILTRATE UNTIL pH 5.8
↓
CALCIUM CITRATE
↓
ADD H_2SO_4 WHILE AT 60°C
↓
FILTER ON ROTARY VACUUM FILTER TO REMOVE $CASO_4$
↓
ACTIVATED CHARCOAL TO DECOLORIZE
↓
CATION AND ANION EXCHANGE RESIN COLUMNS
↓
EVAPORATE TO POINT OF CRYSTALLIZATION AT 36°C
↓
CRYSTALS OF CITRIC MONOHYDRATE SEPARATED IN CONTINUOUS CENTRIFUGES
↓
DRIERS AT 50 - 60°C

Figure 1.28. Recovery and purification of citric acid (Reference 57).

Sodium clavulanate (Figure 1.29) (58) belongs to the β-lactam family and is used to inhibit the β-lactamase bacterial enzyme. After obtaining the supernatant from the fermentation broth, the compound is concentrated by extraction into butanol at acidic pH. This must be done rapidly and at low temperatures to prevent dehydration of clavulanic acid. The back extraction into water at pH 7.0 not only has the advantage of further purifying the sodium clavulanic acid, but also returns it to a more stable condition. Further purification is carried out by gradient elution ion exchange chromatography. The salt is removed by molecular size separation on a gel column. The final purification is achieved by partition chromatography on cellulose, followed by evaporation and freeze drying.

Rather than further concentrating the supernatant containing micrococcal nuclease (Figure 1.30) (59), it is diluted to reduce the conductivity. After acidification with glacial acetic acid, microccal nuclease is adsorbed onto an ion exchange resin. The enzyme-loaded resin is placed in a column from which the nuclease can be eluted. The desired

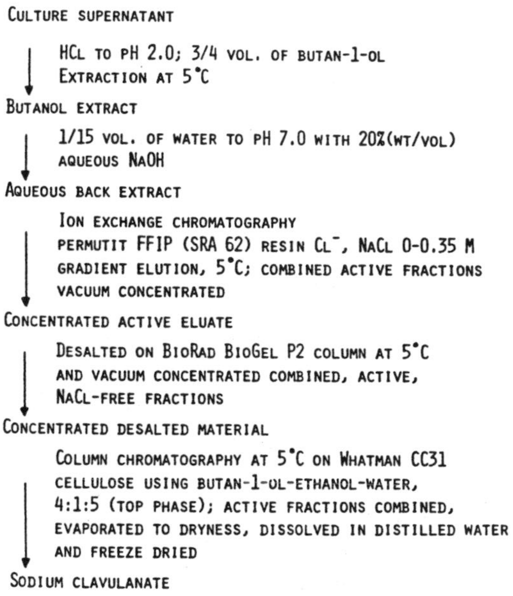

CULTURE SUPERNATANT

↓ HCL TO pH 2.0; 3/4 VOL. OF BUTAN-1-OL
EXTRACTION AT 5°C

BUTANOL EXTRACT

↓ 1/15 VOL. OF WATER TO pH 7.0 WITH 20%(WT/VOL)
AQUEOUS NaOH

AQUEOUS BACK EXTRACT

↓ ION EXCHANGE CHROMATOGRAPHY
PERMUTIT FFIP (SRA 62) RESIN Cl^-, NaCl 0-0.35 M
GRADIENT ELUTION, 5°C; COMBINED ACTIVE FRACTIONS
VACUUM CONCENTRATED

CONCENTRATED ACTIVE ELUATE

↓ DESALTED ON BIORAD BIOGEL P2 COLUMN AT 5°C
AND VACUUM CONCENTRATED COMBINED, ACTIVE,
NaCl-FREE FRACTIONS

CONCENTRATED DESALTED MATERIAL

↓ COLUMN CHROMATOGRAPHY AT 5°C ON WHATMAN CC31
CELLULOSE USING BUTAN-1-OL-ETHANOL-WATER,
4:1:5 (TOP PHASE); ACTIVE FRACTIONS COMBINED,
EVAPORATED TO DRYNESS, DISSOLVED IN DISTILLED WATER
AND FREEZE DRIED

SODIUM CLAVULANATE

Figure 1.29. Isolation procedure for sodium clavulanate (Reference 58).

elution fraction is dialyzed to remove the unwanted salts. Concentration is achieved by ultrafiltration and centrifugation. That product may be used directly or further purified on a gel filtration column followed by freeze drying.

Figure 1.31 (60) shows the recovery process steps used to purify human insulin, human growth and human leukocyte interferon made in recombinant *E. coli*. Such proteins, synthesized from genetically engineered organisms and intended for injection into humans, must be purified to exacting standards. After disruption of the cells to extract the desired protein, a series of precipitation and chromatography steps are applied in each case to achieve the required purity.

There is a limit, placed by economics, on the complexity of an acceptable product recovery process for a given biochemical. In general, the initial fermentation broth concentration of the prepared solute will be inversely proportional to the recovery costs. This is shown in Figure 1.32 (61) where recovery costs have been augmented by the profit margin per kg to give the selling price. The clear

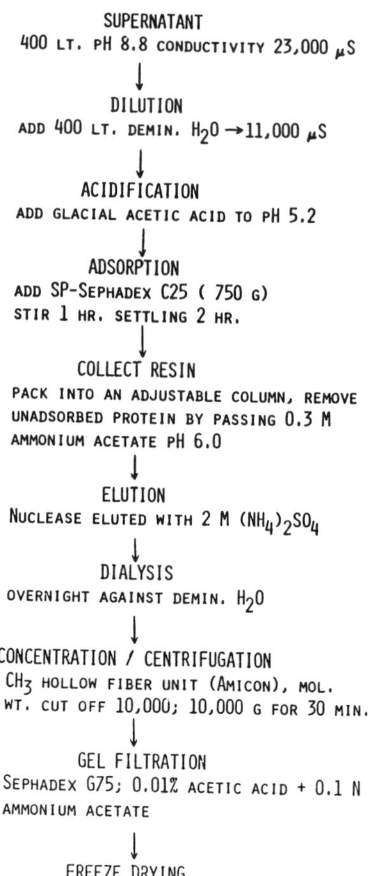

Figure 1.30. Purification procedure for micrococcal nuclease (Reference 59).

trend in this figure illustrates the continuing pressure for technological improvements in fermentation and recovery yields.

Example 1.6

A new enzyme has been prepared by a fermentation process that yields a concentration of 10 mg/L. Assuming a 35% return on sales (ROS), what is the limit on the process and recovery costs to justify commercialization of this product?

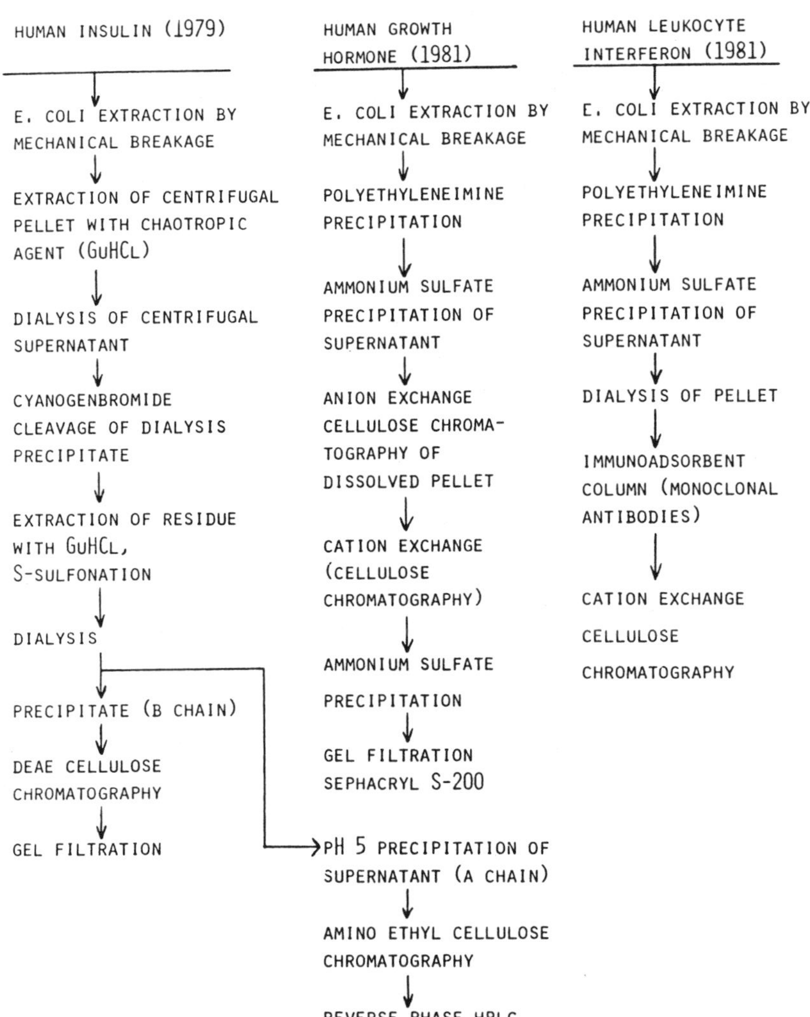

Figure 1.31. Recovery processes for human insulin, human growth hormone and human leukocyte interferon synthesized from *E. coli* (Reference 60).

From Figure 1.32 it is seen that the enzyme would need to have a selling price of about \$10,000/kg, about the same as for research and diagnostic enzymes. Therefore, with a 35% ROS, the maximum the process and recovery costs can be for this enzyme is \$6,500/kg.

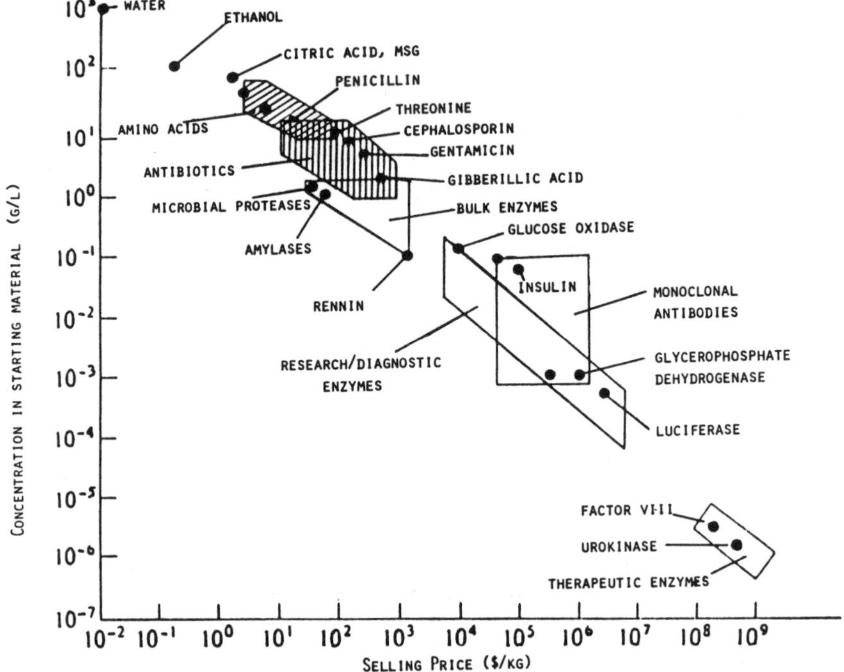

Figure 1.32. Relation between starting product concentration and final selling price of prepared product (Reference 61).

1.4 REFERENCES

1. Lipinsky, E.S., Allen, B.R., Jenkins, D.M., LM Current Research and Development 5:102 (1984)

2. Hawtin, P., The Chemical Engineer (Jan 11, 1982)

3. Dlouhy, P.E., Dahlstrom, D.A., Chem Eng Progr 64(4):116 (1968)

4. Shirato, S., Esumi, S., J Ferment Tech (Japan) 41:87 (1963)

5. Bell, G.R., Hutto, F.B., Chem Eng Prog 54(6):75 (1958)

6. Boss, F.C., "Filtration" In: Fermentation and Biochemical Engineering Handbook (Vogel, H.C., ed.), Noyes Publications, Park Ridge, NJ, p 163 (1983)

7. Purchas, D.B., Industrial Filtration of Liquids, Leonard Hill, London, 2nd Ed. (1971)

8. Savorsky, L., (ed.), Solid-Liquid Separation, Butterworth, London (1977)

9. Tilles, F.M., et al., Theory and Practice of Solid-Liquid Separation, University of Houston Press, Houston, TX, 2nd Ed. (1975)

10. Purchas, D.B., Solid-Liquid Separation Technology, Uplands Press Ltd., Croyden, England (1981)

11. Ambler, C.M., "Centrifugation" In: Handbook of Separation Techniques for Chemical Engineers, (Schweitzer, P.A., ed.), McGraw-Hill Book Company, New York, NY, p 4-55 (1979)

12. Solomons, G.L., Materials and Methods in Fermentation, Academic Press, London, p 299 (1969)

13. Boss, F.C., "Centrifugation" In: Fermentation and Biochemical Engineering Handbook, (Vogel, H.C., ed.) Noyes Publications, Park Ridge, NJ, p 296 (1983)

14. Aiba, S., Humphrey, A.E., Millis, N.F., Biochemical Engineering, Academic Press, New York, 2nd Ed. (1973)

15. Gray, P.P., Dunnill, P., Lilly, M.D., Fermen Technol Today, Proc Int Fermen Symp 4th, p 347 (1972)

16. Bird, R.B., Stewart, W.E., Lightfoot, E.N., Transport Phenomena, John Wiley & Sons, New York, NY (1960)

17. Bennett, R.C., "Evaporation" In: Handbook of Separation Techniques for Chemical Engineers, (Schweitzer, P.A., ed.), McGraw-Hill Book Company, New York, NY, p 2-131 (1979)

18. Freese, H.L., "Evaporation" In: Fermentation and Biochemical Engineering Handbook, (Vogel, H.C., ed.), Noyes Publications, Park Ridge, NJ, p 227 (1983)

19. Newman, H.H., Chem Eng Prog, 64(7):33 (July 1968)

20. Perry, J.H., Chilton, C.H., Chemical Engineers Handbook, McGraw-Hill New York, NY, 5th Ed., sec. 11 (1973)

21. Standiford, F.C., Chem Eng, p 157 (Dec 9, 1963)

22. Mullin, J.W., Crystallization, Butterworth, London (1972)

23. Randolph, A.D., Larson, M.A., The Theory of Particulate Processes, Acad. Press, New York (1971)

24. Shirotsuka, T., Toyokura, K., Mem School Sci Eng Waseda Univ, 30:57 (1966)

25. Nyvlt, J., Industrial Crystallization from Solutions, Butterworth, London (1971)

26. Nyvlt, N., Industrial Crystallization, Verlag Chemie, New York, p 129 (1978)

27. DeVivo, J.F., "Nonadiabatic Drying" In: Fermentation and Biochemical Engineering Handbook, (Vogel, H.C., ed) Noyes Publications, Park Ridge, NJ, p 317 (1983)

28. Keey, R.B., Introduction to Industrial Drying Operations, Pergamon Press, Oxford (1978)

29. Cook, E.M., Chem Engr Prog, (April 1978)

30. Quinn, J.J., Jr., "Adiabatic Drying" In: Fermentation and Biochemical Engineering Handbook, (Vogel, H.C., ed.) Noyes Publications, Park Ridge, NJ 334 (1983)

31. Keey, R.B., Drying Principles and Practice, Pergamon Press, Oxford (1972)

32. Belcher, D.W., Smith, D.A., Cook, E.M., Chem Eng, p 112 (Jan 17, 1977)

33. Dittman, F.W., Chem Eng, p 106 (Jan 17, 1977)

34. Porter, M.C., Michaels, A.S., Chemtech, p 56 (Jan 1971)

35. Cruver, J.E., Marine Technology, 9:216 (1972)

36. Li, N.N., Long, R.B., Henley, E.J., Ind Eng Chem, 57(3):18 (1965)

37. Michaels, A.S., In: Progress in Separation and Purification, Vol 1, (Perry, E.S., ed.), Interscience, New York, p 297 (1968)

38. Cheryan, M., Ultrafiltration Handbook, Technomics, Lancaster, PA (1986)

39. Fane, A.G., Progress in Filtration and Separation, Vol 4 (Wakeman, R.J., ed.) (1986)

40. Trudel, M., Trepanier, P., Payment, P., Process Biochem, 18:2-4,9 (1983)

41. Olson, W.P., Process Biochem, 18:29 (1983)

42. Hood, J.D., Box, S.J., Verrall, M.S., J Antibiot, 32(4):295 (1979)

43. King, C.J., Separation Processes, McGraw-Hill, New York (1971)

44. Brian, P.L.T., Staged Cascades in Chemical Processing, Prentice-Hall, Englewood Cliffs, NJ (1972)

45. Belter, P.A., In: Microbial Technology, (Peppler, H.J., Perlman, D., eds.), Academic Press, New York, Chapter 16, p 403 (1979)

46. Robbins, L.A., In: Handbood of Separation Techniques for Chemical Engineers, (Schweitzer, P.A., ed.), McGraw-Hill, New York, Section 1.9, p 1-256 (1979)

47. Lo, T.C., In: Handbood of Separation Techniques for Chemical Engineers, (Schweitzer, P.A., ed.), McGraw-Hill, New York, Section 1.10, p 1-283 (1979)

48. Todd, D.B., "Solvent Extractions", In: Fermentation and Biochemical Engineering Handbood, (Vogel, H.C., ed.), Noyes Publications, Park Ridge, NJ, p 175 (1983)

49. Kroner, K.H., Hustedt, H., Kula, M.R., Process Biochem, 19:170 (1984)

50. Abramson, H.A., Electrophoresis of Proteins, Hafner, New York (1964)

51. Bier, M., ed., Electrophoresis, Academic Press, New York, Vol I (1959); Vol II (1967)

52. One such mixture is called Ampholine, which is available from LKB Instrument Corporation.

53. Shaw, D.J., Electrophoresis, Academic Press, New York, (1969)

54. Lacey, R.E., "Dialysis and Electrodialysis", In: Handbook of Separation Techniques for Chemical Engineers, (Schweitzer, P.A., ed.), McGraw-Hill, New York, Section 1.14, p 1-449 (1979)

55. Bailey, J.E., Ollis, D.F., Biochemical Engineering Fundamentals, 2nd Ed., McGraw-Hill, New York, p 276 (1986)

56. Wheelwright, S.M., Bio/Technology 5:789 (1987)

57. Sodeck, G., Modl, J., Kominek, J., Salzbrunn, G., Process Biochem, 16(6):9 (1981)

58. Reading, C., Cole, M., Antimicrob Ag Chemother, 11(5):852 (1977)

59. Darbyshire, J., In: Topics in Enzyme and Fermentation Biotechnology, Vol 5, (Wiseman, A., ed.), Ellis Harwood, Chichester, p 147 (1981)

60. McGregor, W.C., Ann N Y Acad Sci, 413:231 (1983)

61. Dwyer, J.L., Bio/Technology, 2:957 (1984)

2

Adsorption

2.1 INTRODUCTION

Adsorption is the phenomenon in which molecules in a fluid phase concentrate on a solid surface without any chemical change. Adsorption takes place due to unsatisfied forces in the surface which attract and hold certain molecules, the adsorbate, from the fluid surrounding the surface of the solid, the adsorbent. The adsorption energy determines the strength with which any given molecule is adsorbed relative to other molecules in the system.

The adsorbed species are considered to form a monolayer on the surface of the adsorbent. The method is generally limited to molecules with molecular weights less than 2,000 since many of the surface sites may be of limited accessibility. Large molecules may be difficult to elute because of the possibility of multiple sites of adsorptive interaction. The concentration of the solute in the adsorption technique is usually less than 1000 ppm. Above that concentration, one should think in terms of chromatographic methods of purification.

Adsorption, essentially a surface effect, must be distinguished from absorption, which implies the penetration of one component throughout the body of a second.

The utilization of adsorption processes for the purification of drinking water, wine and oils by solid adsorbents, particularly charcoal, has been done for centuries. With the expansion of the chemical industry over the last century and with the development of the biotechnology applications in the last decade, the range of substances purified by adsorption has increased enormously.

Adsorption is utilized in fermentation recovery processes: (a) to remove unwanted molecules, such as color bodies, from the solution of the fermentation product; (b) to adsorb the fermentation product to increase its concentration when the product is eluted from the adsorption medium; or (c) to adsorb selectively the solutes in the fermentation produce stream to separate them by chromatography. The first two types of purification will be covered in this chapter and the chromatographic application will considered separately in Chapter 4.

2.2 ADSORPTION THEORY

Despite the long history of adsorption applications, research into the fundamentals of adsorption is only a little over sixty years old. Some of the early experiments with solutions were reviewed by Freundlich (1). A variety of substances, including phenol, succinic acid and simpler aliphatic acids were adsorbed by charcoal from dilute aqueous solution. These experiments measured the adsorption isotherm, which for these dilute solutions can be represented by the curve shown in Figure 2.1.

The curves can be described mathematically by what has become known as the Freundlich equation, even though it had been used earlier by Boedecker (2) and Kuester (3). The equation is:

$$x/m = \alpha c^{1/n} \tag{2.1}$$

where x is the weight of the adsorbate taken up by a weight m of solid adsorbent; c is the concentration of the solution at equilibrium; α and n are constants. The form $1/n$ was used to emphasize that c is raised to a power less than unity.

The Freundlich equation is based upon an approximation which is valid only for dilute solutions. Many attempts have

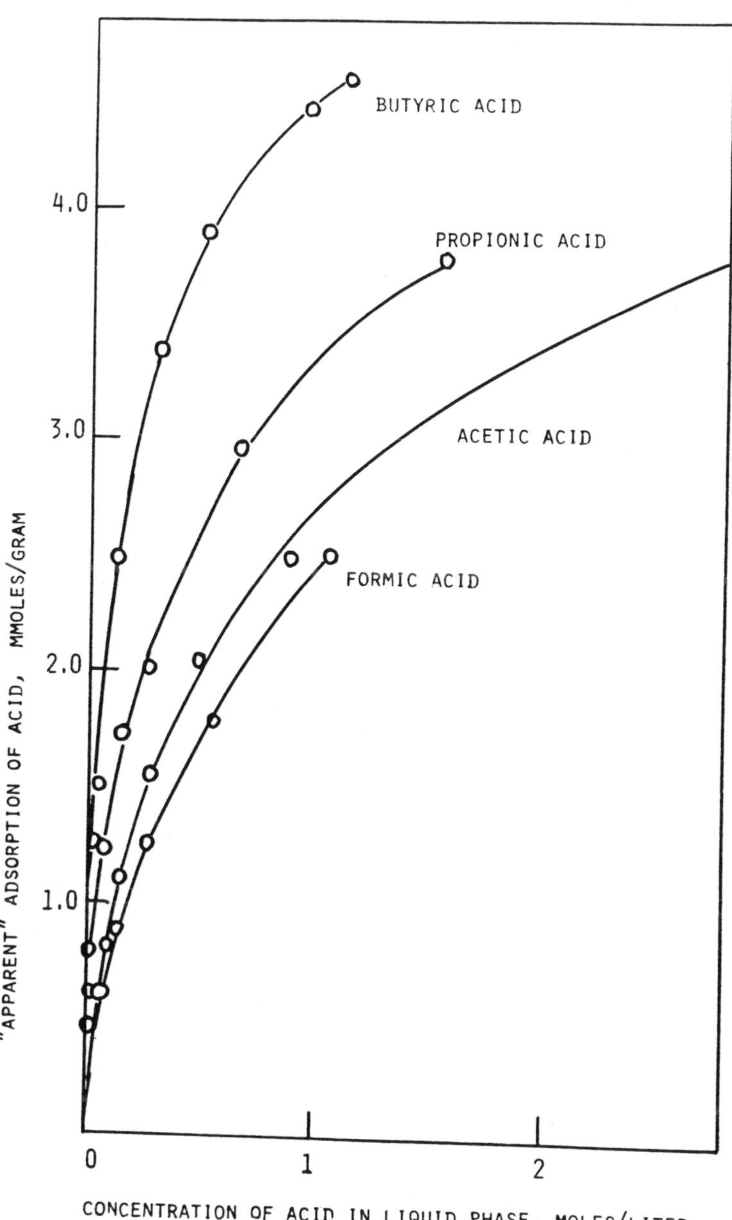

Figure 2.1. Adsorption of organic acids onto activated carbon following the Freundlich equation (Reference 1).

been made to derive a general equation for the adsorption isotherm. These have been reviewed recently by Langmuir (4) and by Jossens and coworkers (5).

Most of the theoretical models for adsorption were developed for gases adsorbed on solids. However, these models have also proven useful for liquid solutes with solid adsorbents. For any specific system of solute, solvent and adsorbent, the adsorbent's surface sites are covered. It is the relative affinity of this surface for the solute and the solvent that determines the efficiency of solute adsorption. Adsorption isotherms are used to represent the equilibrium points of solute adsorbed on the surface as a function of the concentration of solute in solution. At equilibrium, the rate of solute being adsorbed on the surface equals the rate of solute leaving the surface.

The adsorption isotherm is the starting point for any analysis of adsorption processes. The isotherm is an essential part of modeling adsorption kinetics and thus, column or batch processing design, efficiency and economics. The isotherm reveals the degree of purification that might be achieved, the approximate amount of adsorbent required to reach that degree of purity and the sensitivity of the purification process to the concentration of the solute.

2.2.1 Adsorption Isotherm

The relatively simple Freundlich Adsorption isotherm has probably been most widely used precisely because it is so simple and because it was quite successful in correlating data for adsorption on activated carbon over limited concentration ranges. It is useful to examine the derivation of this equation to appreciate its range of application.

Henry (6) has shown that the Freundlich isotherm can be derived from the Gibbs adsorption equation combined with a mathematical description of the free energy of the surface. The surface free energy, σ, when a fraction (θ) of it is covered with a solute can be expressed as:

$$\sigma = \sigma_0(1 - \theta) + \sigma_1\theta = \sigma_0 - (\sigma_0 - \sigma_1)\theta \qquad (2.2)$$

where σ_0 is the surface free energy of the surface covered with the pure solvent and σ_1 is the surface free energy of the surface covered with a monolayer of solute. The fraction of surface covered is the amount of sorbate per unit sorbent (x/m) divided by the monolayer capacity of the sorbent for the sorbate $(x/m)_M$:

$$\theta = (x/m)/(x/m)_M \tag{2.3}$$

Equation 2.2 can then be expressed as:

$$\sigma = \sigma_0 - \frac{(\sigma_0 - \sigma_1)(x/m)}{(x/m)_M} \tag{2.4}$$

The Gibbs adsorption equation forms the mathematical description of the surface excess, Γ, which is the extent by which the surface concentration of a given component exceeds its concentration in the bulk liquid. For dilute solutions, the Gibbs surface excess equation is:

$$\Gamma = -\frac{c\, d\sigma}{RT\, dc} \tag{2.5}$$

where c is the concentration of the solute in solution.

For dilute solutions, x/m is the Gibbs surface excess, Γ, so that:

$$\frac{x}{m} = -\frac{c}{RT} \cdot \frac{d\sigma}{dc} = \frac{c}{RT} \frac{\sigma_0 - \sigma_1}{(x/m)_M} \frac{d(x/m)}{dc} \tag{2.6}$$

which can be integrated to give:

$$\ln \frac{x}{m} = \frac{RT(x/m)_M}{\sigma_0 - \sigma_1} \ln c + \ln k \tag{2.7}$$

By replacing $RT(x/m)_M/(\sigma_0 - \sigma_1)$ by 1/n, Equation 2.7 becomes the familiar form of the Freundlich isotherm:

$$\frac{x}{m} = k\, c^{1/n} \tag{2.8}$$

This derivation points out that the Freundlich adsorption isotherm is limited to those cases where a dilute solution is involved, that is, where x/m = Γ. A second limitation of the Freundlich expression is that it does not predict a linear isotherm in the limit of zero concentration.

The adsorption isotherm derived by Langmuir has also been useful over limited concentration ranges. This isotherm was based upon the following assumptions: 1) all adsorption sites are equivalent, 2) only one layer of solutes or solvent is adsorbed, and 3) there is no interaction between adjacent adsorbed molecules. For liquid systems, one must also assume that all adsorption sites are occupied by either solute or solvent molecules.

The adsorption process for such a system can be written as:

$$A + SY \rightleftharpoons AY + S \tag{2.9}$$

with the equilibrium constant:

$$K_a = \frac{[AY]\,[S]}{[A]\,[SY]} \tag{2.10}$$

In this equation, A designates the solute molecules, S is the solvent molecules and Y the surface sites. Although the equation is valid only for thermodynamic activities, since the sites are assumed to be of equal activity, the ratio of AY to SY is the same as the ratio of the actual concentration of AY to SY in the adsorbent. Thus:

$$\frac{[AY]}{[SY]} = \frac{\theta}{1 - \theta} \tag{2.11}$$

where θ is the fraction of the adsorbent's surface covered by A and $(1 - \theta)$ is the fraction covered by S. Since the concentration of A is usually small compared to the concentration of S, the mole fraction of S in the liquid phase, [1 - A], is approximately 1. Thus, Equation 2.10 can be written as:

$$K = \frac{\theta\,[1 - A]}{(1 - \theta)\,[A]} = \frac{\theta}{(1 - \theta)\,[A]} \tag{2.12}$$

At low concentrations of A and small values of θ, the term θA is small and the ratio θ/A becomes nearly constant. At high concentrations of A in the liquid phase, θ approaches 1 and the concentration of A in the adsorbent comes constant. This is shown in Figure 2.2.

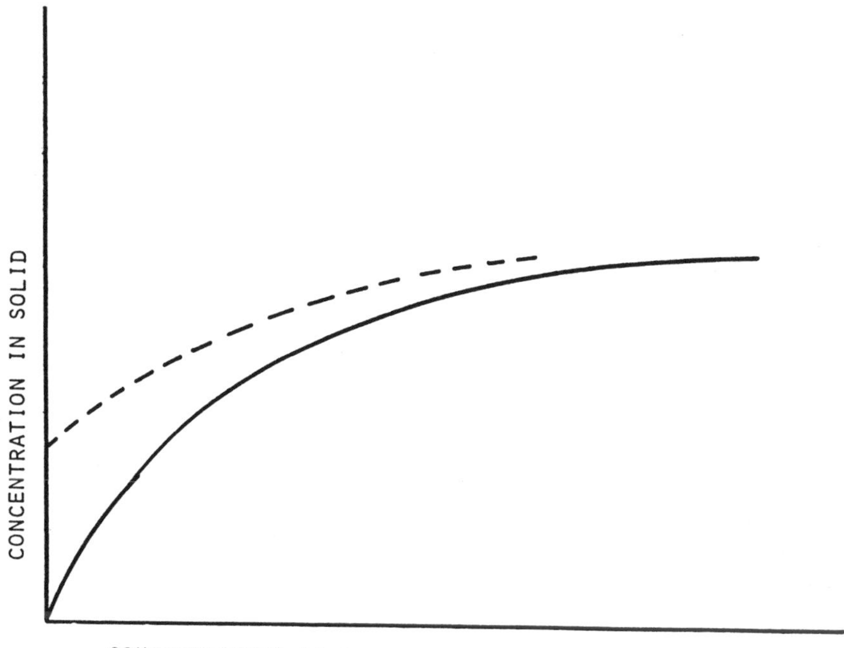

Figure 2.2. General Langmuir adsorption behavior. When the behavior follows the dashed line, the adsorption is too great for product recovery.

By replacing [A] with c and θ with q/b, Equation 2.12 can be rearranged to give the more familiar form of the Langmuir equation:

$$q = \frac{b\ K\ c}{1 + K\ c} \tag{2.13}$$

where b is the number of moles of adsorbate adsorbed per gram of adsorbent in forming a layer one molecule thick on the adsorbent surface and q = x/m, the amount adsorbed per gram of adsorbent.

The assumption concerning the equal activity of surface sites is the most common failing of the Langmuir isotherm. The Langmuir derivation lends itself to modification for those cases where the assumptions are not valid.

Beverloo and coworkers (7) classified the many adsorption equations according to the number of parameters in the model and to the "credibility" of the relation at low and high solute concentration. As measures of credibility, adsorption equilibrium relations should have the following features: (1) it should asymptotically approach a linear equation at low concentration, (2) it should also asymptotically approach a maximum value at high concentrations, and (3) the level of adsorption should not have a maximum value in the range of concentration values. The third requirement should be seen as a limit on the permissible values for the parameters in the adsorption equilibrium equations.

Table 2.1 lists the equations, the number of parameters and their applicability (credibility) at low and high concentrations. It is interesting to note that the much used Freundlich equation has a low "credibility" at low and high concentrations. Beverloo (7) recommended the use of the exponential, the Langmuir, the Volmer and the Toth equations.

The adsorption of solutes that are completely miscible in water, such as organic acids, could not be adequately represented by these isotherm equations. A Chebyshev rational approximation polynomial provided a good fit to the adsorption data over the concentration range 10^{-2} to 10^{-4}M. The Chebyshev Equation used was:

$$\frac{x}{m} = \frac{ac}{1 + bc + dc^2} \tag{2.14}$$

The fitting parameters (a, b and d) are given in Table 2.2 for the temperatures 278°K, 298°K and 323°K when the units on x/m are mmol/g and on C are mol/L.

While most adsorption isotherms follow the type of adsorption-concentration behavior shown in Figure 2.2, other profiles have been observed. Giles (9) classified these according to the system shown in Figure 2.3. The class to which the isotherm belongs is based on the initial slope of the isotherm. The sub-group is assigned according to the shape at higher concentrations.

The S class is obtained if: (1) the solvent is strongly adsorbed; (2) there is strong intermolecular attraction within the adsorbed layer; and (3) the adsorbate has a single point of strong attachment.

Table 2.1: Adsorption Equilibrium Equations(7)

Equation	Name	Parameters	Credibility Low c	Credibility High c
$q = Hc$	Henry	1	+	-
$q = A(c + \frac{1}{B})$	linear	2	-	-
$q = \frac{A}{B}\{1 - \exp(-Bc)\}$	exponential	2	+	+
$q = \frac{Ac}{1 + Bc}$	Langmuir	2	+	+
$c = \frac{q}{A - Bq}\exp(\frac{Bq}{A - Bq})$	Volmer	2	+	+
$c = \frac{q}{A}\exp(\frac{Bq}{A})$	Myers (reduced)	2	+	-
$q = A_f c^n$	Freundlich	2	-	-
$q = \{\frac{(Ac)^n}{1 + (Bc)^n}\}^{1/n}$	Toth	3	+	+
$c = \frac{q}{A}\exp(\frac{Bq}{A})^n$	Myers	3	+	-
$c = \frac{q}{A - Bq}\exp(\frac{Dq}{A - Bq})$	Volmer (extended)	3	+	+
$q = \frac{Ac}{1 + Bc}\{1 - \exp(-Dc)\}$	Langmuir extended	3	+	+
$q = \frac{Ac}{(1 + Bc)(1 - Dc)}$	B.E.T.	3	+	-
$q = \frac{A(Bc)^m}{B\{1+(Bc)^n\}}$ $(0 < n < m)$	Fritz--Schluender	4	-	-
$q = \frac{Ac(1 + Dc)}{(1+Bc)(1+Ec)}$ $(B > D > E)$	Langmuir (extended)	4	+	+

Table 2.2: Chebyshev Approximations Parameters (8)

Organic Acid	T(°K)	a	b	d
acetic	278	0.7100	2.2177	-2.4230
	298	0.4059	-2.6185	19.0550
	323	0.2761	-1.3078	10.4550
propionic	278	2.6665	-4.2880	46.592
	298	1.6768	-1.2595	12.291
	323	1.2356	-1.1009	9.453
butyric	278	13.4945	1.6677	222.072
	298	9.8098	0.4733	144.518
	323	7.5380	12.5698	-137.767
hexanoic	278	193.6	254.7	-6101.9
	298	154.1	227.7	-6454.3
	323	121.4	225.6	-7613.7

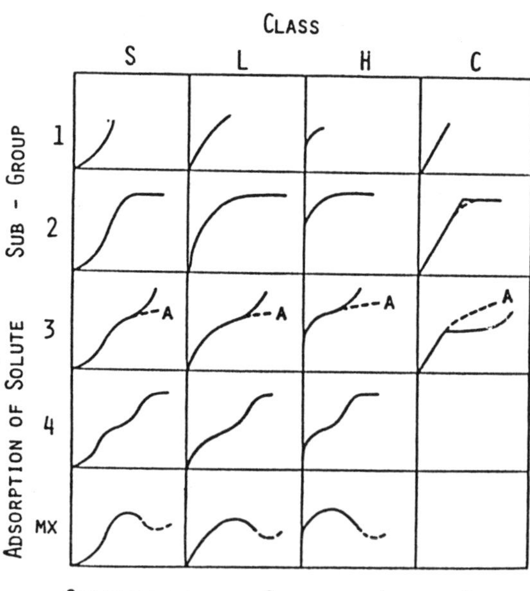

Figure 2.3. Classification system of isotherms (Reference 9).

The L class contains the Langmuir-type of curves. It is found when there is no strong competition from the solvent for sites on the surface. It may also be found if the adsorbate has linear or planar molecules with the major axis adsorbing parallel to the surface.

The H class occurs when there is high affinity between the adsorbate and the adsorbent which is shown even in very dilute solutions. This may result from chemisorption or from the adsorption of polymers or ionic micelles.

The C class indicates constant partition of the adsorbate between the solution and the adsorbent. This type of adsorption mainly occurs with textile fibers, into which the solute penetrates to increasing levels as its concentration in the solution is increased.

The steps which occur for sub-groups 3 and higher are usually interpreted to mean a phase change in the adsorbed layer or the onset of the formation of a second molecular layer after the completion of the first.

Example 2.1

The equilibrium adsorption of two amino acids, phenylalanine and glycine, on a crosslinked polystyrene adsorbent is given in Table 2.3 (10). What are the coefficients that will fit this data to a Freundlich isotherm and to the Langmuir isotherm?

Table 2.3: Equilibrium Adsorption of Phenylalanine and Glycine on Amberlite XAD-2 (10)

Phenylalanine			Glycine		
solution concentration (moles/liter)	amount adsorbed (mg/g)	amount adsorbed (moles/g)	solution concentration (moles/liter)	amount adsorbed (mg/g)	amount adsorbed (moles/g)
0.0112	10.9	6.60×10^{-5}	0.0126	0.60	7.94×10^{-6}
0.0224	19.8	1.20×10^{-4}	0.0251	1.06	1.41×10^{-5}
0.0302	26.1	1.58×10^{-4}	0.1000	4.22	5.62×10^{-5}
0.0355	29.4	1.78×10^{-4}	0.1995	8.41	1.12×10^{-4}

The coefficients for the Freundlich isotherm may be obtained by plotting the log of the amount adsorbed versus the log of the concentration. This is shown in Figure 2.4.

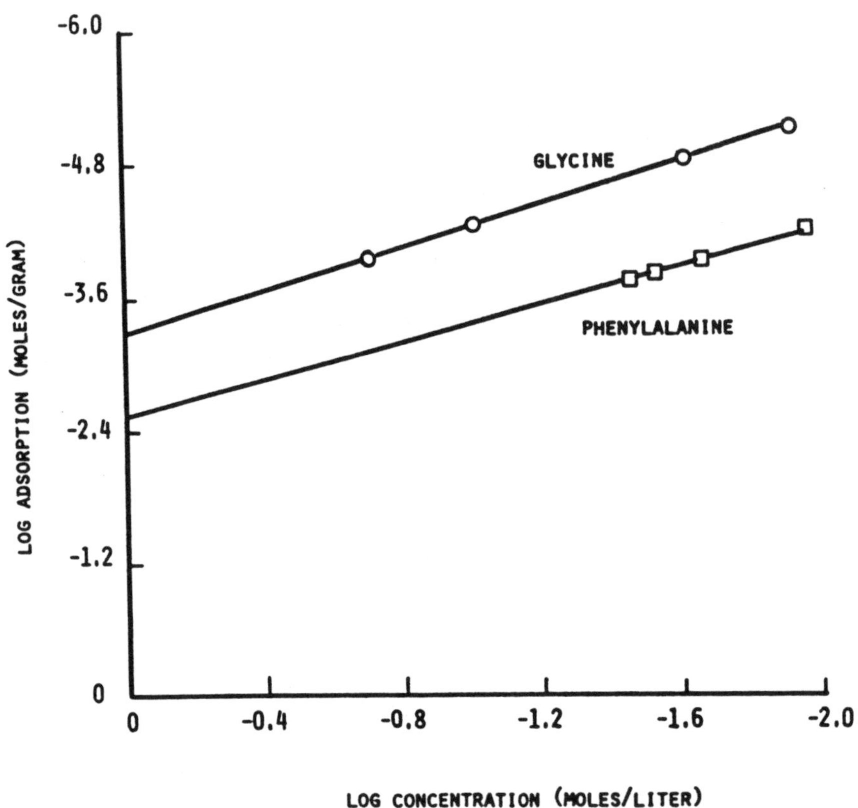

Figure 2.4. Freundich plot for the phenylalanine and glycine data from Table 2.3.

The values for k, from the intercept of the line, are 6.4 x 10^{-5} and 6.6 x 10^{-6} for phenylalanine and glycine, respectively. The values for 1/n, from the slope of the lines, are 0.884 and 0.940 for phenylalanine and glycine, respectively.

The coefficients for the Langmuir isotherm may be obtained by plotting the concentration divided by the amount adsorbed versus the concentration. This is shown in Figure 2.5. The values for b, from the reciprocal of the slope of the lines, are 8.88 x 10^{-4} and 6.48 x 10^{-5} for phenylalanine and glycine, respectively. The intercept of the lines, 1/Kb, may then be used to give the values for K, 7.167 and 11.084 for phenylalanine and glycine, respectively. Note that the slope for the glycine adsorption goes to 0 when the concentration is above 0.025 moles/liter.

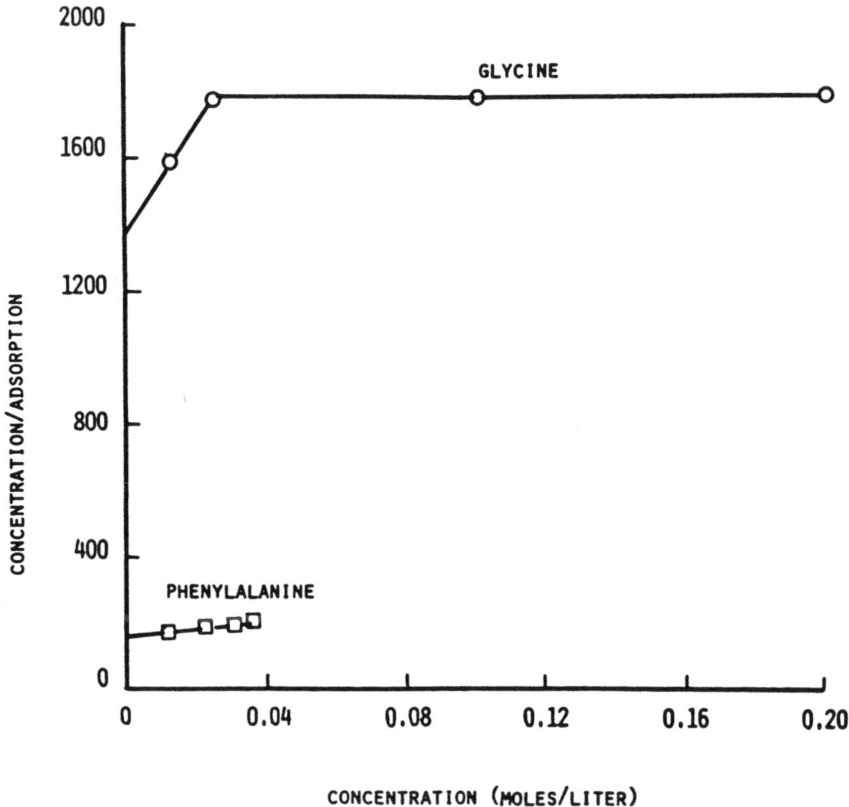

Figure 2.5. Langmuir plot for the phenylalanine and glycine data from Table 2.3.

2.2.2 Adsorption Kinetics

Adsorption of small molecules onto a non-porous solid surface normally takes place rapidly as long as the solution is not viscous. The equilibrium adsorption of butanol from an aqueous solution onto Graphon occurs within ten minutes (11). Larger molecules require a longer time to reach adsorption equilibrium. Stearic acid in a benzene solution requires four hours to reach adsorption equilibrium on iron powder, even though 90% of the equilibrium value had been adsorbed in the first five minutes (12).

Adsorption by porous solids may also be fast in those cases where the liquid-adsorbent slurry is continuously agitated. Adsorption equilibrium with aqueous solutions of mono-, di- and trichloroacetic acid onto a silica gel was attained within 1.5 to 2.5 minutes with continuous agitation, but required 12 to 30 minutes with intermittent agitation. The adsorption equilibrium with charcoal took much longer, indicating the kinetic dependence on the pore size of the adsorbent and on the molecular size of the adsorbate (13).

The adsorption of macromolecules proceeds more slowly. The adsorption of dextran by collodion was still incomplete after 50 hours. The presence of substances in the solution with dextran, such as salts and detergents, altered the adsorption rate (14).

Many models of the adsorption process have been developed which may be solved to generate detailed parametric curves using numerical techniques. These mathematical models are useful if they assist in the design of the pilot studies and in the interpretation of the pilot data so that it may be used in the engineering design of the full scale unit. A model will do this if it can predict the effluent concentration of solutes in complex mixtures and if it can predict the impact of changes in process parameters on the performance of the adsorption unit (15).

Figure 2.6 illustrates the different steps in the adsorption process which must be included or accounted for in the model. First, the solute moves from the bulk solution to the vicinity of the adsorbent material. If the adsorbent is a porous material, the solute diffuses into the fluid contained in the pores. The solute may adsorb on the exterior of the adsorbent particle or on the surface of its pores. Once adsorbed, the solute (now the adsorbate) may diffuse along the surface of the pores. The concept of surface diffusion along the surface of the pores was introduced to explain the effective diffusivities which were determined (16,17) to be higher than molecular diffusivities.

The adsorption process usually occurs much more rapidly than the diffusion through the bulk solution or than the pore or surface diffusion processes. Therefore, it is reasonable that only the diffusion processes are included in mathematical descriptions of the rate of adsorption.

The adsorption rate models reported in the literature have usually been derived or solved using one or more of these simplifying assumptions: external mass transfer resistance has been neglected, pore diffusion has been

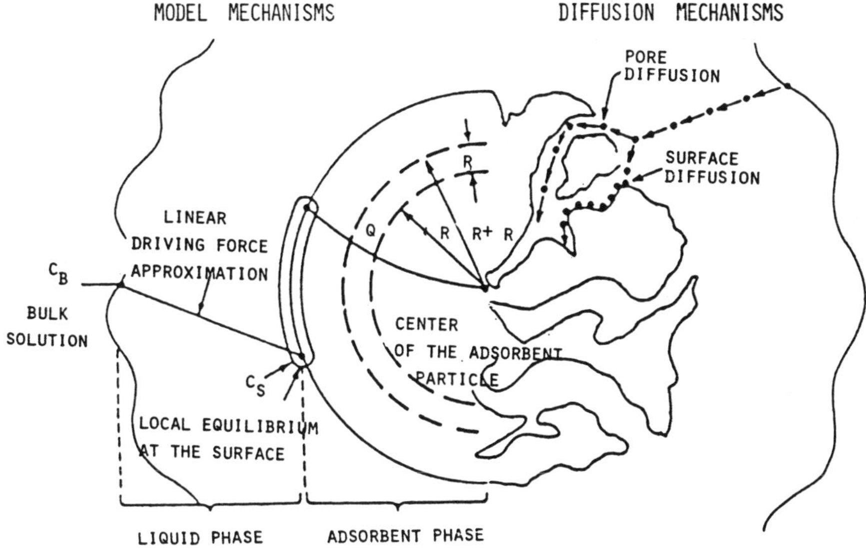

Figure 2.6. Mechanisms and schematic for the homogeneous surface diffusion model for porous adsorbents (Reference 15)

neglected, the accumulation of the solute in the liquid phase in the pores has been neglected, or a linear isotherm was used for mathematical expediency. Recent literature reviews of rate models and their limitations have been prepared by McKay (18) and by Leyva-Ramos and Geankoplis (19).

The simplest systems to be modeled are for adsorbents without porosity (7). In this case, the rate of adsorption will be dependent on the diffusion rate at which the solute is transported from the bulk of the solution to the surface of the adsorbent particles. This is normally viewed as a concentration driven process:

$$N_A = \beta_1 a (c - c_{iA}) = \varepsilon \, d\overline{q_A}/dt \qquad (2.15)$$

where N_A is the rate of adsorption: β_1 is the mass transfer coefficient through the liquid; a is the specific outer particle surface; c is the bulk solution concentration of the solute; c_{iA} is the solute concentration near the adsorbent surface; $\overline{q_A}$ is the average concentration of the solute adsorbed to the adsorbent particle; ϵ is the adsorbent fraction of the total volume; and t is the time.

For "infinite" bath experiments in which the Langmuir adsorption isotherm is applicable, Equation 2.15 can be integrated to give:

$$\frac{\beta_1 a t}{\varepsilon} = \frac{A}{1 + Bc}\left[\frac{B\overline{q}_A}{A} - \frac{1}{1 + Bc} \ln\left|1 - \frac{(1 + Bc)\,\overline{q}_A}{Ac}\right|\right] \quad (2.16)$$

where A and B are the Langmuir parameters.

For "finite" concentrations of the solute in solution, the analytical solution becomes more complex:

$$Y = \tfrac{1}{2}\left\{\ln\left[\frac{\alpha^2 + (\alpha + \beta - 1)z - 1}{2\alpha + \beta - 2}\right] + \frac{\alpha - 3\beta + 3}{((\alpha + 1)^2 + 2\beta(\alpha - 1) + \beta^2)^{1/2}}\right.$$
$$\left.\ln\left[\frac{(3\alpha + \beta - 1)z + \alpha + \beta - 3 + (z - 1)\sqrt{(\alpha + 1)^2 + 2\beta(\alpha - 1) + \beta^2}}{(3\alpha + \beta - 1)z + \alpha + \beta - 3 - (z - 1)\sqrt{(\alpha + 1)^2 + 2\beta(\alpha - 1) + \beta^2}}\right]\right\} \quad (2.17)$$

where

$$Y = \frac{\beta_1 a t}{1 - \varepsilon} \quad (2.18)$$

$$z = c/c_0 \qquad (z = 1 \text{ for } Y = 0) \quad (2.19)$$

$$\alpha = (B c_0)^{-1} \quad (2.20)$$

and

$$\beta = \frac{A \varepsilon}{B c_0 (1 - \varepsilon)} \quad (2.21)$$

A general model for porous adsorbents must include both the mass balance of the solute in the fluid phase surrounding the adsorbent particles and the mass balance of the solute within the particle. The adsorption isotherm equation is used to describe the concentrations of the adsorbate during the adsorption process. The initial and boundary conditions allow simplification of these equations for the determination of several parameters from measurable properties.

Rasmuson (20) specified the mass balance equations and boundary conditions for batch, slurry and fixed bed adsorption operations as follows:

Mass balance in the fluid phase surrounding the adsorbent:

Batch:

$$\frac{dc}{dt} = \frac{-3 N_0 (1 - \varepsilon)}{\varepsilon R} \quad \text{as} \quad \tau \to \infty \tag{2.22a}$$

Slurry:

$$\frac{dc}{dt} = \frac{1}{\tau}(c_0 - c) - \frac{3 N_0 (1 - \varepsilon)}{\varepsilon R} \tag{2.22b}$$

Fixed Bed:

$$\frac{\partial c}{\partial t} + V \frac{\partial c}{\partial z} - D_L \frac{\partial^2 c}{\partial z^2} = \frac{-3 N_0 (1 - \varepsilon)}{\varepsilon R} \tag{2.22c}$$

Mass balance inside the adsorbent particle (identical in each case):

$$\varepsilon_p \frac{\partial c}{\partial t} + N_i = \varepsilon_p D_p \frac{\partial^2 c_p}{\partial r^2} + \frac{2}{r} \frac{\partial c_p}{\partial r} \tag{2.23}$$

where

$$N_i = k_{ads}(c_p - \frac{q_A}{K_A}) = \frac{\partial q_A}{\partial t} + k_r q_A \tag{2.24}$$

Boundary conditions:

Fixed Bed	Batch and Slurry	
$c(0,t) = c_0$	-------	(2.25)
$c(\infty,t) = 0$	-------	(2.26)

Fixed Bed	Batch and Slurry	
$c(z,0) = 0$	$c(0) = c_i$	(2.27)
$\frac{\partial c_p}{\partial t}(0,z,t) = 0$	$\frac{\partial c_p}{\partial r}(0,t) = 0$	(2.28)
$c_p(R,z,t) = c_{p/r=R}$	$c_p(R,t) = c_{p/r=R}$	(2.29)

given by

$$N_0 = \epsilon_p D_p \left(\frac{\partial c_p}{\partial r}\right)_{r=R} = k_f (c - c_{p/r=R}) \quad (2.30)$$

$$c_p(r,z,0) = q_A(r,z,0) = 0 \qquad c_p(r,0) = q_A(r,0) = 0 \quad (2.31)$$

N_0 which couples Equation 2.23 with the appropriate form of Equation 2.22, is the mass flux from the exterior fluid phase to the outer surface of the adsorbent particles.

In these equations, c_p is the concentration of the solute in the pores of the adsorbent, D_L is the longitudinal dispersion coefficient; D_p is the diffusivity of the solute in the pores of the adsorbent; K_A is the adsorption equilibrium constant; k_{ads} is the adsorption rate constant; k_r is the surface reaction of rate; r is the radial distance from the center of the adsorbent particle with radius R; V is the fluid flow velocity; z is the distance along the direction of flow in the adsorption column; and ϵ_p is the void fraction of the adsorbent particle.

The two terms on the left of Equation 2.23 represent the accumulation of the solute in the liquid pore volume and of the solute adsorbed on the pore surface, respectively. The two terms on the right represent the pore volume diffusion and the surface diffusion.

The boundary conditions state that the fixed bed is initially free of solute in the fluid phase and on the adsorbent. For the batch and slurry cases, the solute in the fluid surrounding the adsorbent is initially at a concentration c_i.

These equations were solved analytically using the Laplace transform technique and inversion in the complex plane to give solutions in the form of infinite integrals. The integrals were evaluated numerically for specific values of adsorption parameters to give the breakthrough curves shown in Figure 2.7, 2.8 and 2.9 when $k_{ads} \rightarrow \infty$ and $k_r = 0$; that is, when the adsorption occurs instantaneously when the solute reaches the adsorption site and there are no chemical reactions taking place. These curves are plotted in terms of the change in concentration as a function of the contact time parameter, Y. The variables are expressed in terms of the parameters:

Contact time:

$$Y = \left(\frac{2\, D_p\, \varepsilon_p}{(K_A + \varepsilon_p)\, R^2} \right) t \tag{2.32}$$

Residence time:

$$\delta = \frac{3\, D_p\, \varepsilon_p\, z\, (1 - \varepsilon)}{R^2 \cdot \varepsilon\, V} \tag{2.33}$$

Peclet number:

$$P_e = \frac{zV}{D_L} \tag{2.34}$$

Distribution ratio:

$$D_R = \frac{K_A + \varepsilon_p (1 - \varepsilon)}{\varepsilon} \tag{2.35}$$

Diffusion resistance ratio:

$$\nu = \frac{D_p\, \varepsilon_p}{\beta_1\, R} \tag{2.36}$$

The residence time, δ, has also been called the bed length parameter since it contains the ratio of the bed length to the adsorbent particle size, with adjustments for the void space in the bed and in the particles. The particle diameter is always very small in comparison to the overall bed length; typically 0.5 cm compared with 50 to 100 cm for the packed adsorbent bed length. Therefore, values for δ will typically range from 5×10^{-2} to 1.0 with values of V between 0.001 and 0.21 m/s. The very sharp breakthrough curve of Figure 2.8 was obtained assuming $\delta = 100$ which would mean that $V = 1.0 \times 10^{-5}$, assuming one uses the same particle diameter and bed length as in Figure 2.7.

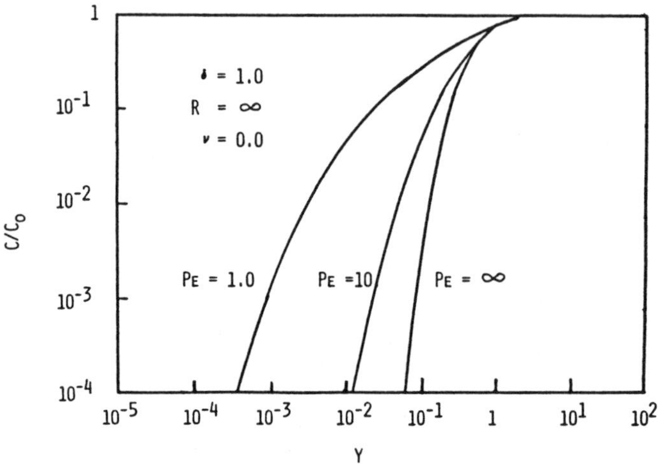

Figure 2.7. Fixed bed breakthrough curves modeled for different Peclet numbers with $\delta = 1.0$ (Reference 20).

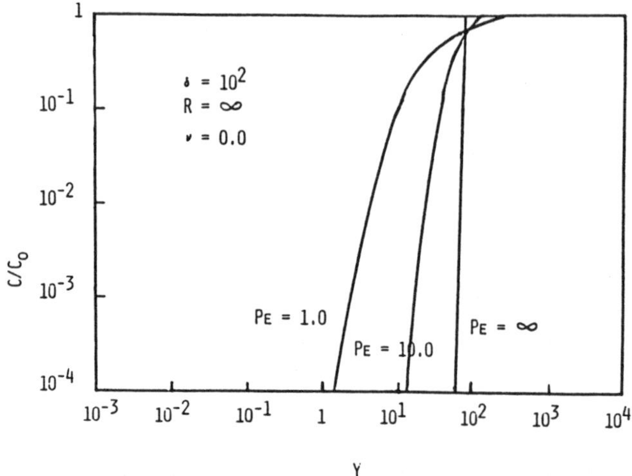

Figure 2.8. Fixed bed breakthrough curves modeled for different Peclet numbers with $\delta = 100$ (Reference 20).

The changes in the Peclet number show the influence of longitudinal or axial dispersion at different flow rates. In this instance, lower values of V correspond to lower Peclet numbers which show shallower breakthrough curves. This behavior is the result of the longitudinal molecular diffusion causing increased spreading of the solute concentration during the increased residence time of the lower flow rates. Longitudinal diffusion is known to arise for Peclet numbers less than 1.0. If longitudinal dispersion can be neglected, then P_e becomes equal to ∞ .

The distribution ratio, D_R, gives a measure of the accumulation of the solute on the adsorbent and in the void spaces both of the bed and of the adsorbent particles. If there is negligible accumulation in the void spaces, $D_R \rightarrow \infty$. Figure 2.9 shows that $D_R = \infty$ and $D_R = 10^6$ have nearly identical breakthrough curves. As the value of δ increases, the breakthrough curves for lower values of D_R will coincide with the curve for $D = \infty$. For example, at $\delta = 1.0$, the breakthrough curve of $D_R = 10^3$ coincides with that of $D_R = \infty$.

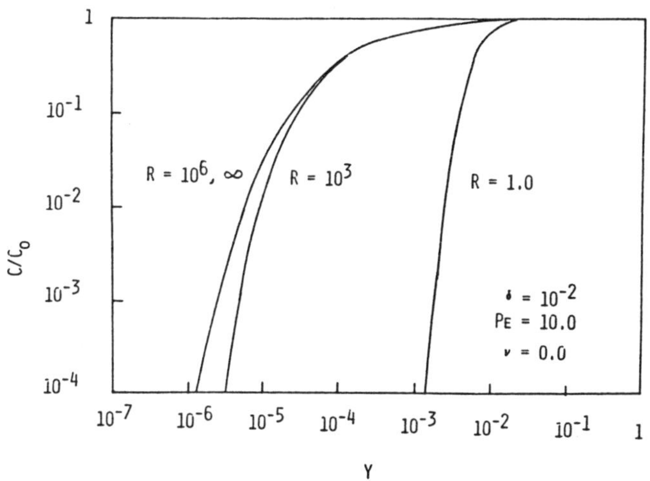

Figure 2.9. Fixed bed breakthrough curves modeled for different distribution ratios (Reference 20).

The parameter ν is the ratio of the external diffusion resistance to the internal diffusion resistance. There are few circumstances for fixed bed operations where ν would be greater than 10^{-2}. Simulation results (20) have shown that the breakthrough curves for those cases when $\nu = 10^{-2}$ are very similar to those when $\nu = 0$.

The experimental values for the pore diffusivity of the solute, D_p, can be determined from $\epsilon_p D/\tau$, where D is the molecular diffusivity and τ is the tortuosity factor of the adsorbent pores. A typical value of τ is 3.6. A typical value for D_p in a system with activated carbon as the adsorbent is 2.5×10^{-10} m^2/s (19). For other systems, the value will generally not vary from this by more than a factor of 2 or 3.

The mass transfer coefficient, β_1, will typically have values around 0.0001 m/s. The range of values will be from 0.0001 up to 0.0002 m/s. The mass transfer coefficient in packed beds at low Reynolds numbers, 0.5 to 900, has been calculated from the equation (21):

$$\beta_1 = 1.15 \frac{D_{mol}}{2R} \left(\frac{2 R \rho u_o}{\mu} \right)^{1/2} \left(\frac{\nu}{D_{mol}} \right)^{1/3} \quad (2.37)$$

where u_o is the solute solution velocity (m/s); μ is the kinematic viscosity and ρ is the density. The molecular diffusivity, D_{mol}, can be calculated from: (22)

$$D_{mol} = 7.4 \times 10^{-8} \frac{T M^{0.5}}{\mu_D V_m^{0.6}} \quad (m^3/sec) \quad (2.38)$$

where V_m is the molecular volume of the solute; μ_D is the dynamic viscosity; T is the temperature (°C) and M is the molecular weight of the solute.

Example 2.2

How long does it take for the relative concentration, c/c_0, to rise to a breakthrough value of 0.90 of the solute concentration in the feedstream, assuming $\delta = 1.0$ and $P_e = 100$? What flow rate is needed, in this instance, when the bed length is one meter?

With an adsorption equilibrium constant equal to 1.5×10^3 m^3/m^3, a particle radius of 0.05 cm, a particle porosity of 0.43 and a pore diffusivity of 2.5×10^{-10} m^2/s, the time to breakthrough is 1.74×10^4 seconds or 4.8 hours. With a void fraction of 0.38 in the bed, the flowrate through a one meter adsorption bed during this 4.8 hours should be 2.1 x 10^{-3} m/s.

For specific cases that are not modeled by the curves in Figures 2.7 to 2.9, McKay (18) has provided a solution that may be used fairly easily to calculate the concentration distribution of a solute within the fluid and solid phase in fixed beds of adsorbents. His derivation was based upon the work proposed by Spahn and Schlunder (16) and by Brauch and Schlunder (23).

The theoretical adsorption rate can be determined using the following equation and experimental values obtained from batch tests:

$$\frac{d\eta}{d\tau} = \frac{3\,(1 - C_h\,\eta)\,(1 - \eta)^{1/3}}{1 - (1 - 1/B_i)(1 - \eta)^{1/3}} \tag{2.39}$$

in which C_h is the column capacity factor; B_i is the Biot number; η is a dimensionless solid phase concentration parameter; and τ is a dimensionless time parameter.

$$C_h = \frac{M\,q_e^h}{V_s\,c_0} \tag{2.40}$$

$$B_i = \frac{\beta_1\,R}{D_{eff}} \tag{2.41}$$

$$\eta = \bar{q}\,/\,q_e^h \tag{2.42}$$

$$\tau = \frac{c_0}{\rho\,q_e^h}\,\frac{D_{eff}}{R^2}\,t \tag{2.43}$$

where M is the mass of adsorbent; V_s is the solution volume; D_{eff} is the effective diffusivity; and q_e^h is the concentration of solute adsorbed at equilibrium.

An analytical solution for τ allows D_{eff} to be calculated:

$$\tau = \frac{1}{6\,C_h}\left[\ln\left(\left|\frac{x^3 + a^3}{1 + a^3}\right|^{2B - 1/a}\right)\right] + \ln\left(\left|\frac{x + a}{1 + a}\right|^{3/a}\right)$$
$$+ \left[\frac{1}{(a\sqrt{3})\,C_h}\left(\tan^{-1}\frac{2 - a}{a\sqrt{3}} - \tan^{-1}\frac{2x - a}{a\sqrt{3}}\right)\right] \tag{2.44}$$

where

$$B = 1 - 1/B_i \tag{2.45}$$

$$x = (1 - \eta)^{1/3} \tag{2.46}$$

$$a = \left(\frac{1 - C_h}{C_h}\right)^{1/3} \tag{2.47}$$

The effective diffusivity describes the concentration versus time decay curve over the whole range of the isotherm for which it applies. In the instance cited by McKay, for an initial solute concentration of 450 mg/dm, the equation was valid over the time interval 5 to 1440 minutes.

The column may be divided into two sections: the section where a constant pattern of adsorption exists and the section which is in the transitional state. The concentration of the solute in solution as a function of position and time can then be given by:

for the constant pattern region:

$$\frac{c_t}{c_0} = \frac{1}{6}\ln\frac{(1 - x_2^3)^{2B}(x_2^2 + x_2 + 1)}{(1 - x_2)^2} - \frac{1}{\sqrt{3}}\tan^{-1}\frac{2x_2 + 1}{\sqrt{3}}$$
$$- \frac{1}{6}\ln\frac{(1 - x_0^3)^{2B}(x_0^2 + x_0 + 1)}{(1 - x_0)^2} + \frac{1}{\sqrt{3}}\tan^{-1}\frac{2x_0 + 1}{\sqrt{3}} \tag{2.48}$$

where:

$$x_0 = \left[1 - \eta(0,\tau)\right]^{1/3} \tag{2.49}$$

$$x_2 = \left[1 - \eta(0,\tau)\ \xi\right]^{1/3} \tag{2.50}$$

for the transition region:

$$\frac{c_t}{c_0} = \frac{1}{6} \ln \frac{(1 - x_2^3)^{2B}(x_2^2 + x_2 + 1)}{(1 - x_2)^2} - \frac{1}{\sqrt{3}} \tan^{-1} \frac{2x_2 + 1}{\sqrt{3}}$$

$$- \tau_1 - \tau - \frac{\pi}{18}\sqrt{3} \qquad (2.51)$$

where

$$\tau_1 = \frac{1}{2} - \frac{B}{3} \qquad (2.52)$$

and ξ is a dimensionless liquid phase concentration parameter.

Example 2.3

Using the Equation 2.48 and 2.51, the breakthrough curves shown in Figure 2.10 were calculated. The flow rate was 2 cm^3/s; the column height was 200 cm; the particle size was 600 microns; the equilibrium adsorbent capacity was 0.2 g/g; and the effective pore diffusion coefficient was 2.0 x 10^{-10}m^2/s. As the initial concentration was changed from 350 mg/dm^3 to 450 mg/dm^3 to 550 mg/dm^3, the shape of the breakthrough changes and is shifted to the left, indicating earlier breakthrough times.

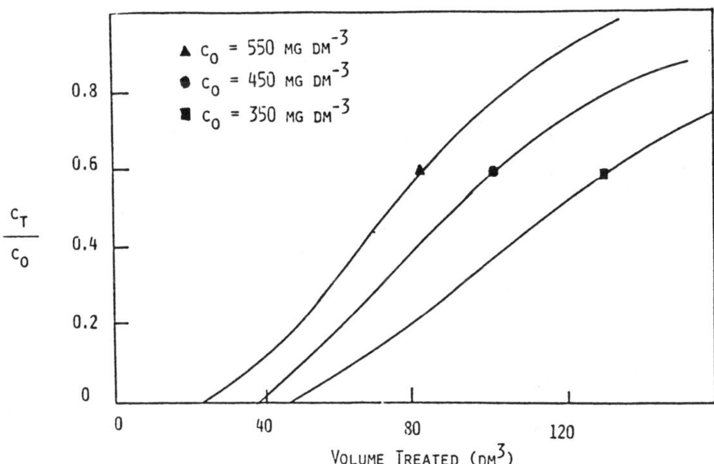

Figure 2.10. Effect of initial solute concentration on the breakthrough profile using a fixed bed height of 0.20 m (Reference 18).

If the maximum allowable concentration of solute in the effluent is 35 mg/dm^3, what is the maximum volume of each feedstream that can be treated under these conditions?

This would represent c_t/c_o values of 0.100, 0.078 and 0.064 for the three feedstreams. Therefore, the volume treated would be 66 dm^3, 49 dm^3 and 30 dm^3. Note that this corresponds to different amounts (23.1 g, 21.6 and 16.5, respectively) of solute being adsorbed with the same amount of adsorbent, same flow rate and same bed height.

While the previous techniques used mass transfer coefficients to model the resistance a molecule encounters as it transfers from the bulk solution until it is eventually adsorbed on the surface of an adsorbent particle, it is also possible to model column behavior using the mass transfer zone approach (24). Figure 2.11 shows a schematic representation of the adsorbate loading on the column as a function of time and position. Prior to the introduction of the feedstream, the adsorbate concentration (q) on the adsorbent is q_0 or zero. Once the feedstream is introduced at time t_1 at a flow rate of u, the adsorption process is underway.

The flow of the feedstream enters the column at a constant concentration, c_0, and continues through the column under steady-state (constant temperature and pressure) conditions. By time t_2, the adsorbate concentration on the first portion of the bed has attained its equilibrium loading, q_e. That portion of the bed whose adsorbate loading equals q_e is called the Equilibrium Zone. The portion of the bed where the adsorbate concentration remains at q_0 is called the Unused Zone. Between these two zones is the Mass Transfer Zone, where the solute is being transferred from the bulk feedstream to the adsorbent. This normally has the "S-shaped" curvature noted at time $t + \Delta t$. Under steady state conditions, this Mass Transfer Zone moves steadily down the column, retaining the same "S-shaped" curvature.

Eventually the leading edge of the Mass Transfer Zone reaches the end of the column. This is termed the breakthrough time, $t = t_b$. Further flow results in the entire bed reaching the equilibrium adsorbate concentration at time $t = t_e$.

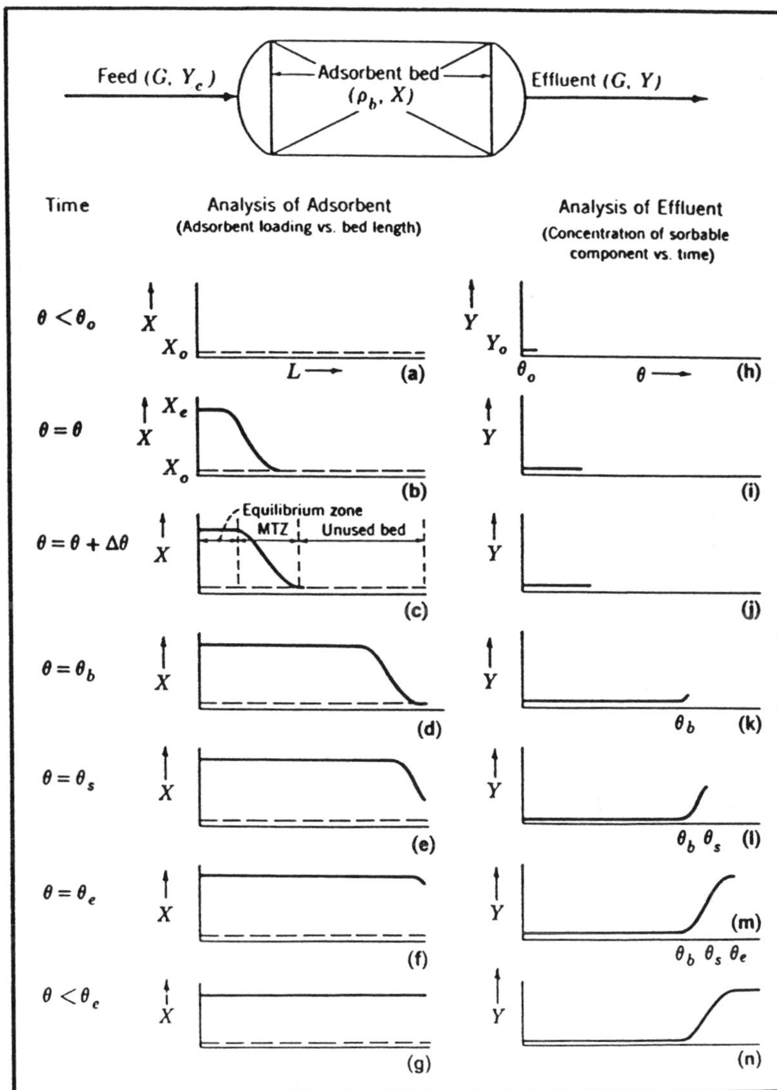

Figure 2.11. Schematic of mass transfer onto adsorbent and solute concentration in the effluent (Reference 24).

Rather than analyze the adsorbate concentration change in the bed, it is more usual to analyze the effluent from the column as a function of time. The initial effluent is at a solute concentration, c_s, equal to zero, at $t = t_1$. It is not until $t = t_b$ that the solute concentration begins to increase.

The breakthrough concentration, however, does not necessarily occur at t_b. It occurs at a more arbitrarily set time, t_s, which may be either the maximum allowable or the minimum detectable concentration of the solute in the effluent.

The effluent concentration goes through the same S-shaped increase as it reaches the equilibrium inlet solute concentration as the adsorbate concentration went through the S-shaped decrease. The relative length of these mass transfer waves represents the relative strength of the resistance to mass transfer.

This approach can be formulated into mathematical equations for the simple estimation of adsorbent column operation. It should be noted that the model requires that the adsorbent bed is uniformly packed; that the feed rate, feed temperature and the adsorbent temperature are constant; that the feed temperature and the adsorbent temperature are the same; that the flow is uniformly fed to the top of the bed with no channeling in the bed; that the fluid does not undergo a phase change; and that no chemical reactions occur in the bed.

Under these conditions, the total amount of solute, m, removed from the feedstream between time t and time t + Δt is given by:

$$m = u\,(C_e - C_0)\,A_B\,\Delta t \qquad (2.53)$$

where A_B is the cross-sectional area of the adsorbent bed.

As m kg of solute is adsorbed, the shape of the mass transfer wave remains the same while its position changes. The Equilibrium Zone has not been increased by a length ΔL:

$$m = (\Delta L)\,A_B\,\rho_B\,(q_e - q_0) \qquad (2.54)$$

Therefore,

$$\Delta L = (u\,\Delta C\,\Delta t)/(\rho_B\,\Delta q) \qquad (2.55)$$

During the steady state operation, the values of u, ρ_B, ΔC and Δq are fixed, so that $u\Delta C/(\rho_B\,\Delta q)$ can be represented by a constant U, which represents the velocity at which the mass transfer wave moves through the bed. While the local mass transfer rates may affect the shape of the front, the velocity of the front under steady state conditions is only a function of the space velocity, the bulk density of the adsorbent and the boundary conditions.

At any time, t, the position of the breakthrough concentration in the bed is given by:

$$L_s = U\,t \tag{2.56}$$

At the breakthrough time, t_b, the equilibrium concentration is a distance L_b from the end of the column, L_0. That is:

$$L_b = U\,t_b \tag{2.57}$$

Likewise, the equilibrium concentration reaches L_0 at the time t_s:

$$L_0 = U\,t_s \tag{2.58}$$

The length of the unused bed, L_u, of the adsorption column is given by:

$$L_u = L_0 - L_b = U\,(t_s - t_b) \tag{2.59}$$

U can be eliminated to give:

$$L_u = L_0\,(1 - t_b/t_s) \tag{2.60}$$

An alternate form of this equation is:

$$L_u = L_0 - \left(\frac{u\,\Delta C}{\rho_B\,\Delta q}\right) t_b \tag{2.61}$$

which allows the estimation of the unused portion of the adsorbent when the adsorption process is stopped at t_b. Figure 2.12 shows the change in the amount of adsorbent used to capacity as the adsorption column is changed relative to the length of the mass transfer zone. For processes which are to remove impurities, the bed length should be between 1.5 and 3 times the size of the mass transfer zone.

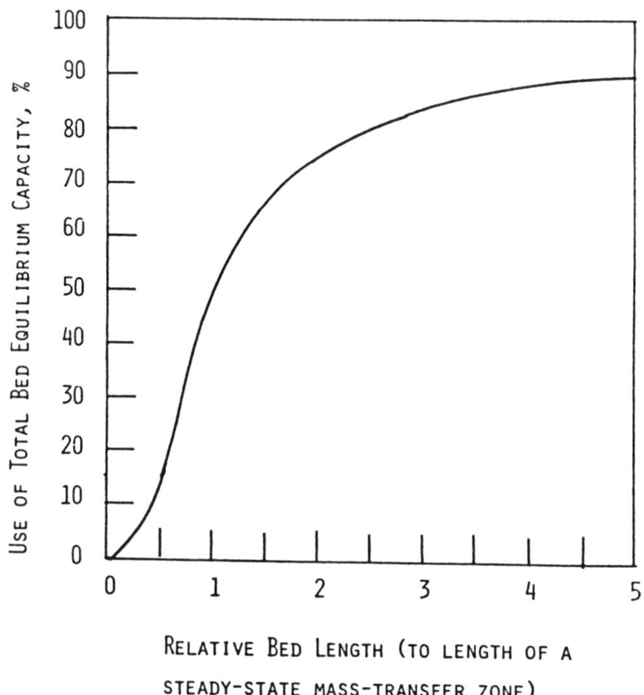

Figure 2.12. Comparison of bed use as a function of increasing bed length (Reference 24).

The particle size distribution of the adsorbent has been shown (25) to be of importance primarily when film diffusion contributes significantly to the rate of adsorption or for short contact times (batch) or short column lengths (fixed bed). Film diffusion contributions are significant mainly when the particle size distribution is skewed toward smaller particles. For particles with Gaussian distributions of moderate standard deviations, the effect is quite small.

Nearly identical breakthrough curves are calculated (26) for spherical, cylindrical and slab adsorbent particles with the same surface-to-volume ratio when the column contact time is either quite short or quite long. Some differences are noted in the intermediate time domain. However, they are of minor importance for engineering considerations of commercial adsorption column design and operation since the calculations always include margins for error.

Example 2.4

Figure 2.13 shows three breakthrough curves that have different lengths in their mass transfer zones. If the allowable concentration of the solute in the effluent is 0.2 C_0, what is the degree of column utilization?

In the first case, the mass transfer zone is quite small so that the column achieves a high degree (~90%) capacity utilization. In the second case, the mass transfer zone is longer and only 60% of the adsorbent bed utilized. The very long mass transfer zone of the third case allows only 30% utilization of the adsorbent bed before breakthrough is reached.

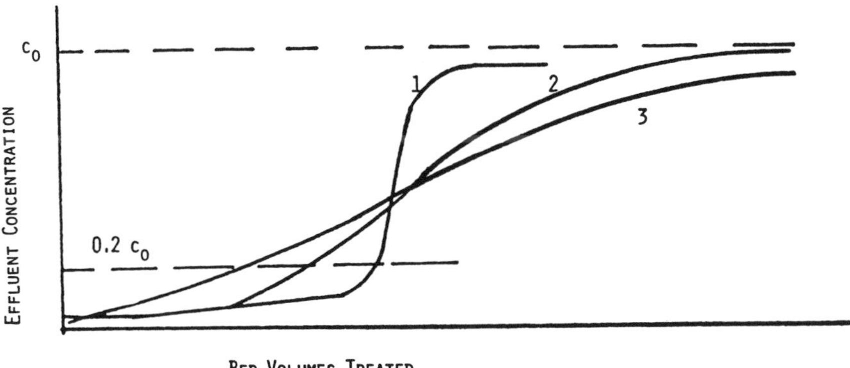

Figure 2.13. Column exhaustion curves for systems with three different lengths of mass transfer zones.

2.3 TYPES OF ADSORBENTS

The usefulness of an adsorbent in a purification process is a function of its composition, the presence and type of functional groups at the surface, its porosity and surface area, its degree of polarity and its relative hydrophilicity/hydrophobicity. The adsorbents which have been and will continue to be most useful in biotechnology purifications are activated carbon, oxides of silicon and aluminum, and crosslinked organic polymers. These materials, their generally accepted surface functionality and surface areas are listed in Table 2.4.

Table 2.4: Surface Structure and Area of Adsorbents

Adsorbent	Structure	Specific Surface Area (m^2/g)
Bone Charcoal	C–O–C	60 - 80
Granular Charcoal	C–O–C	500 - 200
Silica Gel	SI–O–SI	600
Zeolites	AL–O–SI–O–MG	400
Polystyrene	—CH—CH_2—CH—CH_2— (each CH bearing a benzene ring)	100 - 700
Poly(methylmethacrylate)	—C(CH_3)($COOCH_3$)–CH_2—C(CH_3)($COOCH_3$)–CH_2—	100 - 500

Bone chars, natural and synthetic, are sized granules used almost exclusively for decolorizing sugar. Activated carbons are sized granules derived from the pyrolysis of wood or coal. Activated alumina is a synthetic, porous crystalline gel formed into spherical pellets or granules.

Silica gel is a synthetic, porous non-crystalline gel formed into spherical pellets or granules. Zeolite-type aluminosilicates (molecular sieves) are unique in that their synthesis produces a pore volume of almost 100% uniform size. The most common synthetic macroporous polymer beads used in adsorption are copolymers of styrene-divinylbenzene, methacrylate-divinylbenzene or phenol-formaldehyde.

Although it is possible to analyze the chemical constituents of these adsorbents, it is difficult to characterize the chemical nature of their surfaces. In general, polar adsorbents contain surface groups capable of hydrogen bonding with adsorbed materials. This is probably the single most important property of this type of adsorbent. Non-polar adsorbents have no hydrogen bonding properties and London dispersion forces become the most important in the adsorption mechanism. Synthetic polymers may have functional groups to enhance their adsorptive strength.

Since adsorption occurs on the surface of the adsorbent, the total surface area is an important parameter. If the surface of the adsorbent were completely uniform, the amount of surface covered would be a function of the shape, but not the size, of the solute. However, the capacity of activated carbon to adsorb solutes decreases as the solute size increases, indicating the fine pore structure inside the carbon particles.

If the adsorbent did not have pores throughout its structure to provide additional adsorption sites, the maximum amount of solute that could be adsorbed would increase as the particle size decreased. This has the practical limitation caused by excessive pressure drop across a packed column of the adsorbent when the particle size becomes too fine. The surface area of a non-porous adsorbent is shown as a function of particle size in Figure 2.14. Most adsorbents used in industrial purification processes have surface areas greater than 100 m^2/g and particles ranging from 150 to 1500 microns in size. Therefore, a large portion of the adsorbent's surface area must consist of a pore network inside each adsorbent particle. The size and uniformity of the pores affect the adsorption of large molecules. The effective surface area available for adsorption should be expected to decrease as the molecular size of the adsorbed solute increases.

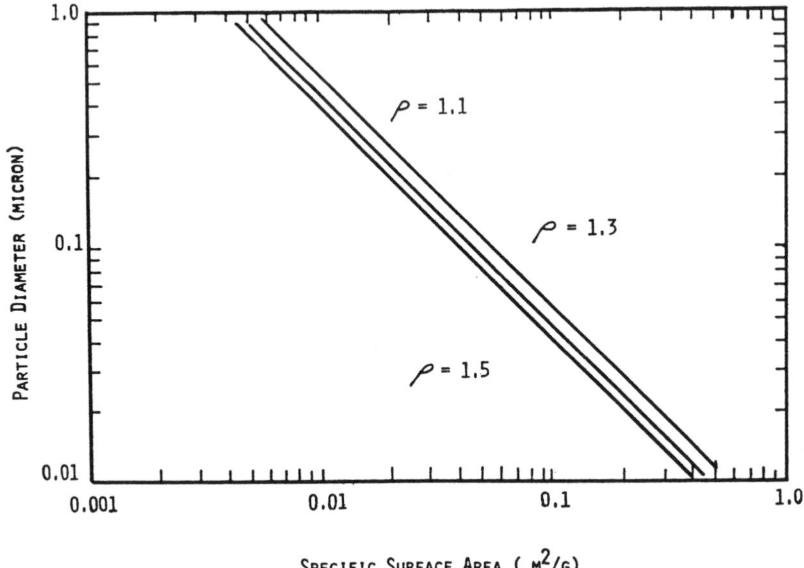

Figure 2.14. The surface area of a non-porous adsorbent as a function of particle size at three different densities.

The surface area of adsorbents is usually measured by degassing the adsorbent under high vacuum at elevated temperatures, followed by the addition of a known quantity of gas, such as nitrogen or helium. The final pressure of the chamber containing the adsorbent is related to the surface area of the adsorbent (S) using the Brunauer, Emmett and Teller (BET) equation (27):

$$S(m^2/g) = \left[\frac{v\,(p^o - p)}{c\,p}\left(\frac{p^o + (c-1)p}{p^o}\right)\right] A\, a_N\, (10^{-10}\ m/A)^2 \qquad (2.62)$$

where the first term is equal to V_m, the volume of gas adsorbed when the entire adsorbent surface is covered with a complete monomolecular layer; v is the volume of gas adsorbed; p is the measured pressured; p^o is the vapor pressure of the gas at experimental temperature; c is a constant; A is Avogadro's number; and a_N is the area occupied by a single gas molecule (16.2 Å^2 for nitrogen).

Abrams (28) has pointed out the danger in relying upon high surface area (as measured by BET) in comparing resins with inorganic or charcoal adsorbents for use in purification processes. As Table 2.5 shows, a gel resin with a surface area of only 1.4 m^2/g had the highest relative adsorption. The gel resin swells in water to form an open-lattice which readily allows diffusion of the organic impurities from the pond water into the particle for adsorption. BET experiments on the dry particle would not have shown these lattice openings. The relatively poor adsorption with the carbon is most likely due to its fine pore size not allowing the organic impurities to the interior adsorption sites. The adsorption performance of the macroporous adsorbent resin is limited by its lack of anion functional groups.

Table 2.5: Influence of Surface Area on Relative Adsorptivity(28)

Adsorbent	Composition	Surface area (m^2/g)	Relative adsorptivity
Nuchar CEE	Carbon	740	1.0
Duolite S-30	Phenol-formaldehyde	128	1.3
ES-140	Styrene-divinylbenzene	110	2.4
Duolite A-7D	Phenol-formaldehyde	24	3.4
Duolite A-30B	Epoxy-amine	1.4	5.0
ES-111	Styrene-divinylbenzene	1.0	2.8

It is generally held that one should use the adsorbent with the highest surface area having a suitable polarity, with consideration for the size of the molecule to be adsorbed. Since the average pore diameter of the adsorbent decreases as its surface area increases, the adsorption of large molecules in a reasonably rapid process necessitates the use of adsorbents with larger pores and thus less surface areas.

2.3.1 Carbon and Activated Charcoal

Commercially available carbons are characterized by having high surface areas which are covered with combined hydrogen and oxygen. These surfaces are likely to consist of reactive polar groups. It was suggested (29) in 1959 that the surface of these carbons may have carbonyl, hydroxyl, carboxyl and lactone groups. Measurements (30) on commercial activated carbons showed that they had carboxylic acid capacities ranging from 0.30 to 0.90 meq/g and phenolic hydroxyl capacities from 0.05 to 0.07 meq/g.

The raw material selected for the preparation of activated charcoal has a definite effect on the adsorbent's properties. Wood, coconut, lignite, bituminous coal, subbituminous coal, petroleum acid and sludge activated coal have been used as raw materials. Table 2.6 shows the test parameter values for activated charcoal prepared from these different materials.

Table 2.6: Characteristics of Activated Carbon Prepared from Different Raw Materials(34)

Raw Material	Iodine Number	Molasses Number	CCl_4 Number	Pore Volume (cc/g)
Wood	1230	470	76	0.57
Coconut	1350	185	63	0.49
Lignite	550	490	34	0.23
Bituminous Coal	900-1000	200-250	60	0.45
Subbituminous Coal	1050	230	67	0.48
Petroleum Acid Sludge Coke	1150	180	59	0.46

The tests used to arrive at these numbers are the industry standards for carbons and activated charcoal. The iodine number (31) can be roughly correlated with the surface area of the charcoal, which is an indication of its capacity to adsorb small to medium molecules. The molasses number (32) can be correlated with the rate of adsorption of large molecules such as occurs in decolorization processes. The carbon tetrachloride number (33) is primarily of interest for those using activated charcoal for vapor phase adsorption. The higher the number, the greater the amount of an organic vapor which can be expected to be adsorbed per unit weight of charcoal. The determination of the pore volume as a function of the pore size of the adsorbent is a useful indication of the molecular size which may be adsorbed by the charcoal. Figure 2.15 shows such a plot for charcoals prepared from different raw materials with butane as the penetrant.

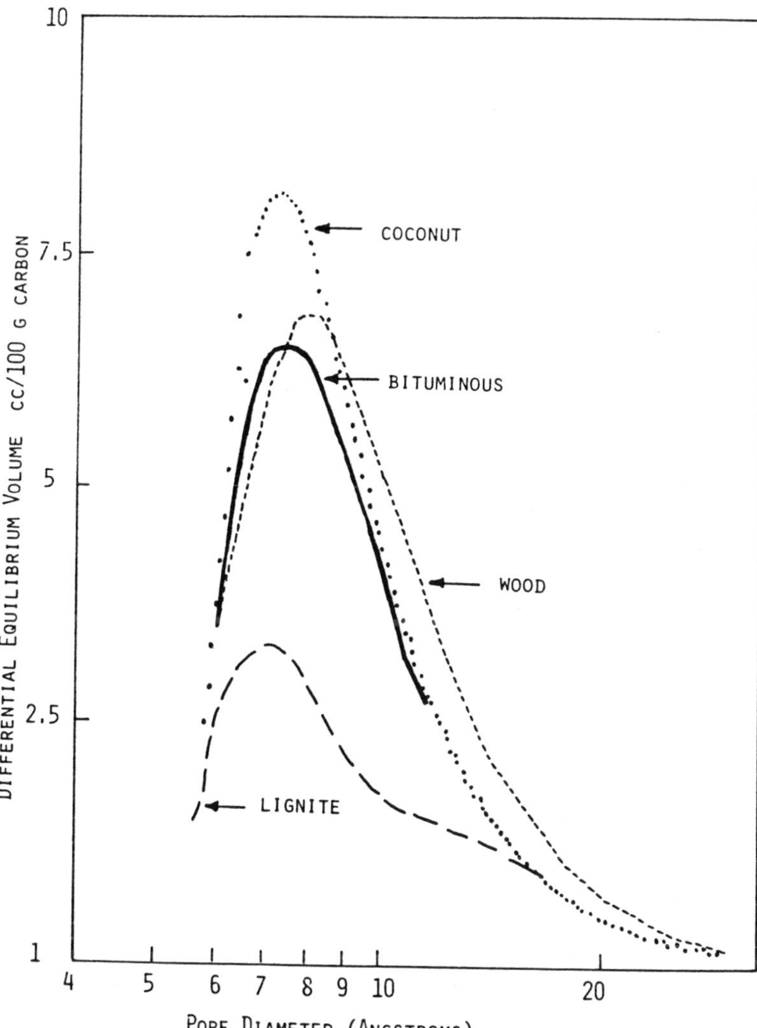

Figure 2.15. Pore volume as a function of pore size for carbons from different sources (Reference 30).

Higher molecular weight dissolved organic molecules adsorb on activated charcoal in preference to lower molecular weight molecules. Also, adsorption is favored for molecules which are non-polar. Organic compounds such as methanol, which are both polar and possess low molecular weight, adsorb very poorly and are easily displaced by other organic compounds. Aromatic, non-polar compounds such as toluene are effectively adsorbed and held by the activated charcoal until it is regenerated.

The pH of the water and the solute's solubility in the solvent are also important in its adsorbability on activated charcoal. The presence of additional solutes will also affect the adsorption characteristics for the removal of a specific contaminant. In general, however, the molecular weight of the solute is the primary factor which controls the feasibility for its adsorption. Molecules with three or more carbon atoms may usually be removed by adsorption on activated charcoal.

There is usually an upper limit of molecular weight above which adsorption decreases with further increases in molecular weight. This happens as the solute approaches polymer size (molecular weight > 600). At this point, the molecule is so large that its diffusion into the internal pore structure of activated charcoal is inhibited.

In general, bituminous coal based granules have a high surface area and small pores which make them suitable for adsorbing lower molecular weight non-polar solutes. Lignite based granules have a much lower surface area but larger pores which make them better for larger molecular weight solutes. Table 2.7 lists several of the common carbon and activated charcoal adsorbents.

New developments continue to be made in the processing of raw materials to obtain modification in the structure of the charcoal so that it will be more suitable for a specific application. Table 2.8 (34) shows the range of values coal-based adsorbents can have. It should be noted that the primary difference between the charcoals used in liquid or vapor applications is in the diffusion rates of the solutes through the charcoals which are possible. In biotechnology purifications, activated charcoal will be used almost exclusively in liquid media.

2.3.2 Silica Gels, Aluminas and Zeolites

Silica gels can be considered to be an amorphous, inorganic condensation polymer of silicic acid. Silica gel, a

Table 2.7: Commercial Activated Carbons

Company	Brand Name	Source	Surface Area (m^2/g)
American Norit Company	Norit (various)	Wood	700 - 1400
Atlas Chemical Industries	Darco		
	S 51	Lignite	500 - 550
	G 60	Lignite	750 - 800
	KB	Wood	950 - 1000
	Hydro	Lignite	550 - 600
Calgon Corporation	Pittsburgh	Bituminous Coal	
	PCC SGL		1000 - 1200
	PCC BPL		1050 - 1150
	PCC RB		1200 - 1400
	PCC GW		800 - 1000
Union Carbide	Columbia	Coconut Shell	
	CXA/SXA		1100 - 1300
	AC		1200 - 1400
	G		1100 - 1150
Westvaco	Nuchar	Pulp Mill Residue	
	Aqua		550 - 650
	C		1050 - 1100
	(various)		300 - 1400
Witco	Witco	Petroleum Acid Sludge Coke	

Table 2.8: Comparison of Activated Carbon Used in Different Applications (34)

Application	Iodine Number	Molasses Number	CCl_4 Number	Pore Volume (cc/g)
Liquid Phase	1080	251	65.6	.472
Vapor Phase	1065	209	63.4	.447
Evaporative Loss	1200	400	90.1	.62

typical polar adsorbent, contains four main types of functionality at its surface:

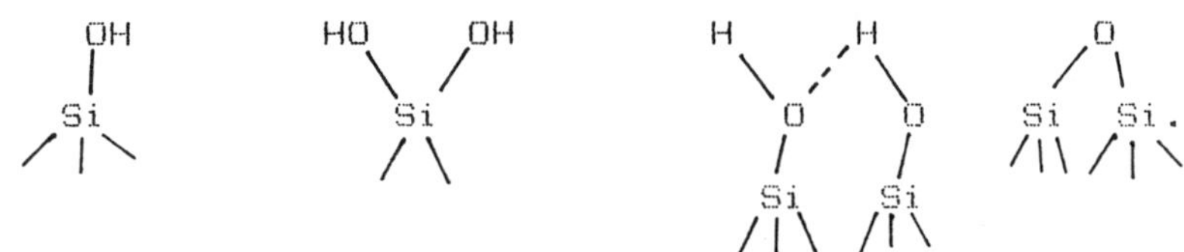

The formation of hydrogen bonds between the surface and the adsorbed compound contributes substantially to the adsorption process. Some surface groups are weakly acidic and therefore can enter into ion exchange reactions at high pH. At high activation temperatures, silica surfaces sinter, becoming less active. This is most likely due to chemical changes in the surface groups.

Most commercial silica gels are either of the narrow pore type (average pore diameter about 2.5 nm, specific surface area about 700 m^2/g and pore volume about 0.4 cc/g or the wide pore type (average pore diameter about 14 nm, specific surface area about 300 m^2/g and pore volume about 1.1 cc/g). Sorbents capable of hydrogen bonding are good eluents for regenerating silica gels.

The surface of the alumina adsorbents contains hydroxyl groups of similar structure to the silica gels. However, the potential for ion exchange reactions is greater with alumina than with silica gel. Calcined alumina is strongly basic and must be acid washed to produce a neutral or slightly acidic surface. The strength of the adsorption on alumina for aliphatic compounds containing various functional groups has been shown (35) to increase in the following order: -S-, -O-, -C≡N, $-CO_2-$, -CO-, -OH, -N=, $-NH_2$, -SO, $-CONH_2$.

Zeolites are characterized by ordered crystal structures. They are hydrated crystalline metal aluminosilicates and possess uniformly small-sized pores leading from the exterior surface to an internal, three dimensional cagework formed of interconnecting silica and alumina tetrahedra. This formulation combines the surface functionality of the silica gel and of the alumina adsorbents, resulting in an adsorbent that has a strong affinity for polar or polarizable molecules.

Since the zeolites have the uniform, small pore size

throughout each particle, most of their surface area is inside the particle. Zeolites are often called molecular sieves because the pores may be small enough to allow certain molecules in solution to pass through the pores while larger molecules are excluded, thus separating one compound from others.

The affinity of zeolites for polar or polarizable molecules lead to their use for the removal of water, carbon dioxide and hydrogen sulfide from liquids (36). Zeolites have been used for the dehydration of chlorinated solvents, refrigerants, acetylene and aromatics. They have been used for the removal of CO_2 from ethylene and liquid propane gas. Butane and liquid propane gas have been desulfurized with zeolites.

Zeolites, along with silica gels and alumina, have been used extensively in the immobilization of enzymes for biotechnology applications (37-39). However, the bulk of the purification applications with zeolites are in the treatment of gas streams. The purification of biological fluid streams themselves is usually best accomplished with an adsorbent with a larger pore size and one which is more stable over a wider pH range.

Table 2.9 shows the properties of several of the commercial silica gels, aluminas and zeolites. New families of zeolites have been developed in the mid-1970's which have a hydrophobic surface. These new materials prefer lipophilic over a more polar compounds. In 1984 Union Carbide (36) announced a new family of microcrystalline porous structures based on aluminum and phosphorus. It is possible that these new materials may find applications in the purification of biotechnology products.

2.3.3 Organic Polymer Adsorbents

Ion exchange resin manufacturers have also produced macroporous resins which may be used as adsorbents in purification applications. Although these resins may possess very weakly acidic or basic groups, in this chapter we will only consider their use as adsorbents, not as ion exchange resins. They are generally used in neutral or slightly acidic solutions where little or no ion exchange activity is displayed.

Table 2.9: Properties of Some Commercial Aluminas, Silica Gels and Zeolites

Tradename	pH	Specific Surface Area (m^2/g)	Mean Pore Diameter (Å)	Source
ICN Alumina A	4.5	155-200	----	ICN Biomedicals
ICN Alumina B	10.0	155-200	----	ICN Biomedicals
ICN Alumina N	7.5	155-200	----	ICN Biomedicals
A-305 CS Alumina	---	325	700	Kaiser Chemicals
Celite Silica	7	---	200	Manville
Matrex Silica	7	500-600	60	Amicon
Matrex Silica	7	---	100	Amicon
Matrex Silica	7	---	275	Amicon
Nova-Pak Silica	7	120	60-100	Waters
Resolve Silica	7	200	60-175	Waters
μ Porasil Silica	7	300	50-300	Waters
Nucleosil 50	7	500	50	Macherey-Nagel
Nucleosil 100	7	300	100	Macherey-Nagel
Zeolite ZSM-5	---	---	5.4 x 5.6	Mobil
Zeolite A	---	---	5	Union Carbide
Zeolite X	---	---	7-8	Union Carbide

The surface active groups were shown in Table 2.4. The more complete chemical structures are shown in Figure 2.16. The non-functionalized polystyrene adsorbents, made from styrene and divinylbenzene, are extremely hydrophobic, with a dipole moment of only 0.3 Debyes. The polymethacrylate structures are more hydrophilic, with dipole moments of approximately 1.4 Debyes. The phenol-formaldehyde resins and the functionalized versions of the polystyrene and polymethacrylate resins are very hydrophilic, with dipole moments greater than 2.0 Debyes. Table 2.10 shows the physical properties of commercial adsorption resins.

Most of the adsorbent resins with functionality have primary, secondary, tertiary or quaternary amine groups since these groups are useful in removing the organic contaminants, most of which are acidic, from sugar extracts and aqueous feed streams. However, the affinity of aromatic acids or salts or aromatic acids for quaternary amines is so great that their adsorption is practically irreversible (40).

In general, the non-polar resins are very effective in adsorbing the non-polar solutes form polar solvents. Likewise, highly polar resins are effective in removing polar

STYRENE-DIVINYLBENZENE

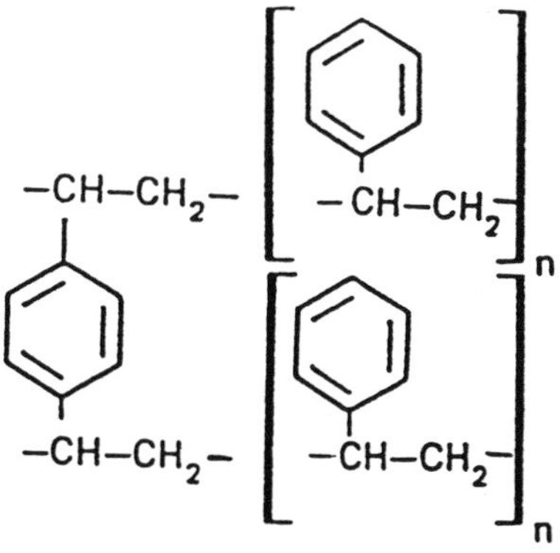

METHYLMETHACRYLATE-DIVINYLBENZENE

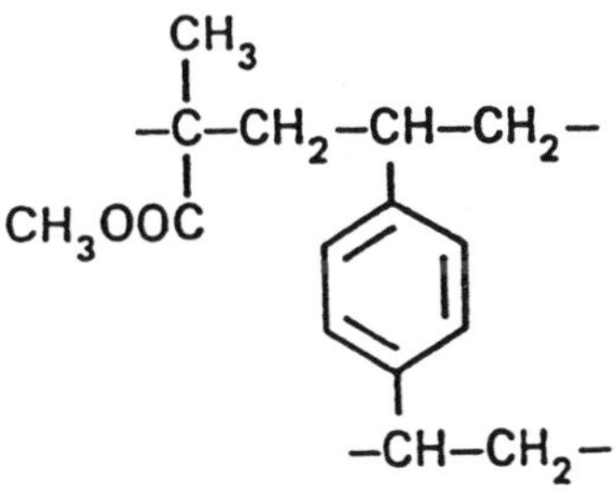

PHENOL-FORMALDEHYDE

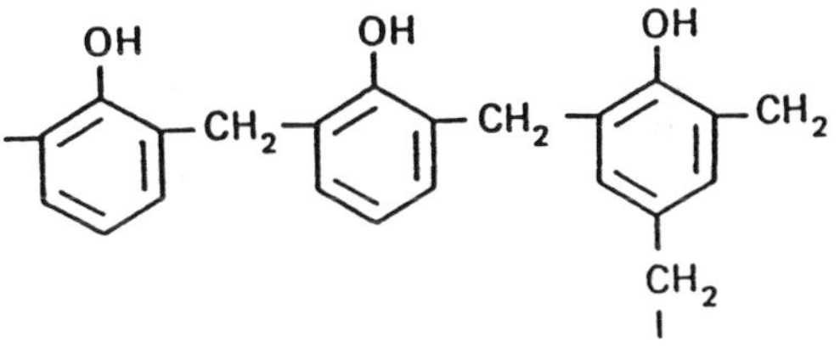

Figure 2.16. Chemical structure of non-functionalized synthetic organic resin adsorbents.

solutes from non-polar solvents. The binding energies of the non-functionalized resin adsorbents are lower than those of activated charcoal for the same organic molecules (41). Enthalpies of adsorption are only -2 to -6 kcal/mole for compounds like butyric acid, sodium naphthalene sulfonate or sodium anthraquinone sulfonate on large pore, non-polar adsorbent resins. Although this decreases the efficiency of adsorption, it also means that the regeneration of the resins is easier to accomplish.

Table 2.10: Physical Properties of Synthetic Organic Resin Adsorbents

Trade Name	Matrix	Functional Group	Physical Form	Capacity to HCl (equiv/l)	Typical Moisture Retention	Surface Area (m^2/g)
Duolite S-761	Phenol-Formaldehyde	Phenolic hydroxyl	Granules	--	48	150 - 300
Duolite ES-33	Phenol-Formaldehyde	Tertiary Amine	Granules	0.7	58	100 - 200
Duolite S-37	Phenol-Formaldehyde	Secondary and Tertiary Amine	Granules	1.4	55	120
Duolite ES-111	Styrene-DVB	Quaternary Amine	Beads	1.0	67	1.0
Amberlite XAD-2	Styrene-DVB	--	Beads	--	--	300
Amberlite XAD-4	Styrene-DVB	--	Beads	--	--	780
Amberlite XAD-7	Carboxylic Ester	--	Beads	--	--	450
Amberlite XAD-8	Carboxylic Ester	--	Beads	--	--	140
XFS-4022	Styrene DVB	--	Beads	--	--	80 - 120
XFS-4257	Styrene DVB	--	Beads	--	--	400 - 600

2.4 LABORATORY EVALUATION OF ADSORBENTS

Several tests (iodine number, molasses number, surface area and pore distribution) were mentioned in the previous section as useful in the characterization of adsorbent materials. While these tests are valuable in assessing the probability that an adsorbent is suitable for a specific purification operation, laboratory evaluations of the selected adsorbents with the fluid to be purified are necessary.

Three main types of experiments are possible for measuring adsorption of solutes from liquids by solid adsorbents. The essence of the first, which is by far the most frequently used, is to bring together known weights of solid and liquid, allow equilibrium to be established at a given temperature, and determine the resulting change in composition of the liquid.

The simplest experimental technique is to seal the liquid and adsorbent into clean, dry test tubes. Freezing the mixture in a carbon dioxide slush bath immediately before the tube is sealed will minimize the loss of volatiles. The tubes are then shaken in a constant temperature bath until equilibrium is reached. The time required for this stage varies considerably depending on the nature of the components of the system being studied.

During the adsorption measurement, the liquid must be in constant motion and the adsorbent must be continuously washed by the liquid. A reciprocating shaker may be used provided the frequency is sufficiently low so that mechanical breakdown of the adsorbent does not occur. For dense solids, tumbling is often more effective. The tumbling rate should be adjusted so that the adsorbent will fall completely through the fluid each half cycle. If the adsorbent is in the powder form, it will be necessary to centrifuge the liquid suspension to separate the adsorbent from the liquid.

A second method (42) of measuring adsorption from a liquid solution on a solid adsorbent is shown in Figure 2.17. In this technique, the liquid containing the solute is circulated over the adsorbent particles while changes in concentration are monitored by either refractometry or spectroscopic changes. The experiment may be carried out in a closed system, with solutions that have been purified and degassed. It also allows a single adsorbent sample to be treated in a reproducible manner with successive experiments. The experimental temperature can be changed over a wide range in a simple experiment with this technique.

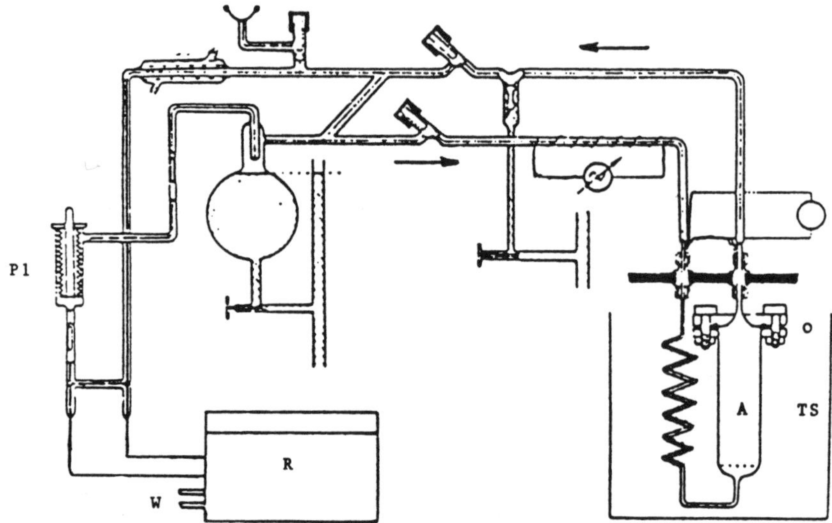

Figure 2.17. Circulation method for determining adsorption from solution. Solution is circulated by pump P1 over adsorbent in cell A in the temperature controlled unit TS and through one arm of a differential refractometer R. Solution of the initial composition is circulated through the reference cell W of the refractometer (Reference 42).

Somewhat related to that technique is a chromatographic method in which a solute at a given mole fraction is passed through a column of the adsorbent particles and the concentration of the effluent is monitored. The concentration of the effluent plotted as a function of time normally gives a curve like the one shown in Figure 2.18. The experiment is complete when the effluent concentration of the solute returns to its concentration in the feed stream. If, beyond a sharp rise, the concentration of the solute in the effluent becomes constant, the system is said to have reached equilibrium. The adsorption isotherm can be plotted from a series of such curves. The method can also be extended to deal with the simultaneous adsorption of two or more solutes. For precise results, it is necessary to correct for the dead volume in the column and for any variation in the molar volume of the solution as its concentration is changed (43).

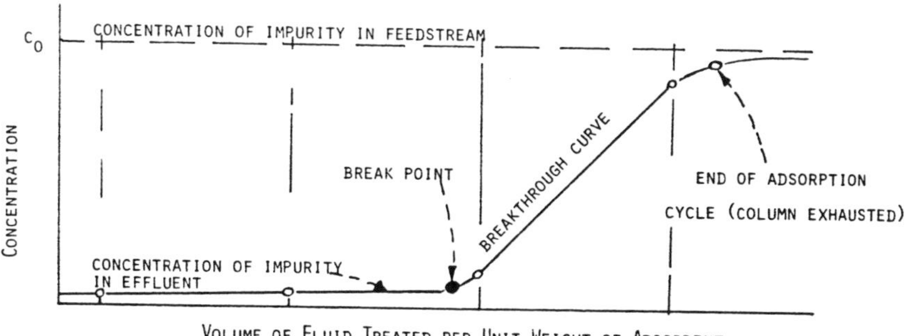

Figure 2.18. Concentration of solute in effluent as a function of time.

Example 2.5

Figure 2.19 shows the concentration of the effluent from an adsorption column using the same amount of adsorbent and the same flow rate but with changes in solute concentration in the feedstream. While the column dimensions (1 cm x 10 cm) were too small for gathering scale-up information, the data can be used to construct the adsorption isotherm shown in Figure 2.20.

Figure 2.21 shows an adsorption measurement technique which is useful for modeling the behavior in a stirred suspension of the adsorbent in the liquid solution with the solute. The relative surface excess is measured and represented by the equation:

$$n_2 = \left(\frac{dn_2}{dm}\right)_{T,x_2^1} \quad (2.63)$$

where dn_2 is the amount of solute 2 which has to be added together with a mass dm of solid adsorbent to maintain a constant bulk liquid concentration. Powdered or particulate adsorbent material can be added in successive amounts along with the solute to maintain a constant mole fraction.

Normally the isotherms obtained should be independent of the ratio of the weights of adsorbent and solution used. If this is not so, the presence of small quantities of impurities should be suspected. There may be other reasons as well. As an example, in the adsorption from aqueous solutions into anodized aluminum, the slight dissociation of

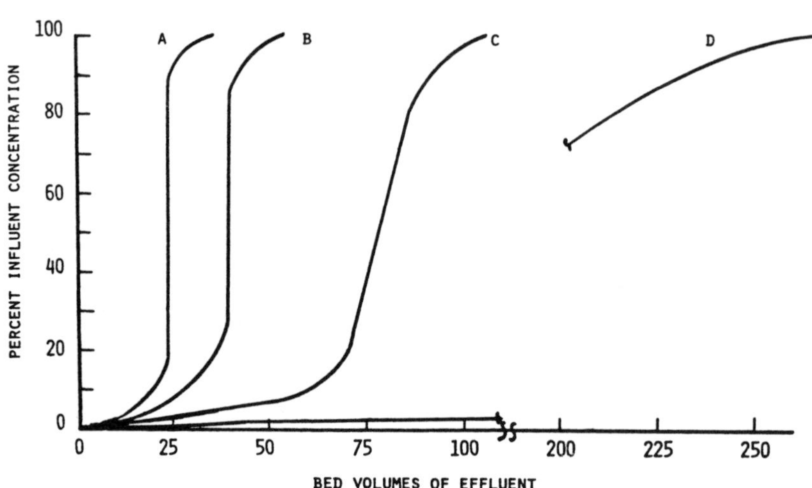

Figure 2.19. Concentration of sodium cholate in effluent as a function of time for different concentrations of the solute in the feedstream.

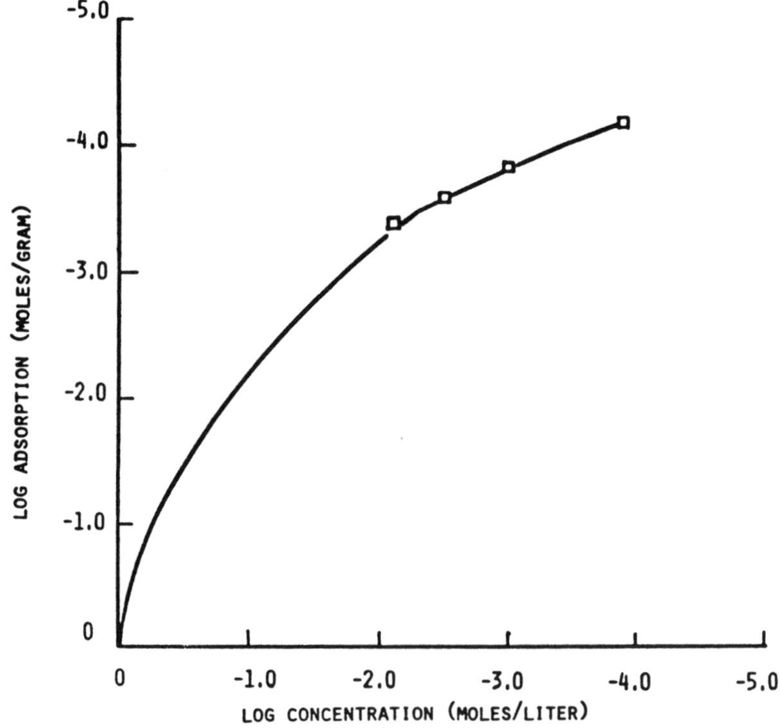

Figure 2.20. Adsorption isotherm for sodium cholate on porous polystyrene-DVB adsorbent, based on data from Figure 2.19.

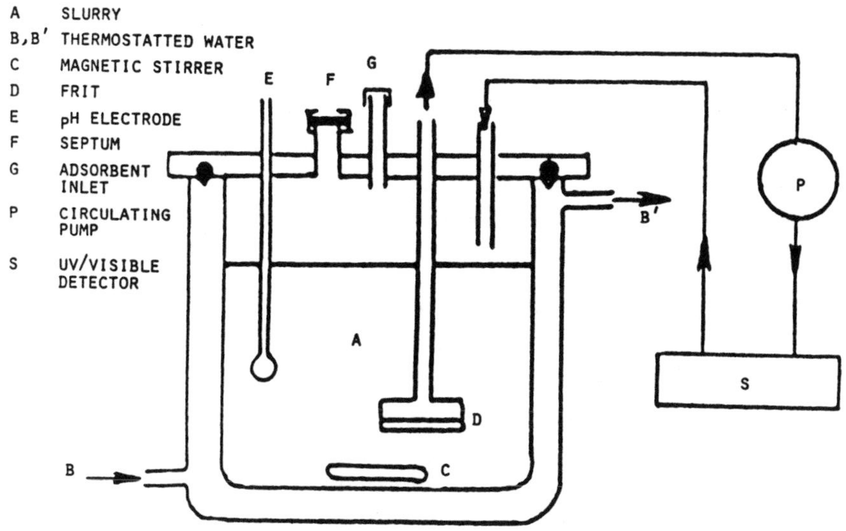

Figure 2.21. Alternate circulation adsorption method in which the adsorbent is used as a suspension rather than in a packed column (Reference 98).

the alumina film in the water causes this type of behavior (44).

In the adsorption of solutes from the liquid phase onto solid adsorbents, substances present in low concentration are often adsorbed preferentially. Therefore, the presence of impurities in the liquid phase may have an effect on the measurement quite out of proportion to the quantity present. A practical check on the presence and effect of all the impurities in the fluid to be treated is valuable in avoiding spurious results. Quantities of adsorbent and, separately, of each impurity should be shaken together for the length of contact time to be used in the adsorption experiments. Then examine the fluid by the analytical technique chosen for the experiment to insure that any change is within the experimental error of the technique.

Many efforts are now focused on ways to correlate and predict the performance of adsorbents in multicomponent systems. The models are often heavily dependent upon experimental data derived using the actual fluid to be treated. Industry has usually simulated full scale column performance by using columns of the same height but with reduced diameters. The time required to carry out a test using such a column is of about the same duration as the full scale operation. Rosene (45,46) has reported column simulation techniques using much smaller columns which permit engineering design in as little as one twentieth of the testing time, with corresponding reductions in the feedstream and adsorbent required for the experiment.

2.5 ADSORBENT OPERATIONS

2.5.1 Pretreatment Operations

Solid adsorbents require some pretreatment before use. The usual treatments are for removal of water vapor or surface impurities or for improving the wetting of the adsorbent surface by the solvent which contains the solute. These pretreatment operations are to insure that a reproducible behavior will be exerted by the surface of the adsorbent during the adsorption process.

When adsorbents must be dried before use, the drying is usually carried out by heating for about two hours at 110 to 120°C. It has been claimed (47) that this temperature is not high enough for complete drying of all adsorbents. However, the use of a higher temperature may result in irreversible changes in the adsorbent, such as sintering, change in crystalline form or change in the functional groups on the surface. The last effect is especially seen for polymer resin adsorbents with strong base functionality. Polymer resin adsorbents are usually only dried before surface or pore characterization experiments and not before use in adsorbent operations.

Many adsorbents, particularly activated carbons, contain impurities which may be leached from the adsorbent to contaminate the solution being purified. The pretreatment needed to overcome this problem must use a solvent at least as powerful in leaching ability as the solution being purified. Activated carbons, which may contain several percent of inorganic oxides or carbonates, are usually washed with a mineral acid to remove these impurities. A volume of 1N sulfuric acid equal to the volume of adsorbent (one bed volume) is passed through the adsorbent at a temperature between 50-60°C. Next, two bed volumes of water, at the same temperature, are passed through the adsorption column.

Finally, ambient temperature water is run through the column until the effluent pH is above 4.0. Figure 2.22 (48) shows the difference between the adsorption which occurs with treated and untreated adsorbents.

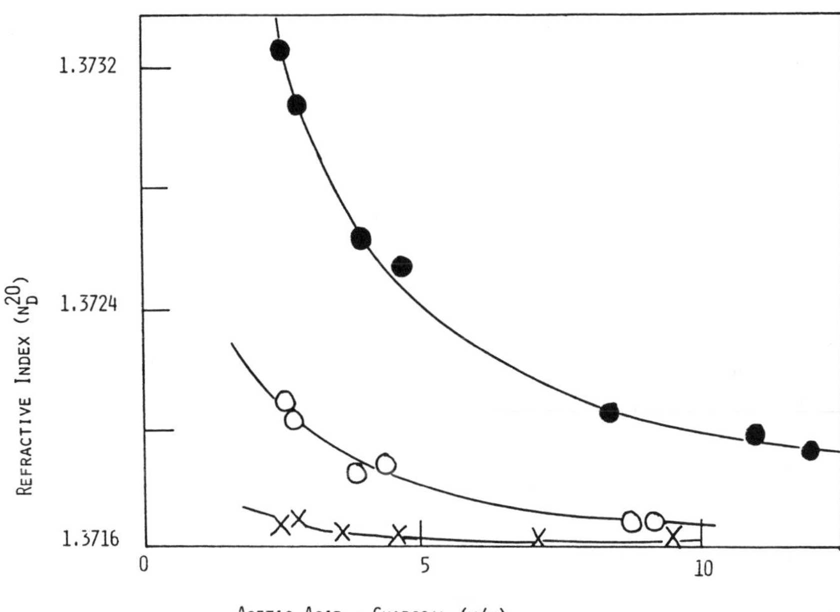

Figure 2.22. Effect of treated vs. untreated charcoals on the refractive index of acetic acid. The refractive index for pure acid is 1.3716 (Reference 48).

The non-polar synthetic resin adsorbents are not easily wetted by aqueous solutions because of their hydrophobic nature. The pretreatment of these resins consists in passing 5 to 7 bed volumes of methanol through the beds of resin, followed by rinsing with 4 to 5 bed volumes of deionized water.

It may also be necessary to pretreat the solution containing the solute before putting it in contact with the adsorbent. This pretreatment is primarily concerned with the removal of excess amounts of suspended solids, oils and greases. The presence of these materials in excess amounts would coat the adsorbent particles, thereby dramatically reducing their adsorptive effectiveness. Suspended solids, including bacteria and yeasts, in amounts exceeding approximately 50 mg/L should be removed prior to applying the fluid to the adsorption column. Oil and greases in concentrations above 10 mg/L should not be applied to adsorbents in either batch or column operations.

Synthetic organic polymer adsorbents are subject to de-crosslinking if oxidative materials are present in the feed solution or the eluant. These should be removed by deaeration or by passing the solution through a sacrificial carbon bed first.

Filtration, oil flotation and chemical clarification are the pretreatment operations commonly employed. The adjustment of the solution pH may also be used to enhance adsorption efficiency. Dissolved organic substances adsorb best at the pH which imparts the least polarity to the molecule. Weak acids, such as phenol, show better adsorptive properties at lower pH values, while amines usually adsorb better at higher pH values.

2.5.2 Batch Operations

In the batch operation, measured quantities of the adsorbent are added to the fermentation broth or filtrate in one or several stages. The adsorbent is left in contact with the fluid for the appropriate length of time, usually with agitation. Then the suspension is screened to isolate the adsorbate-adsorbent complexes. If the object is to recover the adsorbate, the adsorbent is usually transferred to a column through which an eluting solvent is passed. The fractions of the eluate containing the desired product are collected for further purification and/or concentration. If the adsorbate is an unwanted impurity, after being isolated by screening, the adsorbate-adsorbent complex is discarded or regenerated, depending on the economics of the operation.

Example 2.6

An example of this method of purification was given by Ores and Rauber (49). To a fermentation broth of 4.5 liters with 510 units of the antibiotic bacitracin per cc were added 1.5 liters of a porous polystyrene resin. The suspension was stirred together for six hours and then screened to separate the resin. The resin was placed in a column and eluted with 3.9 L of methanol. Water (0.55 L) was added to the eluate which was then distilled at 20°C under vacuum. The result was an 86% recovery of the bacitracin (1.15 L contained 1,740 units/cc).

An important advantage of the agitated slurry systems in adsorptive operations is that the resistance to mass transfer of the solute through the slurry to the adsorbent particles can be lowered by increasing the agitation power input. The mass transfer coefficient in such solutions correlates well with agitation power according to the equation (50):

$$Sh = C_1 \left(\frac{\bar{\epsilon}\, d^4}{\nu^3} \right)^{0.25} Sc^{0.33} \qquad (2.64)$$

where Sh is the Sherwood number; Sc is the Schmidt number; d is the particle diameter; $\bar{\epsilon}$ is the power input per unit mass; ν is the kinematic viscosity; and C_1 is a constant.

Since $\bar{\epsilon}$ is proportional to the third power of the agitator speed for turbulent flow, the mass transfer coefficient is seen to vary as the 3/4 power of the agitator speed. By knowing the mass transfer at one agitator speed, the mass transfer at a second is given by:

$$k_2 = k_1 (n_2/n_1)^{0.75} \qquad (2.65)$$

where k_1 and k_2 are the mass transfer coefficients when the agitator speeds are n_1 and n_2 rpm's, respectively.

The mass transfer coefficients were shown (51) to vary with position and impeller characteristics. Values for the mass transfer coefficient in the plane of the impeller can be described by the equation $15 \times 10^{-5} (nD^2/2r)^{2/3}$ and $11 \times 10^{-5} (nD^2/2r)^{2/3}$ for six bladed disk turbines and curved bladed turbines, respectively. This is shown in Figure 2.23.

2.5.3 Column Operations

Column operations are more common for industrial adsorption than batch operations. The adsorption column holds the adsorbent particles through which the fluid with the solute or solutes flow. Either pressure or gravity can be used as the driving force for fluid flow in a downflow operation. The untreated fluid comes into the top of the column. If the column has a large diameter, the fluid may go through a distributor to insure a minimum amount of unused adsorbent or "channeling" of the fluid. Contact time for the fluid in the bed is determined by the requirements

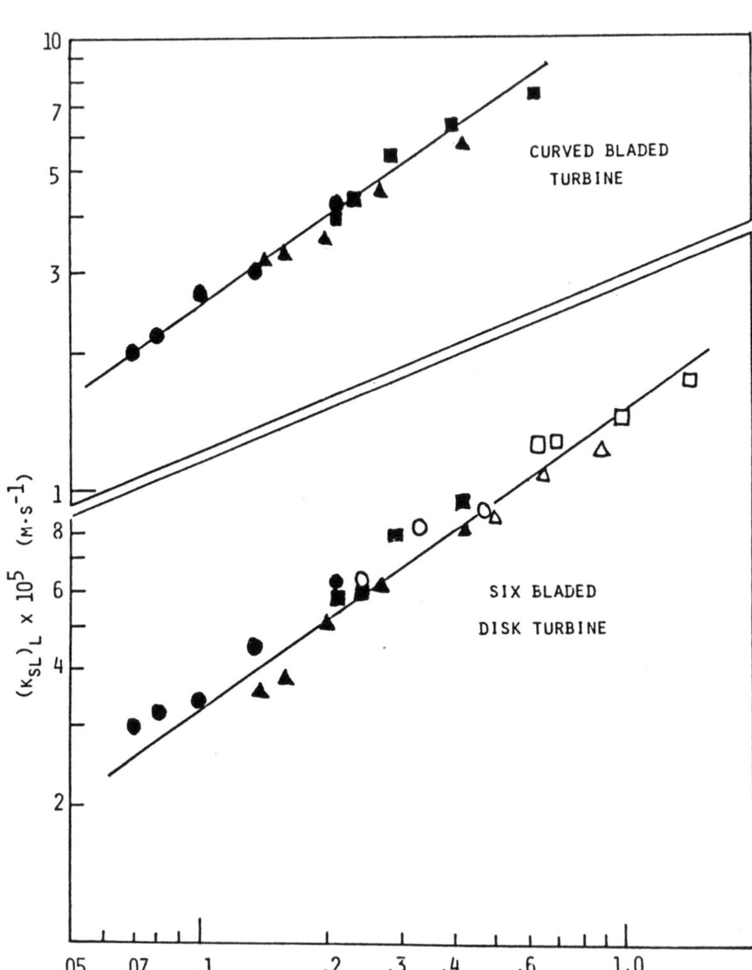

Figure 2.23. Correlation of solid-liquid mass transfer coefficient in the plane of the impeller. The different symbols represent types of contactor and impeller speeds (Reference 51).

for the system. Flow rates or space velocities for fermentation filtrates with small amounts of solute to be adsorbed are usually between 2 l/s-m^3 to 16 l/s-m^3. The treated fluid is collected in a similar distributor device at the bottom of the column and discharged to the next adsorption column or treatment unit. Figure 2.24 (52) shows examples of pressure and gravity flow adsorption units.

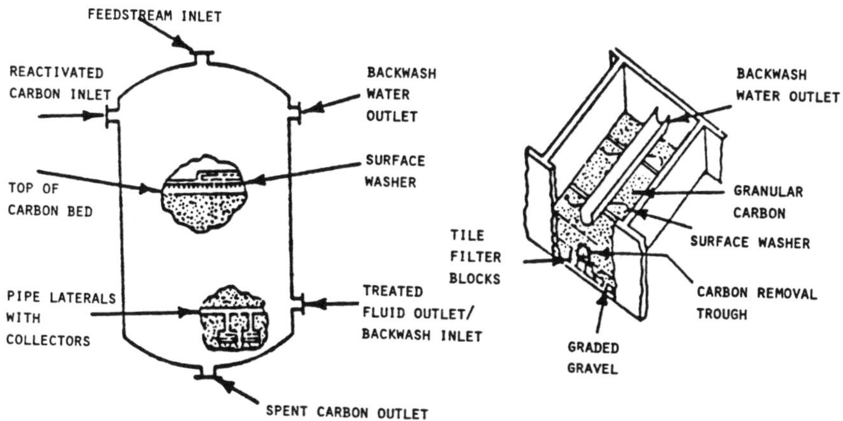

Figure 2.24. Pressure-type and gravity type fixed bed adsorber units (Reference 52).

The shape and length of the adsorption zone in a column determine the number of packed bed adsorbers which should be used to attain the desired degree of solute removal. This is shown in Figure 2.25 (53) for one, two and four columns operating in series. The shape of Curve A indicates that the mass transfer zone is short, that is, saturation of the adsorbent occurs shortly after the allowable concentration of the solute in the effluent is reached. When the effluent has reached the allowable concentration with respect to the solute, it is said to have reached "breakthrough." Curves B and C in this figure show increasingly long mass transfer zones. A second or more adsorption columns are needed to purify adequately the effluent while utilizing more fully the adsorptive capacity of the initial column.

An alternate technique for the purification of a system with an intermediate mass transfer zone using a single packed column is to collect the effluent in a holding tank. Then the overpurified effluent from the earlier part of the cycle can be mixed with the underpurified effluent from the later part to achieve an effluent with the desired purity. This technique is appropriate when impurities are being removed but should not be used when the adsorbate is the desired product.

Another way of mixing underpurified and overpurified effluents without the use of holding tanks is to connect two

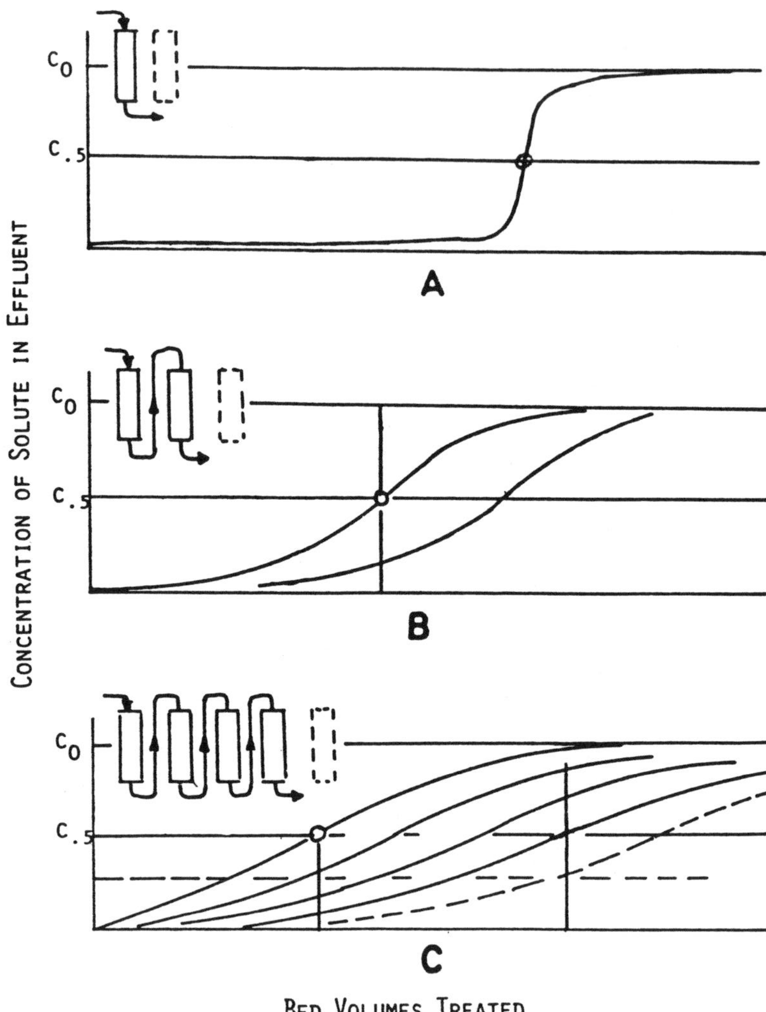

Figure 2.25. Effect of shape and length of the adsorption zone on the number of columns needed to obtain both a high degree of adsorbent utilization and the desired solute removal (Reference 53).

or more adsorption columns in parallel and then begin feeding the untreated fluid to them at evenly spaced intervals. Then, when one column is discharging underpurified effluent late in its exhaustion cycle, a different column would be discharging overpurified effluent since it would be early in its exhaustion cycle. Discharge manifolds need to be arranged such that appropriate effluents are connected to achieve the necessary purity. This packed bed arrangement is shown in Figure 2.26.

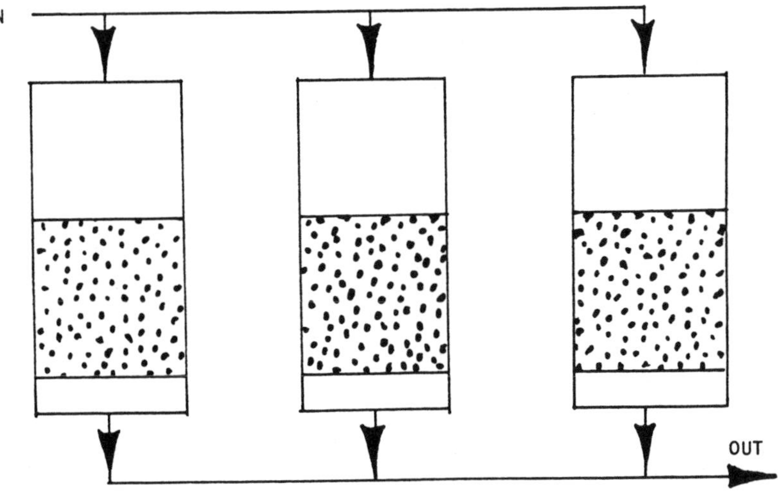

Figure 2.26. Column operation in a down-flow, parallel mode (Reference 54).

There are a number of other column equipment configurations. The three most common types of column adsorber systems are shown in Figure 2.27.

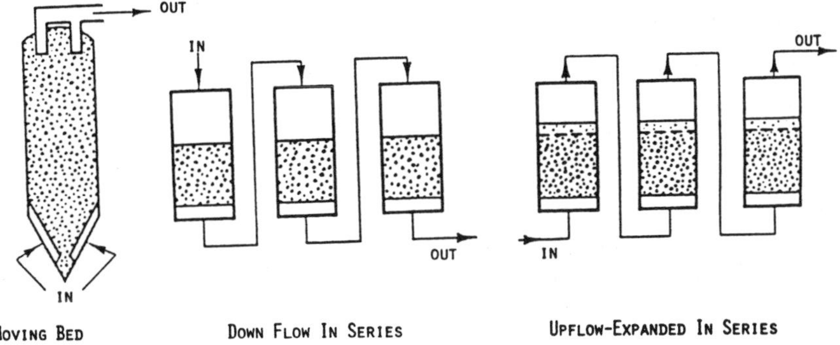

Figure 2.27. Additional adsorption column configurations (Reference 54).

In the packed moving bed adsorber, the untreated fluid flows upward through the bed and is collected and removed at the top of the bed. Periodically, a portion of the adsorbent is removed from the bottom and fresh adsorbent is added to the top of the column. The volume of adsorbent removed at one time is between 2 and 10% of the total adsorbent volume. It is necessary to have a column height to diameter ratio of 3 to 1 in order to facilitate plug flow during the adsorbent removal portion of the cycle. The bottom and the top of these columns are 67° and 35° cones, respectively. Peripheral ring headers are located in these cones for feeding the untreated fluid and for removing the effluent. Well screens, or septs, are submerged in the top of the adsorbent to retain the adsorbent particles while passing the treated fluid. However, the movement of the adsorbent in these systems can add to operational problems and will often result in the loss of friable adsorbents due to attrition.

As was mentioned during the section on pretreatment, packed column operations require that the amount of suspended solids in the feedstream should not exceed 50 mg/L. Packed columns also require that the adsorbent particle size be greater than 300 microns so that the pressure drop through the column is not excessive. Fluidized or semi-fluidized beds of adsorbents may be used to circumvent these limitations.

The feedstream in fluidized systems enters at the bottom of the column at a flow rate 10% to 100% greater than in packed bed systems. The adsorbent bed expands in proportion to the flow rate and in inverse proportion to the particle size of the adsorbent. Fluidized bed operations take advantage of small adsorbent particle size while avoiding the problem of excessive pressure drop which would occur in packed beds with similar sized particles.

Fluidized beds also provide good contact between the fluid and the adsorbent because of the high degree of mixing and high mass transfer coefficients which the small particles allow. This is shown in Table 2.11 (55) where the effect of particle size on the mass transfer rate is illustrated. However, because of the high degree of mixing, plug flow does not occur with this technique. The lack of plug flow and the short residence time mean that fluidized adsorbent operations do not remove all of the solute. This limits the use of fluidized beds to applications which require some removal of impurities but not the complete removal of solutes.

Table 2.11: Effect of Carbon Adsorbent Particle Size on Mass Transfer Rate with a 200 ml/min Flowrate(55)

Solute	Particle Size (micron)	Mass Transfer Rate (min^{-1})
Astrazone Blue	150 - 250	0.77
Astrazone Blue	250 - 355	0.58
Astrazone Blue	355 - 500	0.58
Astrazone Blue	500 - 710	0.41
Teflon Blue	150 - 250	1.06
Teflon Blue	250 - 355	0.98
Teflon Blue	355 - 500	0.58
Teflon Blue	500 - 710	0.40

Semi-fluidized beds combine the features of both fluidized beds and packed beds in a single column. Hsu and Fan (56) have evaluated the usefulness of semi-fluidized beds in biochemical processing. In this technique, the fluid bed is compressed by placing a screen or sieve in the column to form two distinct regions of packed and fluidized particles. The time before breakthrough is comparable for packed beds and semi-fluidized beds; it is much longer than for fluidized beds. However, the flow rate that is possible for the same pressure drop is much greater for the semi-fluidized bed than for the packed bed. This is shown in Table 2.12. (57) Design of the feedstream inlet systems for semi-fluidized and fluidized adsorption columns is critical since the distributors must be able to pass suspended solids, but must retain the adsorbent particles.

Table 2.12: Effect of Fluidization on Flow Rate for Specific Pressure Drops(57)

		Hydraulic Loading Rate (m/min)			
			Degree of Fluidization		
Adsorbent	Pressure Drop	Packed Bed	25%	50%	75%
F400 Carbon	0.3	.13	.20	.28	.55
	0.6	.24	.39	.55	1.05
APC Carbon	0.2	.12	.15	.23	.35
	0.4	.23	.27	.40	.57
	0.6	.38	.43	.60	.92

Tavares and Kunin (58) have reported on the use of a very shallow bed of finely divided adsorbent particles to remove impurities from fluids. This technique combines filtration and purification in a single unit operation. A powder is applied to a retaining screen so that the adsorbent thickness is between 0.5 and 1 cm. The feedstream is passed through several of these adsorbent layers in series to achieve the desired purity. Once the adsorptive capacity has been completely utilized, the layer is removed and discarded. This technique relies upon the increased efficiency of the fine size of the adsorbent, the savings in regeneration costs and the lower capital investment associated with pre-coat filter equipment to compete with typical adsorbent column systems.

Example 2.7

Figure 2.28 shows the effect of changes in bed height for fixed column adsorption. In this case, the particle size was 600 microns; the concentration of the solute in the feedstream was 500 mg/dm^3; the flow rate was 2 cm^3/s; the equilibrium adsorption was 0.2 g/g and the effective pore diffusivity was 2 x 10^{-10}m^2/s.

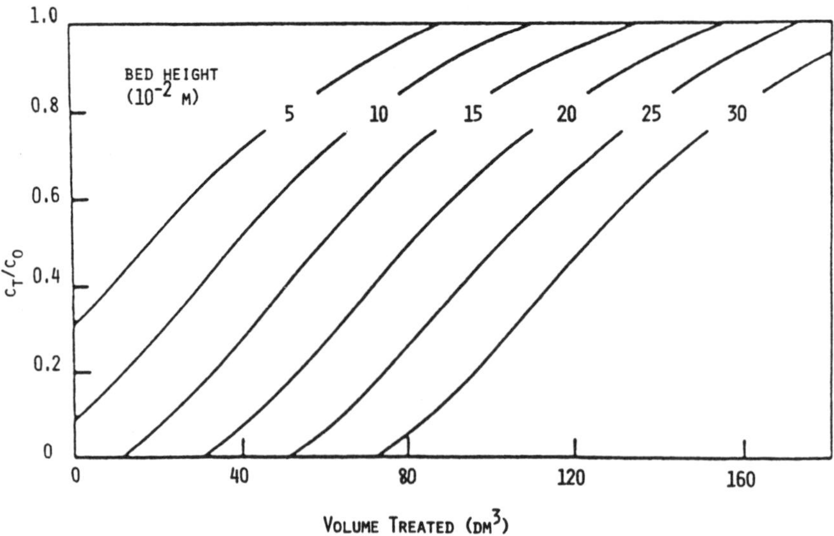

Figure 2.28. Effect of changes in bed height for fixed column adsorption (Reference 18).

If it were required that the solute concentration in the effluent remain below 0.1 C_0 and there was 50 dm^3 of fermentation broth to be treated, the column bed height required would be 25 cm.

2.5.4 Regeneration Operations

The adsorbent may be used once and then discarded. More often, the used adsorbent is regenerated for reuse. When adsorbent is regenerated, two or more beds of the adsorbent are used with a manifold valving arrangement which allows continuous processing of the treated fluid.

As was seen with the adsorption isotherm in the earlier section, adsorption loading increases as the concentration increases and decreases with increases in temperature. The regeneration or desorption portion of the operating cycle utilizes some variation of this behavior. Regeneration may be accomplished by a thermal process, a displacement adsorption process, an inert stripping process or a combination of these processes.

Thermal regeneration is a very common practice with adsorbents, particularly with granular activated carbon, which have been used for the removal of organic species. Temperatures between 870-980°C are normally used in rotary kilns or multiple-hearth furnaces. Typically, the adsorbent spends 5 minutes in a drying zone, 10 minutes in a charring zone and 15 minutes in a reactivation zone. The oxygen content of the atmosphere is held at a low level to allow oxidation of the adsorbed organic impurity rather than of the activated carbon. Carbon losses per cycle are normally between 2 to 10%, depending on the size of the regeneration system. Large systems are designed to operate continuously which allows more efficient control for lower losses.

Schematics of a multiple-hearth furnace and a rotary kiln are shown in Figure 2.29 and Figure 2.30, respectively (59). The multiple-hearth furnace is said to have better control of temperature and residence time, better control of heat input, more uniform regeneration, lower temperature gradient and less space requirements than the rotary kiln units. The primary disadvantage of the multiple-hearth furnaces are their higher capital costs and, in some cases, higher carbon losses compared to the rotary kiln furnaces.

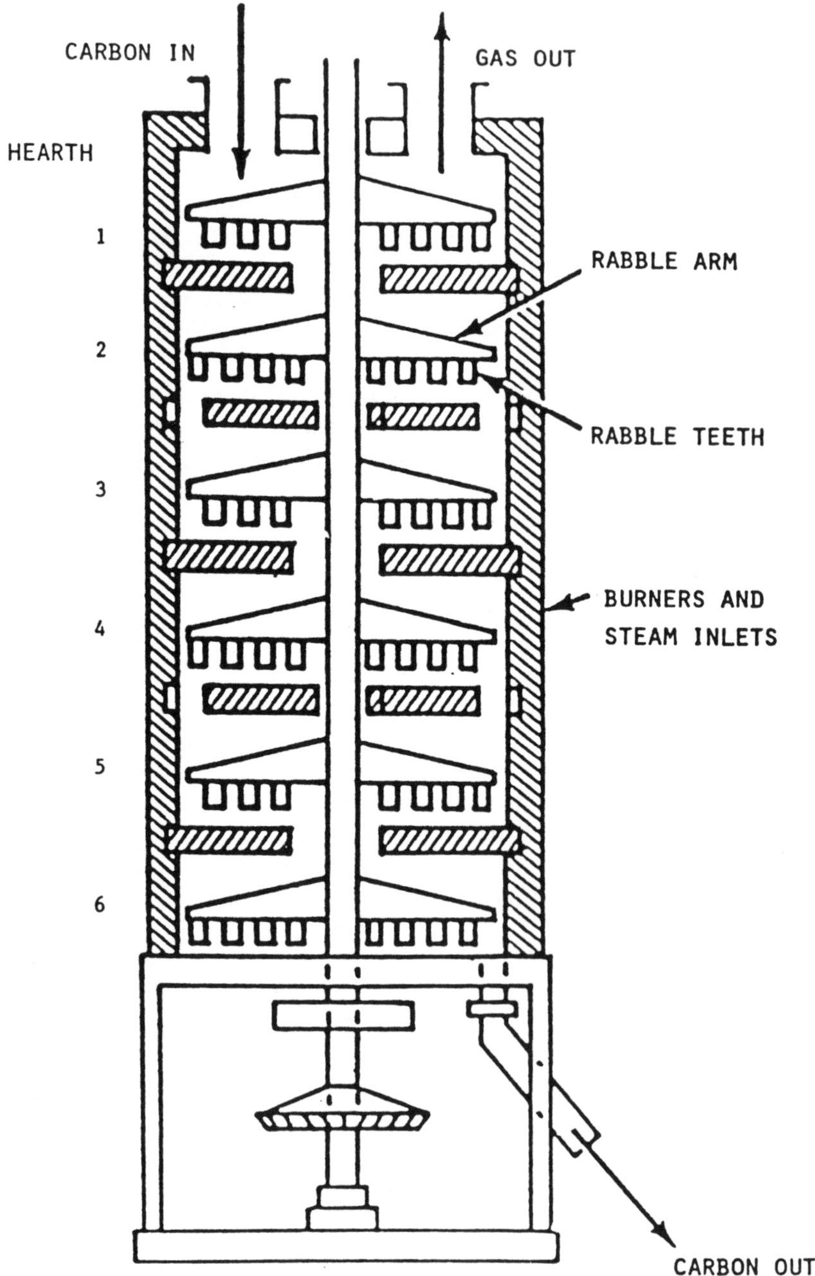

Figure 2.29. Multiple hearth furnace (Reference 59).

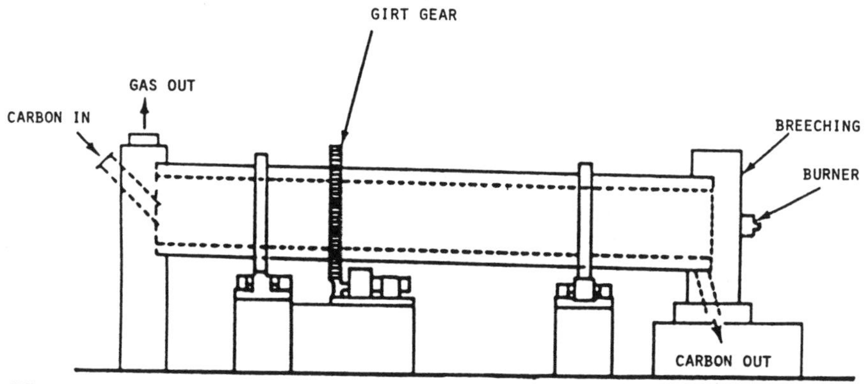

Figure 2.30. Rotary tube regeneration furnace (Reference 59).

Whichever thermal regeneration equipment is used, it is necessary to transport the adsorbent between the columns and regeneration equipment by water slurry. Various centrifugal or diaphragm pumping arrangements can be used. Alternately, hydraulic or pneumatic pressure could push the slurry. The bends in the piping should have a wide radius to reduce adsorbent attrition. About 0.5 to 3.0 kg of adsorbent may be transported with one liter of water, with the exact amount depending upon distance and elevation considerations.

The thermal regeneration of activated carbon also has an effect on the surface area of the adsorbent. As Figure 2.31 (60) shows, even five regenerations cause a significant reduction in surface area. This occurs because the ash that is picked up during the adsorption phase of the cycle plugs small pores and is not burned off during regeneration. The high temperatures of regeneration are also likely to result in the formation of inorganic metal oxide residues which build up as ash in the carbon. An acid wash must be used to remove this ash to reduce fouling of the adsorbent.

Thermal regeneration at more moderate temperatures by passing a purge gas or steam through the adsorbent column may be used in those applications where the adsorbent has been used for solvent recovery or for drying applications. The highest temperature reached in such regenerations is usually less than 200°C for carbon and 300°C for molecular sieves (61). This type of regeneration is more appropriately

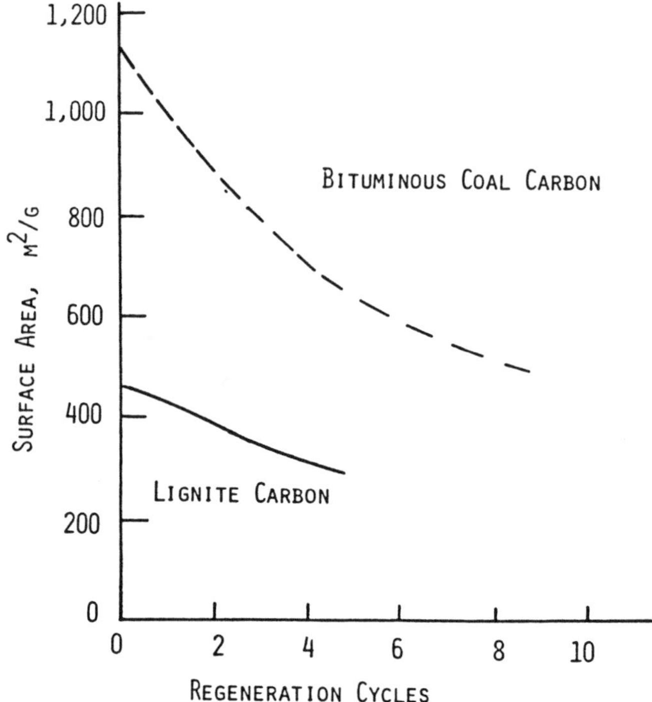

Figure 2.31. Decrease in surface area with use and regeneration (Reference 60).

called inert stripping. The regeneration of the hydrophobic zeolite, ZSM-5, which had been used to adsorb ethanol from a water-ethanol mixture, made use of this technique (62). The ethanol was displaced from the adsorbent with the inert gas CO_2, which was later separated from the ethanol by reducing the pressure.

In the semi-continuous adsorption system shown in Figure 2.32 (63), a hydrophilic zeolite is used to remove water from an ethanol-water mixture. Here, after a column has been completely utilized for water adsorption, the column is drained of liquid which is returned to the feedstock tank since the solute has not been removed from this liquid. Then hot air is blown through the column to remove the ethanol-rich phase which clings to the adsorbent particles. After this is removed, the temperature of the bed of adsorbent is increased to about 260°C as the water vapor is driven from the micropores of the adsorbent. When no more water vapor comes from the column, cool air is passed through the column to cool it to 80°C before the next adsorption cycle begins.

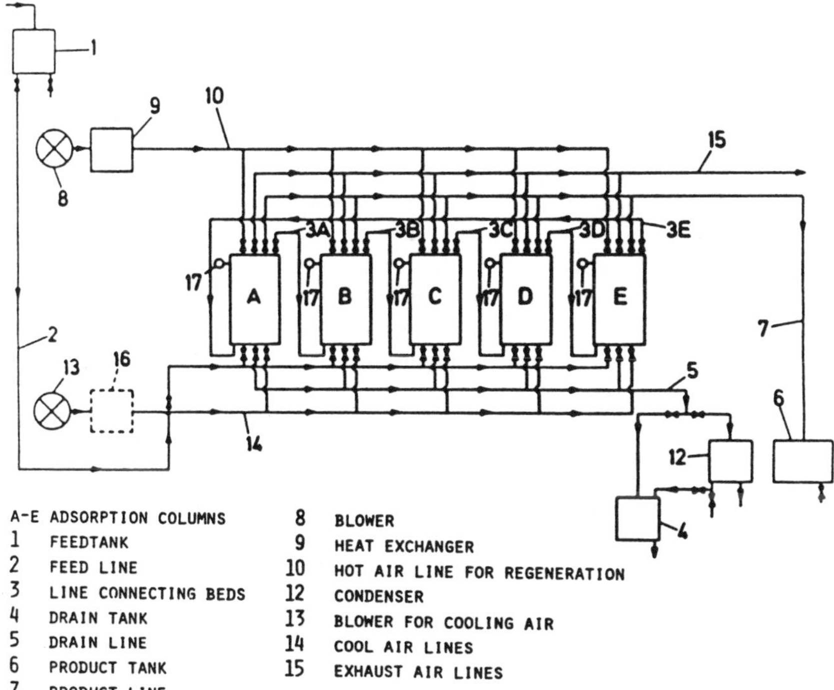

Figure 2.32. Semicontinuous adsorption system described in Reference 63).

When the adsorbent is a synthetic organic polymer, regeneration of the adsorbent is usually carried out by the elution of the adsorbate with a solvent that has a lower solubility parameter than the adsorbate. It is also necessary that the adsorbate be soluble in the eluting solvent so that the adsorbed molecules may dissolve rapidly after the solvent diffuses to the adsorption site.

Methanol is the most commonly used eluting regenerant. It will usually remove the adsorbate from the resin in a minimum effluent volume and is relatively inexpensive. Other low molecular weight alcohols or ketones might also be used alone or in combination as eluting regenerants.

The elution regeneration of granular carbon has been

examined with very little commercial success. The binding energies of activated carbons are much higher than those of adsorbent resins for the same organic molecules (42). While high temperature thermal methods are necessary to overcome the high binding energies for activated carbon adsorbents, the smaller binding energies for resin adsorbents allow the use of the eluting regeneration method.

Several advantages have been pointed out (64) for the elution regeneration method for resin adsorbents compared to the high temperature thermal regeneration required for most carbon adsorbents. These need to be considered when choosing an adsorbent system. Since the resin is regenerated in the same column as the adsorptive process occurred, the elution method does not require investment in the furnace, quenching or conveyance equipment. Nor are there adsorbent losses due to handling, to thermal oxidation or to irreversible fouling by the formation of inorganic ash as occurs during thermal regeneration.

Example 2.8

Table 2.13 shows the solubility parameters of several solvents used in regeneration. Methanol is almost always the eluting solvent of choice since it requires a minimum of volume to remove the adsorbate from the adsorbent and is relatively inexpensive.

Table 2.13: Common Regeneration Solvents for Synthetic Adsorbents

Solvent	Solubility ($Cal^{0.5}/cm^{1.5}$)
2-butanone	9.3
Acetone	9.9
2-propanone	10.0
1-butanol	11.4
1-propanol	11.9
Ethanol	12.7
Methanol	14.5
Water	23.2

Wallis and Bolton (65) have shown that it is also possible to regenerate activated carbon biologically. The *Pseudomonas putida* culture will regenerate carbon which has adsorbed phenol by maintaining the solution concentration of phenol at a low level and by converting the desorbed phenol to less toxic substances. The rate of regeneration was on the order of 25 mg phenol removed per gram of carbon per hour when the carbon was originally loaded with 40 to 65 mg phenol per gram carbon. The rate is much slower, about 1 mg phenol per gram carbon per hour, when the original amount adsorbed is below 40 mg/g.

This approach is limited to those adsorbates which can be converted to other less adsorbable compounds by a biological organism. Prolonged exposure or multiple regeneration reduces the carbon's regenerated capacity for adsorption. This probably occurs because of microbial fouling of the adsorbent's pores or adsorption of microbially produced compounds.

2.6 PROCESS CONSIDERATIONS

2.6.1 Design Factors

The properties of both the feedstream and the adsorbent determine the sizing and overall design of an adsorption system. The aim is to produce a system in which the flow during adsorption is under substantially constant temperature and constant pressure conditions through the adsorbent bed.

The properties of the fluid that need to be known for design purposes are (66):

(1) Chemical composition (c_i, mg/L), including the content of dissolved inorganic solids. It is important that the pH also be known because adverse effects may occur with some adsorbents at the incorrect pH. The composition will dictate any pretreatment of the feedstream which may be necessary.

(2) Fluid density (ρ, g/cc) at the operating temperature and pressure.

(3) Fluid viscosity (μ, cp) at the operating temperature and pressure.

The properties of the adsorbent that need to be known for design are:

(1) The bulk density (ρ_b, g/cc) of the adsorbent and the void fraction (ϵ) of a packed bed of the adsorbent particles. The way in which the adsorbent is loaded in the bed and its particle size can affect the void fraction in the bed.

(2) The particle size distribution and the mean particle size (r_p, mm). It is also necessary to know if the particles undergo volume changes during the course of the adsorption-regeneration cycle.

(3) The pore size distribution, which would permit the elimination from consideration those adsorbents which have too much of their surface area along pores whose diameter will not admit the solute to be adsorbed. The pore size distribution is usually shown as a plot of the increment of pore volume divided by the increment of pore radius versus the pore radius.

(4) Hardness, which indicates the attrition that may occur during the handing of the adsorbent.

There are a few relationships which must also be determined before designing the adsorption unit:

(1) The equilibrium adsorption isotherm as a function of the concentration, temperature and pH conditions of interest for the operation.

(2) Regeneration efficiency under the conditions of the operation. This should include any loss in adsorption capacity or losses in adsorbent so that replacement quantities and frequencies for the adsorbent are known.

(3) The length of the mass transfer zone should be determined in laboratory experiments at the flow rates of interest.

The equilibrium adsorbate concentration, q_e, from the isotherm data and the residual adsorbate concentration, q_r, on the regenerated adsorbent determine the adsorbate gradient across the mass transfer zone. This gradient is also the amount of adsorbate which will be loaded per unit weight of adsorbent each cycle.

With this information, the design of the process consists of the following steps (67):

(1) With an assumption concerning the cycle time desired, usually 8-24 hours, the quantity of solute, m_a, in the feedstream that is to be removed during the adsorption phase of the operating cycle is calculated.

(2) The amount of adsorbent material, m_p, is then calculated from the equation:

$$m_p = m_a / (q_e - q_r) \tag{2.66}$$

(3) The amount of adsorbent material, m_u, that cannot be used because of the length of the mass transfer zone is calculated from:

$$m_u = \frac{(\pi D^2)}{4} \rho_b \left(L_0 - \frac{\bar{u} \Delta c}{\rho_b \Delta q} \right) t_b \tag{2.67}$$

(4) The total amount of adsorbent material, m_c, required for the adsorbent cycle is, therefore:

$$m_c = m_p + m_u \tag{2.68}$$

(5) This amount is modified each cycle by some loss in adsorbent capacity and by attrition or physical loss of adsorbent.

The process requires that this quantity be added incrementally at the beginning of each cycle or that the adsorbent initially in the column be augmented to compensate for these losses for a given number of cycles, n_c. In the general case, the column must be designed to accommodate the following quantity of adsorbent:

$$m_t = m_p + m_u + n_c \Delta m \tag{2.69}$$

(6) The pressure drop that this quantity of resin would have, both during the adsorption phase and the regeneration phase of the operation, is then calculated for various bed configurations. A bed configuration that avoids excessive pressure drop yet achieves the desired adsorption is chosen as the adsorption column.

(7) Auxiliary equipment is sized to meet the requirements of the adsorption column. Controls for insuring the constant feed temperature and pressure, for measuring the pressure drop across the column and for monitoring the effluent concentration and temperature are also included.

2.6.2 Pressure Drop

The flow velocity, the pressure drop across the column and the bed diameter are interrelated. Setting a value or limit for one of these parameters places limits on the other two. Pressure drop limits usually are the basis for fixing operational flow velocities and selecting the bed diameter.

The pressure drop, ΔP, through a packed column of adsorbent is calculated using equations developed for pressure drop in pipes (66):

$$\Delta P/L = 2\ f\ G^2/D_p g\rho \tag{2.70}$$

where D_p is the mean particle diameter; f is the dimensionless friction factor; g is the gravitational constant;

G is the mass velocity; L is the depth of the adsorbent bed; and ρ is the fluid density under operating conditions.

The Reynolds Number, $N_{Re}(= D_pG/\mu)$, can be used to obtain the friction factor. When the flow through the bed is in the laminar region $N_{Re} < 10$), $f = 480/N_{Re}$. In the transition flow zone ($10 < N_{Re} < 200$), the friction factor is given by the value in Figure 2.33. When N_{Re} is greater than 200, flow is in the turbulent zone and the friction factor is given by:

$$f = 60.3/(N_{Re})^{0.339} \tag{2.71}$$

This pressure drop only describes the changes across the bed of adsorbent. Additional considerations are the amount of freeboard volume (liquid level above the adsorbent bed) and the pressure drop through the bed supports and retaining particles. These additional factors can increase the total pressure drop by an additional 50%. Safety requirements add another factor of 2.5 to the designed pressure rating of the column.

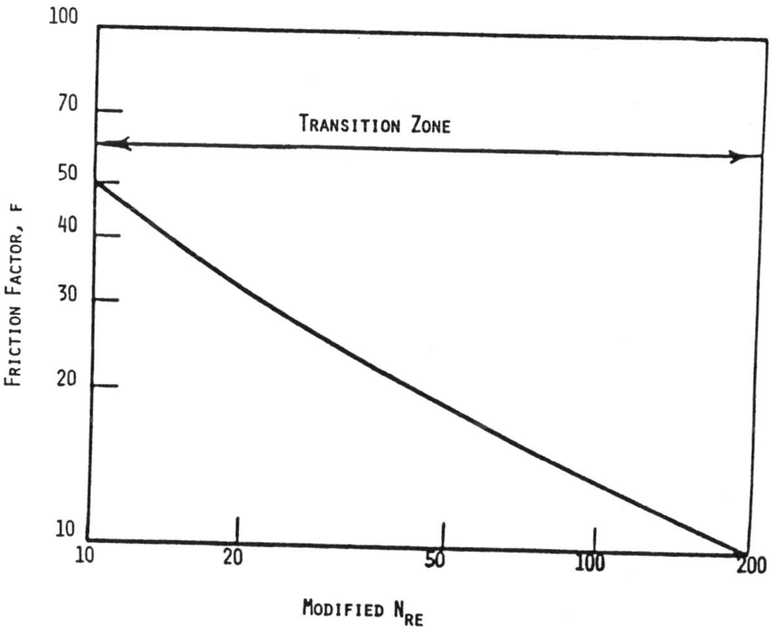

Figure 2.33. Change in friction factor, f, as a function of the Reynolds number, for use in Equation 2.70 (Reference 66).

2.6.3 Computer Design

A computer program (68) was written for the HP-67 to calculate the breakthrough curve for adsorption processes described by the Hiester and Vermeulen (69). Unfortunately this program may result in numerical overflow problems. Tan (70) has written a program for the HP-41 which eliminates the overflow problems and another program which can be used for design calculations.

To generate a breakthrough curve, the concentration of the solute is calculated for a series of throughput values, while holding the separation factor and the number of transfer units constant. The program listing is shown in Figure 2.34 and the user instructions are shown in Table 2.14, along with parameter identification.

That program forms a subroutine of the design program which also requires the listing shown in Figure 2.35. Three types of design problems can be solved with this program:

(1) Given the throughput ratio (T), separation factor (R) and breakthrough limit (X), the length parameter (N) and the run time (NT) can be calculated.

(2) Given a run time, separation factor and breakthrough limit, the length parameter can be calculated.

(3) Given a length parameter, separation factor and breakthrough limit, the run time can be calculated.

The user instructions are shown in Table 2.15. The parameters are the same as in Table 2.14.

2.7 ADSORPTION APPLICATIONS

2.7.1 Decolorization

The earliest applications of adsorption were the use of bone char in the decolorization of raw sugar. The cane sugar industry continues to use activated carbon in its commercial purification and decolorization processes. Barton and Knebel (71) have shown that a blended carbon enhances carbon's intrinsic properties to provide improved levels of purification in the processing of sugar cane.

```
LBL "FIXBED"
LBL B
" N = ?"
AVIEW
STOP
STO 11
" T = ?"
AVIEW
STOP
STO 12
" R = ?"
AVIEW
STOP
STO 13
RCL 11
RCL 12
*
STO 10
CF 12
FIX 3
" N = "
ARCL 11
AVIEW
" T = "
ARCL 12
AVIEW
" R = "
ARCL 13
AVIEW
RTN
LBL J
STO 02
X<>Y
STO 01
XEQ "JFC"
E↑X
RTN
LBL A
CF 00
XEQ "AA"
RCL 17
RCL 00
+
RCL 14
-
230
X>Y?
X<>Y
E↑X
1
+
1/X
STO 20
RCL 16
RCL 00
+
RCL 14
-
E↑X
CHS
1
+
*
STO 19
BEEP
ADV
SF 12
FIX
"Y = "
ARCL 19
AVIEW
"X = "
ARCL 20
AVIEW
CF 12
ADV
RTN
LBL "AA"
CF 01
CF 02
RCL 11
STO 00
RCL 13
*
STO 01
ST- 00
RCL 10
STO 02
ST- 00
RCL 13
*
ST+ 00
RCL 01
RCL 02
*
50
FS? 00
CLX
X<=Y?
SF 01
XEQ "JFC"
STO 14
RCL 13
RCL 10
*
STO 01
RCL 11
STO 02
XEQ "JFC"
STO 15
XEQ "PHI"
STO 16
RCL 15
-
E↑X
CHS
1
+
LN
RCL 15
+
STO 17
RTN
LBL "JFC"
CF 08
RCL 01
X=0?
RTN
CHS
STO 05
RCL 02
X=0?
SF 08
RDN
FS? 08
RTN
FS? 01
GTO "APJ"
LBL "EXJ"
CLX
STO 06
STO 07
230
RCL 01
X>Y?
GTO "APJ"
LN
STO 03
RCL 02
LN
STO 04
1
STO 08
STO 09
RCL 01
CHS
E↑X
STO 05
LBL "J1"
RCL 06
RCL 04
*
RCL 02
-
RCL 07
-
E↑X
ST- 08
1
ST+ 06
RCL 06
LN
ST+ 07
RCL 03
RCL 06
*
RCL 01
-
RCL 07
-
E↑X
RCL 08
X<0?
GTO "J2"
*
X=0?
GTO "J1"
ST+ 05
RCL 05
/
RCL 09
X<=Y?
GTO "J2"
X<>Y
STO 09
1 E-5
X<=Y?
GTO "J1"
LBL "J2"
RCL 05
X=0?
GTO "J3"
LN
RTN
LBL "J3"
RCL 01
CHS
RTN
LBL "APJ"
RCL 01
SQRT
STO 03
RCL 02
SQRT
STO 04
XEQ "ERF"
STO 05
FS? 02
GTO "J4"
CF 02
XEQ "RES"
RCL 05
-
E↑X
1
+
LN
RCL 05
+
LBL "J4"
2
LN
-
RTN
LBL "ERF"
3
STO 08
RCL 03
RCL 04
-
STO 09
STO 05
STO 06
X=0?
GTO "ZO"
X<0?
GTO "ZN"
X>Y?
GTO "E2"
LBL "E1"
RCL 09
X↑2
2
*
ST* 05
RCL 08
ST/ 05
2
ST+ 08
RCL 05
ST+ 06
RCL 06
/
1 E-6
X<=Y?
GTO "E1"
1
RCL 09
X↑2
CHS
E↑X
RCL 06
*
2
*
PI
SQRT
/
-
LN
RTN
LBL "ZN"
ABS
X<=Y?
GTO "E1"
SF 02
2
LN
RTN
LBL "E2"
1
STO 08
RCL 09
/
STO 05
STO 06
LBL "E3"
RCL 09
X↑2
2
*
CHS
ST/ 05
RCL 08
ST* 05
2
ST+ 08
RCL 05
ST+ 06
RCL 06
/
ABS
1 E-3
X<=Y?
GTO "E3"
RCL 06
PI
SQRT
/
LN
RCL 09
X↑2
-
STO 05
RTN
CLX
STO 05
RTN
LBL "RES"
RCL 09
X↑2
CHS
RCL 03
RCL 04
*
SQRT
RCL 04
+
PI
SQRT
*
LN
-
FS? 02
1 E-3
RTN
LBL "PHI"
FS? 01
GTO "APH"
CLX
STO 06
STO 07
STO 09
RCL 01
RCL 02
+
STO 03
200
X<=Y?
GTO "EC"
RCL 01
RCL 02
*
X=0?
GTO "EE"
LN
STO 04
RCL 03
CHS
E↑X
STO 09
LBL "PI"
1
ST+ 06
RCL 06
LN
ST+ 07
RCL 06
RCL 04
*
2
RCL 07
*
-
RCL 03
-
E↑X
ST+ 09
RCL 09
/
1 E-5
X<=Y?
GTO "PI"
RCL 09
LN
RTN
LBL "EC"
CHS
RTN
LBL "EE"
RCL 04
LN
RTN
LBL "APH"
RCL 09
X↑2
CHS
RCL 03
RCL 04
*
PI
*
SQRT
2
*
LN
-
RTN
.END.
```

Figure 2.34. HP-41C, Program "FIXBED" calculates breakthrough curves for fixed bed sorption processes (Reference 70).

Table 2.14: User Instructions for HP-41C Program "FIXBED"(70)

Step/Procedure	Enter/Keystroke	Printer Display
1. Allocate memory	XEQ ALPHA SIZE ALPHA 025	
2. Load program (7 card sides)		
3. Set in user mode	USER	
4. Start	GTO ALPHA FIXED ALPHA	
5. To calculate J:		
By exact series expansion	CF 01	
By approximations	SF 01	
Enter u, v	u↑v	
Calculate J	J	J
6. To calculate X,Y:		
Start	B	N=?
Enter N	N R/S	T=?
Enter T	T R/S	R=?
Enter R	R R/S	N, T, R
Calculate	A	
(Time delay; wait for beep)		Y, X

Table 2.15: User Instructions for HP-41C Program "DESIGN"(70)

Step/Procedure	Enter/Keystroke	Printer/Display
1. Allocate memory	XEQ ALPHA SIZE ALPHA 025	
2. Load "Fixbed" (7 card sides)		
3. Go to end	GTO	
4. Load "Design" (7 card sides)		
5. Set in user mode	USER	
6. Start	GTO ALPHA DESIGN ALPHA	
7. Input R and X	C	R=?
Enter R	R R/S	X=?
Enter X	X R/S	R, X
8. To calculate N, given T:		
Start	F	T=?
Enter T	T R/S	T
(Time delay; a melody is played after each iteration)		N
9. To calculate N, given NT:		
Start	H	NT=?
Enter NT	NT R/S	NT
(Time delay; a melody is played after each iteration)		N
10. To calculate NT, given N:		
Start	G	N=?
Enter N	N R/S	N
(Time delay, a melody is played after each iteration)		NT

```
01 LBL "DESIGN"
02 LBL C
03 SF 00
04 "R = ?"
05 AVIEW
06 STOP
07 STO 13
08 "X =?"
09 AVIEW
10 STOP
11 STO 20
12 1/X
13 1
14 -
15 STO 21
16 ln
17 1
18 RCL 13
19 -
20 /
21 STO 22
22 FIX 3
23 CF 12
24 " R = "
25 ARCL 13
26 AVIEW
27 FIX 6
28 " X = "
29 ARCL 20
30 AVIEW
31 RTN
32 LBL H
33 " NT = ?"
34 AVIEW
35 STOP
36 STO 10
37 FIX 2
38 " NT = "
39 ARCL 10
40 AVIEW
41 RCL 10
42 RCL 13
43 *
44 STO 11
45 RCL 13
46 1
47 X>Y?
48 GTO "H1"
49 RCL 10
50 RCL 22
51 +
52 STO 11
53 RCL 22
54 X<>Y
55 /
56 1
57 +
58 1/X
59 RCL 13
60 X<=Y?
61 GTO "H1"
62 RCL 10
63 RCL 13
64 /
65 ST+ 11
66 RCL 11
67 2
68 /
69 LBL D
70 STO 11
71 LBL "H1"
72 SF 10
73 CLX
74 STO 19
75 LBL "HI"
76 RCL 11
77 FS? 04
78 STOP
79 XEQ "AA"
80 XEQ "PCN"
81 XEQ "DEN"
82 XEQ "TES"
83 FS? 06
84 GTO "HO"
85 RCL 06
86 RCL 07
87 *
88 ST- 11
89 ABS
90 RCL 11
91 /
92 .001
93 X>Y?
94 GTO "HO"
95 BEEP
96 FS? 10
97 GTO "HI"
98 RCL 11
99 RCL 12
100 *
101 STO 10
102 GTO "HI"
103 LBL F
104 " T = ?"
105 AVIEW
106 STOP
107 STO 12
108 CF 12
109 FIX 3
110 " T = "
111 ARCL 12
112 AVIEW
113 RCL 22
114 1
115 RCL 12
116 -
117 /
118 STO 11
119 RCL 12
120 *
121 STO 10
122 1
123 RCL 13
124 X<=Y?
125 GTO "TI"
126 RCL 21
127 SQRT
128 RCL 13
129 X↑2
130 *
131 LBL I
132 STO 11
133 RCL 12
134 *
135 STO 10
136 LBL "TI"
137 CF 10
138 CLX
139 STO 19
140 GTO "HI"
141 LBL "HO"
142 BEEP
143 BEEP
144 ADV
145 SF 12
146 FIX 2
147 "N ="
148 ARCL 11
149 AVIEW
150 RTN
151 LBL "PCN"
152 RCL 17
153 RCL 00
154 +
155 RCL 14
156 -
157 RCL 21
158 LN
159 -
160 STO 18
161 RTN
162 LBL "DEN"
163 RCL 16
164 RCL 17
165 -
166 E↑X
167 RCL 16
168 RCL 00
169 +
170 RCL 14
171 -
172 RCL 13
173 LN
174 +
175 E↑X
176 +
177 1
178 +
179 RCL 13
180 -
181 STO 05
182 RTN
183 LBL "TES"
184 CF 06
185 RCL 19
186 20
187 X=Y?
188 OFF
189 1
190 ST+ 19
191 .005
192 RCL 18
193 X=0?
194 SF 06
195 X=0?
196 RTN
197 ABS
198 X<=Y?
199 SF 06
200 LASTX
201 /
202 STO 07
203 RCL 18
204 RCL 05
205 /
206 ABS
207 STO 06
208 RTN
209 LBL G
210 " N = ?"
211 AVIEW
212 STOP
213 STO 11
214 FIX 2
215 CF 12
216 " N = "
217 ARCL 11
218 AVIEW
219 CLA
220 RCL 21
221 LN
222 RCL 11
223 X<=Y?
224 OFF
225 1
226 RCL 13
227 X>Y?
228 GTO "GB"
229 -
230 SQRT
231 *
232 STO 10
233 RCL 11
234 RCL 22
235 -
236 X>0?
237 STO 10
238 GTO "G1"
239 LBL "GB"
240 RCL 11
241 RCL 13
242 SQRT
243 2
244 *
245 /
246 LBL E
247 STO 10
248 LBL "G1"
249 CLX
250 STO 19
251 STO 23
252 RCL 11
253 STO 24
254 LBL "GI"
255 RCL 10
256 FS? 04
257 STOP
258 XEQ "AA"
259 RCL 19
260 X>0?
261 RCL 18
262 STO 08
263 XEQ "PCN"
264 CF 07
265 RCL 10
266 RCL 08
267 *
268 X>0?
269 SF 07
270 XEQ "DNT"
271 XEQ "TES"
272 FS? 06
273 GTO "GE"
274 RCL 07
275 X<0?
276 GTO 00
277 RCL 23
278 RCL 10
279 X>Y?
280 STO 23
281 GTO 01
282 LBL 00
283 RCL 24
284 RCL 10
285 X<=Y?
286 STO 24
287 LBL 01
288 RCL 24
289 RCL 23
290 -
291 ABS
292 STO 04
293 FS? 07
294 GTO 03
295 RCL 06
296 X<=Y?
297 GTO 02
298 LBL 03
299 RCL 04
300 2
301 /
302 STO 06
303 LBL 02
304 RCL 06
305 RCL 07
306 *
307 ST+ 10
308 ABS
309 RCL 10
310 BEEP
311 /
312 .001
313 X<=Y?
314 GTO "GI"
315 LBL "GE"
316 BEEP
317 BEEP
318 ADV
319 SF 12
320 FIX 2
321 " NT = "
322 ARCL 10
323 AVIEW
324 RTN
325 LBL "DNT"
326 XEQ "PSI"
327 RCL 13
328 LN
329 ST+ 05
330 RCL 05
331 RCL 17
332 -
333 E↑X
334 RCL 05
335 RCL 00
336 -
337 RCL 14
338 -
339 X>0?
340 CHS
341 E↑X
342 +
343 1
344 +
345 RCL 13
346 -
347 CHS
348 STO 05
349 RTN
350 LBL "EXP"
351 225
352 X>Y?
353 X<>Y
354 E↑X
355 RTN
356 LBL "PSI"
357 FS? 01
358 GTO "APS"
359 CLX
360 STO 06
361 STO 07
362 STO 08
363 LBL "AS"
364 RCL 06
365 RCL 04
366 *
367 RCL 07
368 -
369 RCL 02
370 -
371 1
372 ST+ 06
373 RDN
374 RCL 06
375 LN
376 ST+ 07
377 RDN
378 RCL 06
379 RCL 03
380 *
381 RCL 01
382 -
383 RCL 07
384 -
385 +
386 E↑X
387 X=0?
388 GTO "APS"
389 ST+ 08
390 RCL 08
391 /
392 1 E-3
393 X<=Y?
394 GTO "AS"
395 RCL 08
396 STO 18
397 RTN
398 LBL "APS"
399 RCL 09
400 X↑2
401 CHS
402 RCL 01
403 RCL 02
404 *
405 .25
406 Y↑X
407 2
408 *
409 PI
410 SQRT
411 *
412 X≠0?
413 LN
414 -
415 E↑X
416 RCL 01
417 RCL 02
418 /
419 SQRT
420 .25
421 RCL 02
422 /
423 -
424 *
425 X>0?
426 LN
427 STO 05
428 RTN
429 .END.
```

Figure 2.35. HP-41C "DESIGN" performs three types of design calculations for fixed bed sorption processes (Reference 70).

In measuring the performance of adsorbents in sugar decolorization operations, the transmission through a cell at 470 nm for a given dried solids concentration is compared to the same dried solids concentration of sucrose. Asai (72) has shown that contact time is the controlling parameter (Figure 2.36) and that several columns are necessary to obtain the desired degree of color removal at reasonable flow rates (Table 2.16).

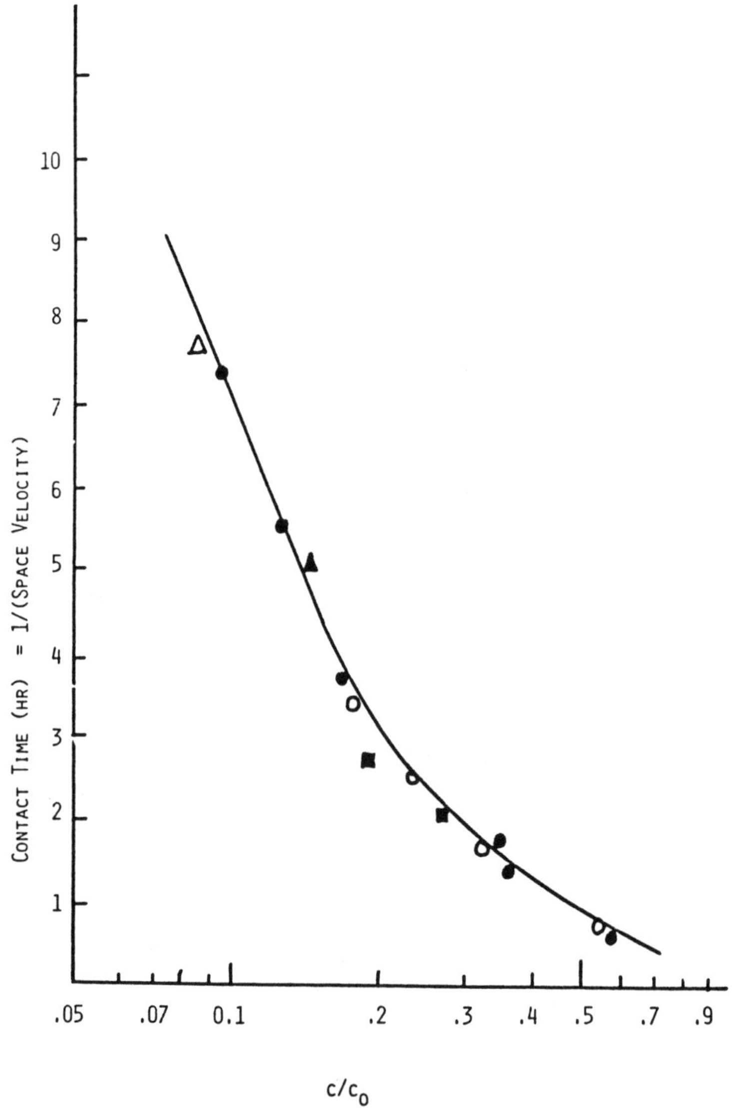

Figure 2.36. The effect of contact time on the residual colored impurities in sugar solutions (Reference 72).

Table 2.16: Effect of Multiple Column Treatment on Effluent Quality (C/Co)(72)

Experiment Number	Flow Rate (cm/min)	C/Co of Effluent			
		Effluent from Column 1	Effluent from Column 2	Effluent from Column 3	Effluent from Column 4
A-1	1	0.38	0.20	0.14	0.11
A-2	2	0.56	0.33	0.24	0.18
A-3	3	0.60	0.37	0.27	0.19

As an alternative to using charcoal based adsorbents to remove color, Abrams (73) has shown that phenol-formaldehyde resins may be used in treating dilute or concentrated sugar solutions. Duolite S-30 removed 80-100% of the color from decationized second carbonation juice of beet sugar. This resin was effective in adsorbing the carmel and melanoidin colorants when the feed solution has a pH between 1 and 6. Regeneration of the resin adsorbent was accomplished by eluting the colorants with 1 to 4% sodium hydroxide. This was followed by rinsing with dilute acid before the next adsorption cycle began.

This type of resin has also been used for the removal of brown color from wine in Japan (74) and for decolorizing fermentation broths. Adsorbents from phenol-formaldehyde amine resins are particularly useful in removing flavonoid pigments which are frequently present in the extracts of plant materials (75). These resins decolorize these types of feed streams in the pH 4-7 range.

The use of carbon and other adsorbents in decolorizing fermentation broths is now so common that it is only casually mentioned as part of the purification process.

2.7.2 Antibiotics

Macroporous polymeric adsorbents have been used to recover and purify many antibiotics. The performance of a polystyrene resin with a surface area of 300 m^2/g and an average pore diameter of 100 Å is shown in Table 2.17 (76). Methanol was used for the elution of the adsorbed antibiotic.

Activated carbon has been shown (77) to remove Cephalosporin C effectively from its fermentation broth. Elution was carried out with dilute aqueous solvents. A

Table 2.17: Performance of Amberlite XAD-2 as an Adsorbent for Antibiotics (76)

Antibiotic	Molecular Wt.	Conc. in Feedstream (ppm)	Saturation Capacity (mg/ml)	Elution: Peak (ppm)	Elution: Elution Required (Bed Volumes)
Tetracycline	444	1000	44	30,000	2
Tetracycline	444	100	7.4	16,000	2
Oxytetracycline	460	100	7.6	17,600	2
Oleandomycin	688	100	40	20,000	2

difficulty with using carbon is its lack of specificity for the Cephalosporin C.

Voser (78) developed a scheme for purifying Cephalosporin C using a macroporous polymeric adsorbent. The concentration of Cephalosporin C in the filtrate from the fermentation broth was 2.1 g. This filtrate was acidified with sulfuric or oxalic acid to adjust the pH to 2.3. The filtrate was fed through a column of a styrene-divinylbenzene resin (XAD-2) at a rate of one bed volume per hour. The height of the resin bed was 76 cm. The elution was carried out with one-half bed volume of deionized water, followed by two and one-half bed volumes of 10% aqueous isopropanol solution. Regeneration of the adsorbent was accomplished with one bed volume of 50% aqueous methanol adjusted to 1 N with NaOH, followed by a water rinse.

This technique removed 65 to 70% of the impurities during the water wash and the first third of a bed volume of the isopropanol elution. The rest of the isopropanol elution contained 95% of the Cephalosporin C and 20 to 25% of the impurities from the filtrate. The rest of the product and impurities were removed during the regeneration. When the flow rate was increased to 2 bed volumes per hour, the Cephalosporin C yield dropped to 85.2% while the impurities remaining with the product increased to 30 to 35%.

When carbon (Pittsburgh CAL) was used as the adsorbent (79), the elution required washing with four times as much deionized water (two bed volumes) and using almost twice as much elution solution (4.5 bed volumes of 50% aqueous acetone). Only the first half bed volume could be discarded without a significant reduction in product yield. This means that a much larger volume of eluate had to be further treated to obtain product of the desired purity when carbon was the adsorbent.

The distribution coefficients for different penicillins and cephalosporins between hydrophobic adsorbents and aqueous methanol solutions have been determined as a function of pH (80). These are shown in Table 2.18 and illustrated in Figure 2.37.

Table 2.18: Distribution Coefficients in 10% Aqueous Methanol for Penicillins and Cephalosporins on XAD-4 as a Function of pH

Antibiotic	1.3	2.3	3.7	5.5	6.0	6.8	7.6
Penicillin G, K^+	-	1,150	751	80.2	81.1	76.4	62.3
Penicillin V	-	-	2,590	108	106	96.6	-
7-amino cephalosporin	-	12.1	12.8	15.2	-	15.3	16.2
7-aminodesacetoxi-cephalosporin	11.1	5.1	6.9	6.25	-	5.34	4.42
sodium cephazoline	-	-	-	103	98.8	85.5	97.5

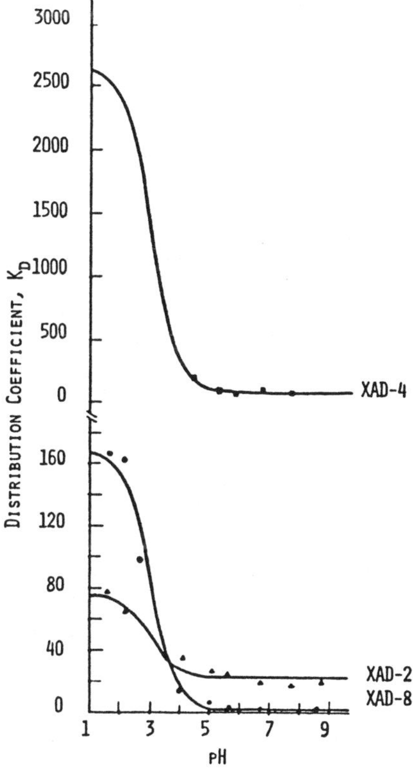

Figure 2.37. The effect of pH on the distribution coefficient for cephalosporin antibiotics on polymer adsorbents (Reference 80).

Chu and Jabuke (81) have shown the effect of antibiotic type and concentration on adsorption by polymeric polystyrene adsorbents (XAD-2). When 20 g of adsorbent was added to a 100 mL solution of streptomycin (5mg/mL), all of the antibiotic was adsorbed on the resin, for a loading of 25 mg per gram of resin. When the solution is changed to lincomycin (1.95 mg/mL), the loading level may be increased to 31.4 mg per gram of resin but only 80% of the antibiotic is adsorbed from solution. Even higher loading levels are possible (57.5 mg per gram of resin) by increasing the lincomycin concentration to 4 mg/mL; however, the amount of antibiotic remaining in solution increases to almost 90%. Neomycin was poorly adsorbed (7.5 mg per gram of resin) from a solution with a 5 mg/mL antibiotic concentration, so that only 15% of the antibiotic was removed from solution.

In all of these cases, the elution was carried out by washing the loaded resin with hot water followed by three bed volumes of 97% aqueous methyl ethyl ketone to recover 99% of the adsorbed antibiotic.

The antibiotic A-4696 has been recovered and purified using two different grades of activated carbon (82). Pittsburgh (12 x 40) carbon was used for the initial recovery, while Darco G-60 was used for the further purification. The elution used an acidified acetone-water (1:1) solution as the eluant for each step. When sulfuric acid was used as the acidifying agent, the sulfate ions had to be removed from the eluate by precipitation as barium sulfate. The yield was reduced by 25% during the purification process.

2.7.3 Enzymes

An alkaline protease (Mol. Wt. 28,000) was recovered from a fermentation broth using a macroporous acrylate non-functionalized adsorbent (83). The effluent concentration of the enzyme is shown as a function of column throughput in Figure 2.38. The resin adsorbed 73% of the enzyme, with a loading of 776,200 AZDU units per gram of resin. After a rinsing with water to remove non-adsorbed enzymes and fermentation impurities, an ethanol-water (1:1) solution with 5.7% NaCl was used to eluate the protease. As the figure shows, good concentration of the enzyme was obtained.

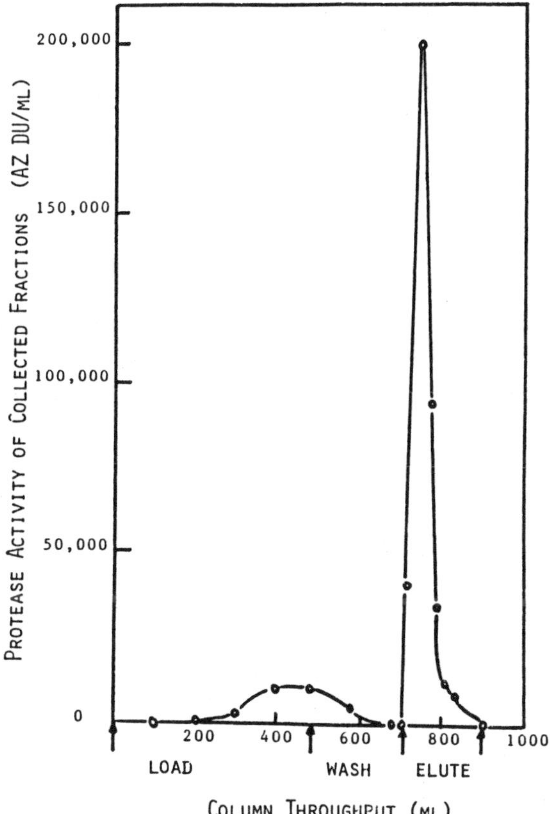

Figure 2.38. Recovery of alkaline protease from fermentation broth with Amberlite XAD-7 (Reference 83).

Volclay KWK, a western bentonite clay adsorbent, has been used to recover lipolytic enzymes from the fermented growth product of *Mucor* microorganisms (84). The adsorption step was carried out by adjusting the growth product to a pH of about 10.0 with 2N NaOH and thoroughly admixing with the adsorbent at ambient temperature for about one hour. The adsorbent material was then separated from the non-adsorbed rennet enzyme material by filtration using 2.0% filteraid. The desired lipolytic enzyme was then eluted from the adsorbent by adjusting the pH to 4.0 with 2N HCl and the solution was evaporated to one sixth the original fluid volume. The recovery was about 100%. The sensitivity of the recovery to the pH adjustment was demonstrated by the yield dropping to only 22.5% when the pH was adjusted to 11.0 rather than 10.0 during the adsorption process. Likewise, when the first pH adjustment was to pH 5.0 instead of 10.0, there was 0% lipolytic enzyme recovered but rather 72.7% of the rennet was recovered.

2.7.4 Microorganism Adsorption

Adsorbents may also be used for the adsorption of microorganisms. While ion exchange resins have been used for much of the reported work, comprehensively reviewed by Daniels (85), charcoal, silica gel and alumino silicates have been used for bacteria, fungi and virus adsorption. Specific examples are listed in Table 2.19.

Table 2.19: Microorganism Adsorption on Charcoal and Silicates (85)

Microorganism	Adsorbent	Capacity (cells/g)
Bacillus subtilis	Charcoal	50×10^{10}
Escherichia coli	Charcoal	50×10^{10}
Escherichia coli bacteriophage T_4	Activated carbon	1.6×10^{12}
Escherichia coli bacteriophage T_2	Activated carbon	3×10^{8}
Escherichia coli bacteriophage MS_2	Bituminous coal	4×10^{6}
Escherichia coli bacteriophage f_2	Activated carbon	$2\text{-}46 \times 10^{5}$
Poliovirus type 1	Activated carbon	6.5×10^{6}
Poliovirus type 1	Silicates	7-9 10^{5}

This type of adsorption of suspended cells on solid surfaces is an additional method of control in isolating components and purifying mixtures. Adsorbents can change the rate of reproduction, morphology of cells, duration of growth cycles, oxygen consumption, carbon dioxide production and the character and amount of metabolites (86). The specific applications of microbial adsorption are listed in Table 2.20 from the review by Daniels (87).

Table 2.20: Applications of Microbial Adsorption(87)

1. Concentration of Cells

   Stimulation or suppression of growth

   Morphology

   Growth cycles

   Oxygen consumption

   Carbon dioxide production

   Metabolic studies of adsorbed vs unadsorbed cells

   Reactants and/or products

   Control of pH

2. Separation of Cells from Suspension

   Nonselective removal

   Selective removal of pathogens

3. Resolution of Cellular Mixtures

   Grossly different microorganisms

   Similar microorganisms

   Distinct species or cell types

2.7.5 Organic Acids and Amines

Charcoal adsorbents have been used to remove organic acids and amines from aqueous solutions. The bed volumes of the feedstream are given in Table 2.21 (88). The amounts of regenerant, 1.0N NaOH, required are also listed. Since charcoal does not have the fixed ion exchange groups which ion exchange resins have, it does not exhibit the ion exclusion (Donnan) effect, causing slower regeneration, especially for the more hydrophobic solutes.

From Table 2.21 it can also be seen that within a given homologous series, a point is reached where the higher members are not easily removed from the charcoal. Also within a series, the extent of adsorption is inversely

Table 2.21: Sorption of Organic Acids and Amines on Charcoal (88)

Solute	Feed Concentration (%)	Breakthrough Point (BV)	Regeneration Requirement (BV)
Formic acid	0.1	4.0	<2
Acetic acid	1.0	6.3	<2
Acetic acid	0.1	21	<2
Oxalic acid	0.1	22	<2
Malonic acid	0.1	36	<2
Propionic acid	0.1	42	2.5
Valeric Acid	0.1	116	13
Caprylic acid	0.1	295	>100
Ethylenediamine	0.1	9.3	<2
Ethylamine	0.1	11	<2
Piperazine	0.1	23	<2
Diethylenetriamine	0.1	24	<2
n-Propylamine	0.1	34	5
n-Amylamine	0.1	95	22
Benzylamine	0.1	123	74

proportional to the water solubility of the compound. The adsorption of monofunctional organic acids and amines is always greater than for polyfunctional molecules of the same molecular weight. Substitution of hydrophilic groups (-OH) lowers the quantity of solute adsorbed and increases the speed of regeneration. Conversely, substitution of hydrophobic groups (-Cl) increases the quantity of solute adsorbed and decreases the efficiency of regeneration.

Paleos (89) studies the adsorption of organic acids and phenols on hydrophobic (styrene-based) and hydrophilic (methacrylate-based) adsorbents. As Figure 2.39 shows, XAD-6 with its higher surface polarity adsorbs more acetic acid but less butyric acid than XAD-7. For both resins, as the chain length of the acid is increased, the amount of adsorption increases, consistent with Traube's Rule.

The adsorption of substituted phenols increases for the hydrophobic adsorbent XAD-4 as the substitution causes the phenol to become more hydrophobic. This is shown in Figure 2.40.

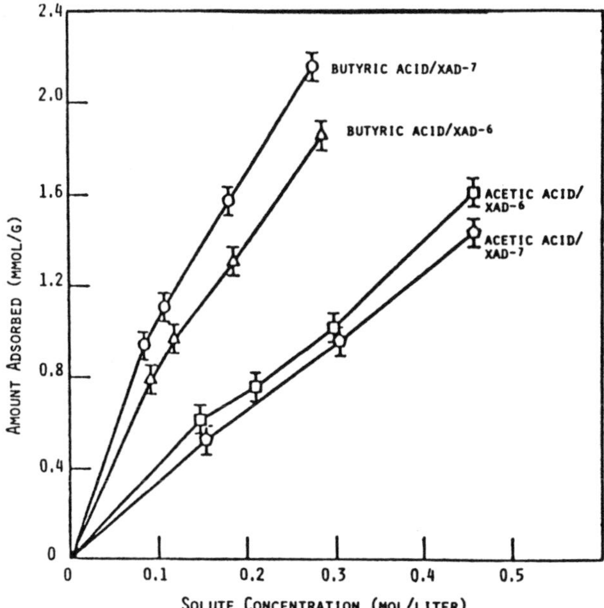

Figure 2.39. Adsorption of acetic acid and butyric acid from aqueous solution onto hydrophobic (XAD-6) and hydrophilic (XAD-7) adsorbents (Reference 89).

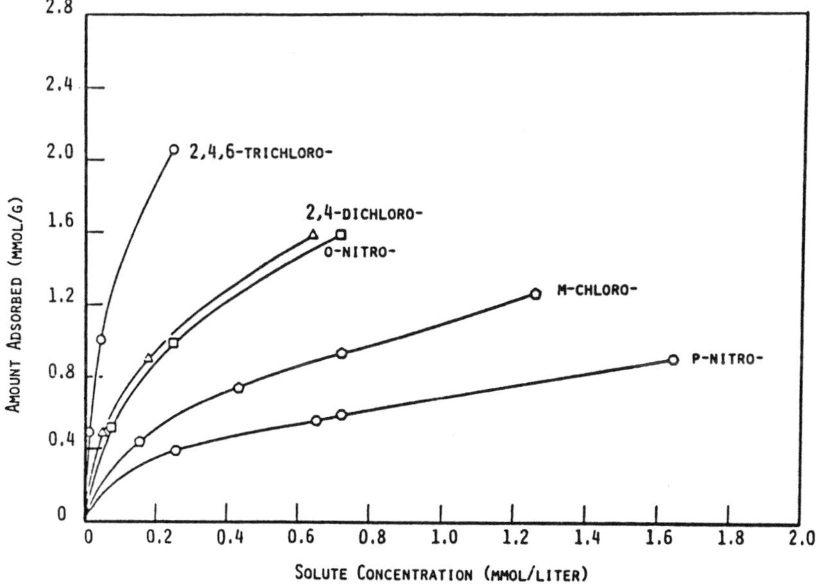

Figure 2.40. Adsorption of substituted phenols from aqueous solutions on Amberlite XAD-4 at 25°C (Reference 89).

2.7.6 Vitamin Adsorption

Porous crosslinked styrene adsorbents were used to examine the effect of porosity, crosslinking, surface area, vitamin concentration and flow rates on the adsorption of Vitamin B-12 (90). The degree of crosslinking, porosity, surface area and equilibrium adsorption are shown in Table 2.22. The porosity and the surface area can be seen to be controlling the amount of vitamin adsorption. This is emphasized by the lack of adsorption seen with the non-porous adsorbent (No. 7).

Table 2.22: Adsorption of Vitamin B-12 on Styrene Adsorbent

Adsorbent	Crosslinking	Porosity (cm^2/g)	Surface Area (m^2/g)	Vitamin B-12 Adsorbed (mg/g)
1	15	1.89	130	3.10
2	20	1.80	174	3.25
3	25	1.77	210	3.00
4	15	0.99	65	2.45
5	20	0.89	232	3.50
6	25	1.10	342	3.85
7	8	0.05	nil	0.00

The decrease in adsorption for adsorbent No. 3, even though it has a higher surface area than No. 2, is most likely due to an unfavorable distribution of the pore sizes which do not admit the vitamin molecule.

The effect of vitamin concentration is shown in Figure 2.41 and the effect of flow rate is shown in Figure 2.42. After a certain vitamin concentration is reached (about 75 mg/dm^3), there is no additional increase in adsorption with further concentration increases. As expected, increasing the flow rate in column operations is seen to decrease the amount of feedstream treated before breakthrough. At a flow rate of 4.5 bed volumes per hour, 71.25 bed volumes of vitamin B-12 solution is treated before breakthrough. This is eluted from the adsorbent with 2 bed volumes of methanol, for a concentration increase of 35.62 times.

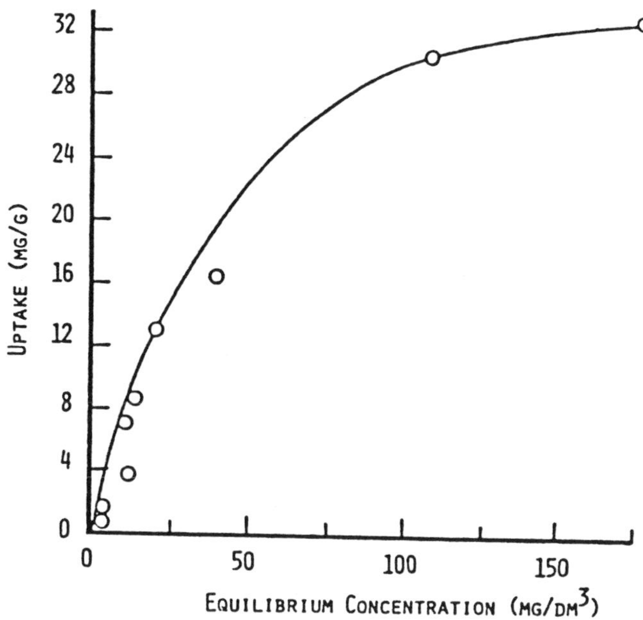

Figure 2.41. The effect of the concentration of the equilibrated solution of vitamin B-12 on its uptake by a porous polymeric adsorbent at 30°C (Reference 90).

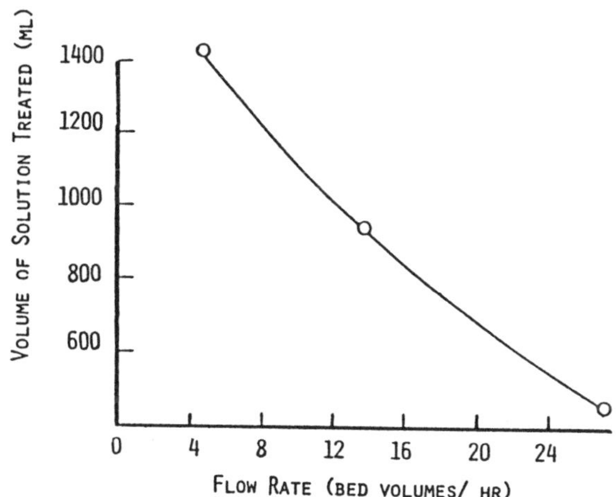

Figure 2.42. The effect of flow rate on the adsorption of vitamin B-12 by the adsorbent used in Figure 2.41 (Reference 90).

Kennedy (84) has examined the adsorption of vitamin B-12 from a fermentation broth using commercial crosslinked styrene adsorbents and weak acid ion exchange resins. Table 2.23 compares the adsorption and recovery with these two types of materials. The styrene adsorbent has a much higher capacity for vitamin B-12 and can be eluted more efficiently than the acrylic ion exchange resin. The concentration of vitamin B-12 in the feedstream was 15 ppm and the flow rate was 40 L/min-m^3. Methanol was again used as the eluting fluid.

Table 2.23: Adsorption and Elution of Vitamin B-12

	Adsorption		Elution	
	Capacity at Breakthrough (mg/ml)	Capacity at Saturation (mg/ml)	Elution Concentration (ppm)	Methanol Required (Bed Volumes)
Amberlite IRC-50	0.03	0.14	150	5
Amberlite XAD-2	3.5	5.2	7200	2

2.7.7 Miscellaneous Product Purifications

Glucuronides and sulfates have been recovered from aqueous solution (91) using hydrophobic polymeric adsorbents. The loadings were typically 10 μg per gram of resin at flow rates up to 0.4 bed volumes per minute. An important aspect of this work was the discovery of a suitable solvent, acetone-water, for the elution of these compounds (Table 2.24), with nearly quantitative recovery. The smaller pore size and larger surface area of XAD-4 allowed more complete rinsing of inorganic salt contaminants without loss of the adsorbed compounds than did XAD-2 with its larger pores and smaller surface area.

Table 2.24: Elution of Glucuronides and Sulfates from XAD-4 with 200 ml of Eluant (91)

Compound	Percent Recovered with	
	Methanol	Acetone - Water (1:1)
p-Nitrophenyl sulfate	2.2 ± 2.6	86.8 ± 1.3
p-Nitrocatechol sulfate	0	0
p-Nitrophenyl glucuronide	29.2 ± 6.8	91.0 ± 3.7
Androsterone glucuronide	9.5 ± 3.5	78.7 ± 12.9

Carcinogenic polycyclic aromatic compounds, such as benz(a)pyrene, in concentrations as low as 15 ppb were removed from crude normal paraffins (10 to 25 carbon atoms) using an upflow adsorption operation through alumina/silica adsorbents (92). The impurity concentration was reduced to less than 1 ppb using a flow rate of 5.2 liters/hour at 50°C. Table 2.25 shows the impurity levels obtained for various amounts of treated fluid per unit of adsorbent. These treated normal paraffins are acceptable for use as starting materials in food petroleum fermentation products.

Table 2.25: Residual Benz(a)Pyrene in Normal Paraffins After Alumina and Alumina/Silica Treatment(92)

Adsorbent	Average Surface Area (m^2/g)	Particle size(mesh)	Amount of paraffins treated/unit adsorbent	Residual benz(a)pyrene(ppb)
*Alumina	200	50	20-30	< 1
*Alumina containing 10% silica	400	50	14-32	< 1
Silica	730	10	20-30	5
Alumina	200	100	20-30	7

*Examples of preferred treatment

2.7.8 Ethanol Adsorption

Walsh and coworkers (93) developed a process for separating ethanol from dilute aqueous fermentation media. In this process, a portion of the fermentation media is passed over activated carbon which adsorbs the ethanol and a small portion of the water. When the adsorbent has reached its adsorption limit, the influx of fermentation medium is halted and a carrier gas, such as CO_2 from the same fermentation process, is used to desorb the adsorbed ethanol and water. The desorbed vapors are passed through a cellulose adsorbent to remove the water, yielding a pure ethanol stream.

This process allows the concentration of ethanol in the fermentor to remain at 6% for minimum inhibition of the fermentation process. The adsorbed phase on the activated carbon concentrates this to approximately 50% by weight ethanol. After this stream emerges from the cellulosic adsorbent, the ethanol content is 95%.

While this technique appears technically feasible with the type of process shown in Figure 2.43, it has not yet been shown to be economically feasible.

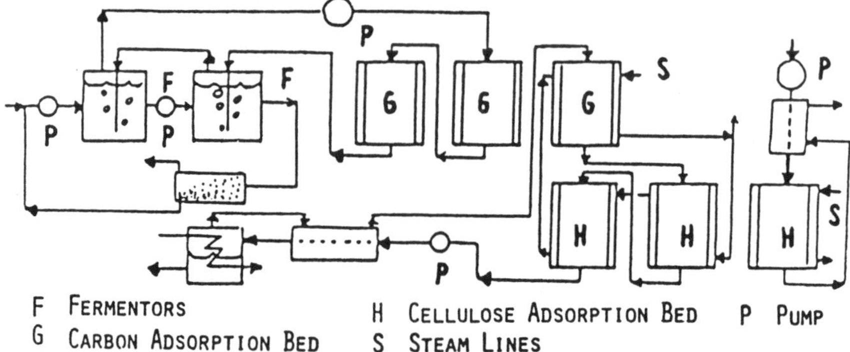

Figure 2.43. Proposed Fermentation-Stripping-two stage adsorption process from Reference 93.

The zeolite adsorbent, ZSM-5, has also been reported to allow the adsorption ethanol from a fermenting mixture so that the ethanol concentration remains below the limit where it becomes toxic to the fermenting microorganism (94). This material has an adsorption capacity of 65 to 85 mg of ethanol per gram of zeolite. The ethanol concentration of an aqueous ethanol solution (5 to 10% ethanol) is reduced by 57 to 86% when the solution is treated with ZSM-5.

These authors suggested the use of butane in a "pressure swing" regeneration. That is, after the elution of ethanol with butane, the pressure is reduced to remove the butane from the ethanol and the butane is recovered for the next elution of ethanol. When the zeolite loses a significant portion of its adsorptive efficiency after several cycles of use due to the accumulation of organic matter in its pores, the efficiency can be recovered by burning the organic matter away at temperatures of 450-650°C.

Feldman (95) uses crosslinked poly(4-vinylpyridine) to remove low concentrations of ethanol from aqueous streams. A maximum adsorption of 0.33 g ethanol per gram of resin was obtained with solutions containing 10.1% ethanol. However, when the concentration of ethanol is above 5%, the time for equilibrium adsorption becomes quite long. Table 2.26 compares the percent adsorption of ethanol and other potential fermentation products on this polymer with that on two different carbon adsorbents. The usefulness proposed was in keeping the ethanol level below the limit that reduces the productivity of the microorganisms by continually adding and removing this type of adsorbent and using the CO_2

generated by the fermentation reaction to strip the adsorbed ethanol from the adsorbent.

Table 2.26: Adsorption of Fermentation Products on Poly(4-Vinyl-Pyridine) Compared with Granular and Powdered Charcoal

	Percent Adsorption at 25°C		
	On PVP	On PCB Granular Charcoal	On PGL Powder Charcoal
Water	33	45	57
Methanol	92	38	48
Ethanol	80	35	49
Isopropanol	40	34	45
Tert-Butanol	19	35	49
Acetic Acid	295	81	90

2.7.9 Adsorption as a Fermentation Aid

The removal of volonomycin from a fermentation broth of *S. paulus*, UC 5142 with a polymeric adsorbent has been shown to protect the *S. paulus* from its antibacterial effect and allows its synthesis to occur at higher levels (96). This is shown in Figure 2.44.

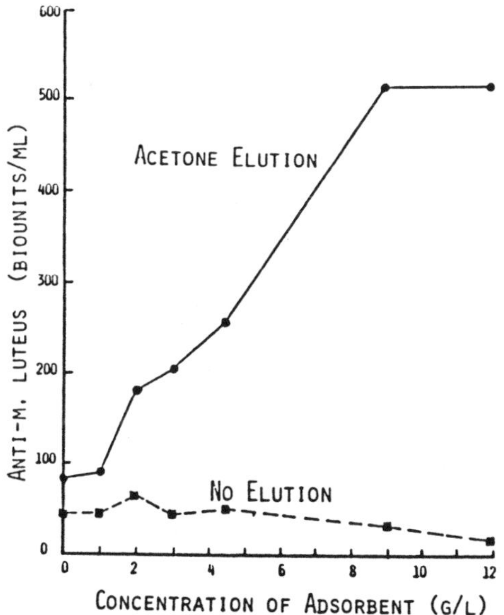

Figure 2.44. Effect of removal of volonomycin from XAD-2 by acetone treatment on the fermentation production of anti-*M. Luteus* (Reference 96).

Adsorbents have also been added to fermentation broths to enhance the microbial production on monoterpenes such as geraniol, citronellol, nerol and linalool by adsorbing excreted metabolites that would otherwise inhibit product concentration (97). These effects are shown in Table 2.27.

Table 2.27: Effect of Presence of Amberlite XAD-2 on Terpenoid Product Concentration (97)

Terpenoid	Product Yields (mg/L) at 162 Hour Point	
	Normal Fermentation	With XAD-2
Geraniol	334	1227
Citronellol	60	198
Linalool	1	46
Nerol	1	28
Total	395	1499

2.8 REFERENCES

1. Freundlich, H., Colloid and Capillary Chemistry, Methuen, London (1926)

2. Boedecker, C., J Landwirtschaft, 7:48 (1859)

3. Kuester, F.W., Annalen, 283:360 (1894)

4. Langmuir, J., J Am Chem Soc, 40:1361 (1981)

5. Jossens, L., Prausnitz, J.M., Fritz, W., et al., Chem Eng Sci, 33:1097 (1978)

6. Henry, D.C., Phil Mag, [6]44:689 (1922)

7. Beverloo, W.A., Pierik, G.M., Luyben, K., In: Fundamentals of Adsorption (Myers, A.L., Belfort, G., eds.) Engineering Foundation, New York, p 95 (1984)

8. Hasanain, M.A., Hines, A.L., Ind Eng Chem Processes Des Dev, 20:621 (1981)

9. Giles, C.H., MacEwan, T.H., Nakhwa, S.N., et al., J Chem Soc, 3973 (1960)

10. Gustafson, R.L., Albright, R.L., Heisler, J., et al., Ind Eng Chem Prod Res Develop, 7:107 (1968)

11. Young, G.J., Chessick, J.J., Healey, F.H., J Phys Chem, 60:394 (1956)

12. Greenhill, E.B., Trans Fraaday Soc, 45:625 (1949)

13. Swearingen, L.E., Dickenson, B.N., J Phys Chem, 36:534 (1932)

14. Rothman, S., Mandel, J., McCann, F.R., Weissberg, S.G., J Colloid Sci, 10:338 (1955)

15. Crittenden, J.C., Hand, D.W., In: Fundamentals of Adsorption (Myers, A.L., Belfort, G., eds.) Engineering Foundation, New York, p 185 (1984)

16. Spahn, H., Schleunder, E.U., Chem Eng Sci, 30:529 (1975)

17. Neretrieks, I., Chem Eng Sci, 31:465 (1976)

18. McKay, G., Chem Eng Res Des, 62(7):235 (1984)

19. Leyva-Ramos, R., Geankoplis, C.J., Chem Eng Sci, 40(5):799 (1985)

20. Rasmuson, A., In: Fundamentals of Adsorption (Myers, A.L., Belfort, G., eds.) Engineering Foundation, New York, p 461 (1984)

21. Kataoka, T., Yoshida, H., Ueyama, K., J Chem Eng (Japan), 5(2):132 (1972)

22. Gorg, D.R., Rutliven, D.M., A I Ch E J, 19:852 (1973)

23. Brauch, V., Schleunder, E.U., Chem Eng Sci, 30:539 (1975)

24. Lukchis, G.M., Chem Eng, 111 (June 11, 1973)

25. Rasmuson, A., Chem Eng Sci, 40(4):621 (1985)

26. Rasmuson, A., Chem Eng Sci, 40(7):1115 (1985)

27. Brunauer, S., Emmett, P.H., Teller, E., J Am Chem Soc, 60:309 (1938)

28. Abrams, I.M., Ind Eng Chem Prod Res Dev, 14(2):108 (1975)

29. Smith, R.N., Quart Rev Chem Soc, 13:287 (1959)

30. Guisti, D.M., Conway, R.A., Lawson, C.T., J Water Poll Control Fed, 46(5):947 (1974)

31. Calgon Test Method TM-4

32. Calgon Test Method TM-3

33. ASTM D-3467-76

34. Carrubba, R.V., Urbanic, J.E., Wagner, N.J., et al., A I Ch E Symp Ser, 80(233):76 (1984)

35. Snyder, L.R., J Chromatog, 23:388 (1966)

36. Anderson, R.A., Sherman, J.D., A I Ch E Symp Ser, 80(233):118 (1984)

37. Strandberg, G.W., Smiley, K.L., Biotechnol Bioeng, 14:509 (1972)

38. Messing, R.A., U.S. Patent No. 3,868,304 (1975)

39. Lee, Y.Y., Fratzke, A.R., Wun, K., et al., Biotechnol Bioeng, 18:389 (1976)

40. Abrams, I.M., Chem Eng Progr, 65(97):106 (1969)

41. Gustafson, R.L., U.S. Patent No. 3,531,463 (Sept 1979)

42. Ash, S.G., Brown, R., Everett, D.H., J Chem Thermodynamics, 5:239 (1973)

43. Wang, H., Duda, J.L., Radke, C.J., J Colloid Interface Sci, 66:153 (1978)

44. Giles, C.H., Mehta, H.V., Rahman, S.M.K., et al., J Appl Chem, 9:457 (1959)

45. Rosene, M.R., Lutchko, J.R., Deithom, R.T., et al., Presentation at ACS National Meeting, Miami, FL, September 1981

46. Rosene, M.R., ACS Presentation, March 1981

47. Berthier, P., Kerlan, L., Courty, C., Compt Rend, 246:1851 (1958)

48. Blackburn, A., Kipling, J.J., J Chem Soc, 4103 (1955)

49. Ores, B., Rauber, C., U.S. Patent No. 3,741,949 (June 26, 1973)

50. Mathews, A.P., A I Ch E Symp Ser, 79(230):18 (1983)

51. Patil, V.K., Joshi, J.B., Sharma, M.M., Chem Eng Res Des, 62:247 (1984)

52. Jordon, D., "Carbon Decolorization", In: Fermentation and Biochemical Engineering course, The Center for Professional Advancement, New Brunswick, NJ (1983)

53. Fornwalt, H.J., Hutchins, R.A., Chem Eng, 179 (April 11, 1966)

54. Hager, D.G., Industrial Water Engineering, 11(1):14 (1974)

55. McKay, G., Chem Eng Res Des, 61(1):29 (1983)

56. Hsu, E.H., Fan, L.T., "A Novel Porous Medium Filter", In: Proceedings of World Filtration Congress III, Philadelphia, PA, p 420 (September 13-17, 1982)

57. Mathews, A.P., Fan, L.T., A I Ch E Symp Ser, 79(23):79 (1983)

58. Tavares, A., Kunin, R., Publ Tech Pap Proc Annu Meet Sugar Ind Technol, 40:242 (1981)

59. Hutchins, R.A., Chem Eng Prog, 71(5):80 (1975)

60. DeJohn, P.B., Chem Eng, 113 (April 28, 1975)

61. LeVan, L.D., Friday, D.K., In: Fundamentals of Adsorption (Myers, A.L., Belfort, G., eds.) Engineering Foundation, New York, p 295 (1984)

62. Dessau, R.M., Haag, W.O., U.S. Patent No. 4,442,210 (April 10, 1984)

63. McCaffrey, D.J.A., Rogers, P.E., Great Britain Patent No. 2151501A (July 24, 1985)

64. Stevens, B.W., Kerner, J.W., Chem Eng, 84 (Feb 3, 1975)

65. Wallis, P.A., Bolton, E.E., Presented at Oak Ridge National Lab Meeting (August 1984)

66. Johnston, W.A., Chem Eng, 87 (Nov 27, 1972)

67. Lukchis, G.M., Chem Eng, 83 (Aug 6, 1973)

68. Tan, H.K.S., Chem Eng, 117 (March 24, 1980)

69. Hiester, N.K., Vermeulen, T., In: Perry's Chemical Engineers Handbook, Sec. 16, 5th Ed, McGraw-Hill, NY (1973)

70. Tan, H.K.S., Chem Eng, 57 (Dec 24, 1984)

71. Barton, W.F., Knebel, W.J., 41st Annual Meeting, Sugar Industry Technologists, Inc.

72. Asai, S., Proc Res Soc Japan Sugar Refin Technol, 16:1 (1965)

73. Abrams, I.M., Sugar Y Azucar, 31 (May 1971)

74. ( ), J Ferment Technol Japan, 46(3):153 (1968)

75. Abrams, I.M., Ind Eng Chem Prod Res Devel, 14(2):108 (1975)

76. Kunin, R., Amber-hi-lites Porous Polymers as Adsorbents - A Review of Current Practice, Rohm and Haas Co. Brochure

77. Abraham, E.P., Newton, G.G.F., U.S. Patent No. 3,093,638 (1963)

78. Voser, W., U.S. Patent No. 3,725,400 (April 3, 1973)

79. Pines, S., U.S. Patent No. 3,983,108 (September 28, 1976)

80. Salto, F., Diez, M.T., Prieto, J.G., An Frac Vet Leon Univ Oviedo, 25(1):31 (1979)

81. Chu, D.Y., Jabuke, H.K., U.S. Patent No. 3,515,717 (June 2, 1970)

82. Hamill, R.L., Stark, W.S., DeLong, D.C., U.S. Patent No. 3,952,095 (April 20, 1976)

83. Kennedy, D.C., Ind Eng Chem Prod Res Develop, 12(1):56 (1973)

84. Moskowitz, G.J., Como, J.J., Feldman, L.I., U.S. Patent No. 3,899,395 (Aug 12, 1975)

85. Daniels, S.L., In: Adsorption of Microorganisms to Surfaces (Bitton, G., Marshall, K.C., eds.) J. Wiley & Sons, NY, p 7 (1980)

86. Zuyagintsev, D.G., Nauch Dokl Vyssh Shk, Biol Nauki, 3:97 (1967)

87. Daniels, S.L., Dev Ind Microbiol, 13:211 (1972)

88. Sargent, R.N., Graham, D.L., Ind Eng Chem Proc Des Dev, 1(1):56 (1962)

89. Paleos, J., J Coll Interf Sci, 31(1):7 (1969)

90. Anand, P.S., Dasare, B.D., J Chem Tech Biotechnol, 31:213 (1981)

91. White, J.D., Schwartz, D.P., J Chromatog, 196:303 (1980)

92. Funakubo, E., Matsuo, T., Taira, T., et al., U.S. Patent No. 3,849,298 (Nov 19, 1974)

93. Walsh, P.K., Liu, C.P., Findley, M.W., et al., In: Fundamentals of Adsorption (Myers, A.L., Belfort, G., eds.) Engineering Foundation, NY, p 667 (1984)

94. Dessau, R.M., Haag, W.O., U.S. Patent No. 4,442,210 (April 10, 1984)

95. Feldman, J., U.S. Patent No. 4,450,294 (May 22, 1984)

96. Marshall, V.P., Little, M.S., Johnson, L.E., J Antibiot, 34(7):902 (1981)

97. Schindler, J., Ind Eng Chem Prod Des Dev, 21:537 (1982)

98. Everett, D.H., In: Fundamentals of Adsorption (Myers, A.L., Belfort, G., eds.) Engineering Foundation, NY, p 1 (1984)

3

Ion Exchange

3.1 INTRODUCTION

Separation and purification operations with ion exchange resins involve the reversible interchange of ions between a functionalized insoluble resin (the ion exchange material) and an ionizable substance in solution. While Thompson (1) reported in 1850 the first ion exchange applications which used naturally occurring clays, ion exchange resins have only been used in biochemical and fermentation product recovery for the last few decades (2,3). In these early studies, biochemicals such as adenosine triphosphate (4), alcohols (5), alkaloids (6), amino acids (7), growth regulators (8), hormones (9), nicotine (10), penicillin (11), and vitamin B-12 (12) were purified using ion exchange resins.

Ion exchange applications intensified following the work of Moore and Stein (13), which showed how very complex mixtures of biochemicals, in this case amino acids and amino acid residues, could be isolated from each other using the ion exchange resin as a column chromatographic separator. Specific chromatographic applications of ion exchange resins will be covered in Chapter 4. This chapter will concentrate on the theory, characterization and process applications of ion exchange resins.

In biotechnology applications, ion exchangers are important in preparing water of the necessary quality to enhance the desired microorganism activity during fermentation. Downstream of the fermentation, ion exchange resins may be used to convert, isolate, purify or concentrate the desired product or by-products.

3.2 THEORY

The important features of ion exchange reactions are that they are stoichiometric, reversible and possible with any ionizable compound. The reaction that occurs in a specific length of time depends on the selectivity of the resin for the ions or molecules involved and the kinetics of that reaction.

The stoichiometric nature of the reaction allows resin requirements to be predicted and equipment to be sized. The reversible nature of the reaction, illustrated as follows:

$$R\text{-}H^+ + Na^+Cl^- \rightleftharpoons R\text{-}Na^+ + H^+Cl^- \qquad (3.1)$$

allows for the repeated reuse of the resin since there is no substantial change in its structure.

The equilibrium constant, K, for equation 3.1 is defined for such monovalent exchange by the equation:

$$K = \frac{\left[R\text{-}Na^+\right]\left[H^+Cl^-\right]}{\left[R\text{-}H^+\right]\left[Na^+Cl^-\right]} \qquad (3.2)$$

In general, if K is a larger number, the reverse reaction is much less efficient than the forward reaction and requires a large excess of regenerant chemical, HCl in this instance, for moderate regeneration levels.

The equilibrium constant defined in Equation 3.2 compares to K_D, the equilibrium constant defined in Chapter 2 for adsorptive processes:

$$K_D = c_i/c_0 \qquad (3.3)$$

where K_D is the ratio of the solute concentration within (c_i) and surrounding (c_0) the adsorbent material.

Starobinietz and Gleim (14) distinguished between the amount of molecular adsorption and ion exchange that occurred with a strong base anion resin and a series of fatty acids. The overall adsorption of carboxylic acids on the resin was determined from the change in its solution phase concentration and the amount of ion exchange from the amount of counterions displaced from the resin into the solution. As Figure 3.1 shows, there is a steady increase in molecular adsorption with hydrocarbon increase in chain length. However, the amount of ion exchange decreases from formic through acetic to propionic acid and increases again for the higher acids. This type of distinction is not often measured, particularly when biomolecules with multiple modes of interaction are under investigation.

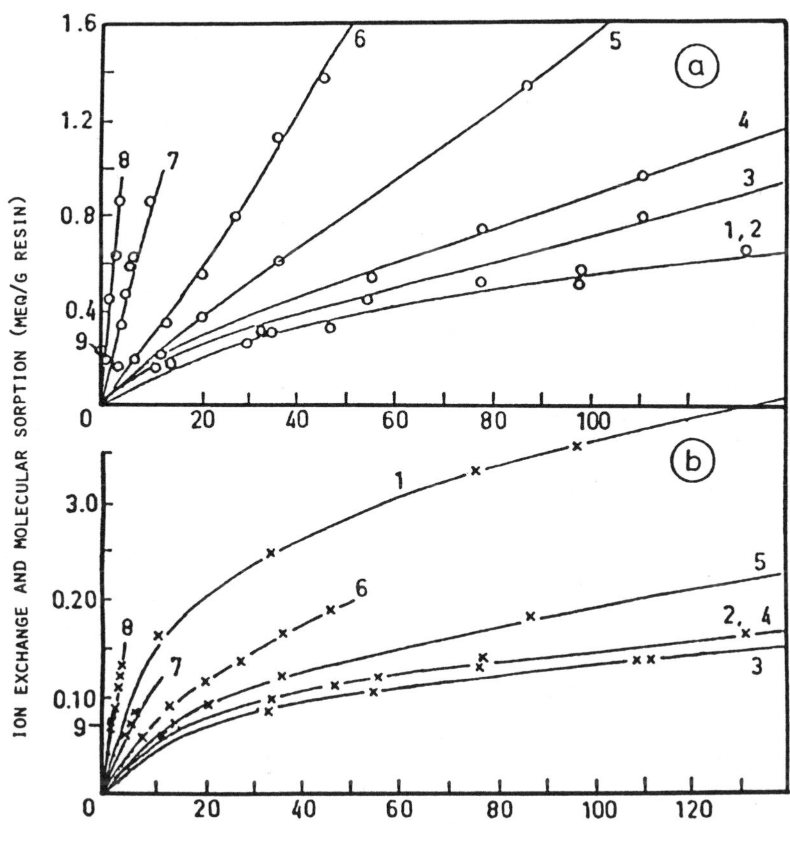

Figure 3.1. Adsorption isotherms of fatty acids on Dowex 1X2 in chloride form: (a) molecular sorption; (b) ion exchange. 1. formic; 2. acetic; 3. propionic; 4. butyric; 5. valeric; 6. caproic; 7. heptanoic; 8. caprylic; 9. pelargonic acids (Reference 14).

With proper processing and regenerants, the ion exchange resins may be selectively and repeatedly converted from one ionic form to another. The definition of the proper processing requirements is based upon the selectivity and kinetic theory of ion exchange reactions.

3.2.1 Selectivity

For ion exchange resins, the selectivity is developed in terms of the Donnan potential, E_D. The Donnan potential is the difference in the electric potential between the ion exchange resin, ϕ_r, and the solution, ϕ_s. This may be related to the chemical potential or activity by the equation:

$$E_D = \phi_r - \phi_s = \frac{1}{z_i F}\left(RT \ln \frac{a_{i,s}}{a_{i,r}} - \pi v_i\right) \tag{3.4}$$

where z_i is the charge on the ion, F is the Faraday constant, $a_{i,s}$ and $a_{i,r}$ are the activities of ion "i" in solution and the resin, respectively, and v_i is the partial molar volume of ion "i". The partial molar volume is assumed to be the same in the liquid and resin phase.

When ion "B", which is initially in the resin, is exchanged for ion "A", the selectivity is represented by:

$$\ln K_B^A = \frac{\pi}{RT}\left(|z_A| v_B - |z_B| v_A\right) \tag{3.5}$$

The selectivity which a resin has for various ions is affected by many factors. These factors include the valence and size of the exchange ion, the ionic form of the resin, the total ionic strength of the solution, crosslinkage of the resin, the type of functional group and the nature of the non-exchanging ions. The ionic hydration theory has been used to explain the effect of some of these factors on selectivity (15).

According to this theory, the ions in aqueous solution are hydrated and the degree of hydration for cations increases with increasing charge and decreasing crystallographic radius, as shown in Table 3.1 (16). It is the high dielectric constant of water molecules that is responsible for the hydration of ions in aqueous solutions. The hydration potential of an ion depends on the intensity of the charge on its surface. The degree of hydration of an ion increases as its valence increases and decreases as its hydrated radius increases. Therefore, it is expected that the selectivity of a resin for an ion is inversely proportional to the ratio of the valence/ionic radius for ions of a given radius. In dilute solution the following selectivity series are followed:

$$Li < Na < K < Rb < Cs$$

$$Mg < Ca < Sr < Ba$$

$$Al < Sc < Y < Eu < Sm < Nd < Pr < Ce < La$$

$$F < Cl < Br < I$$

Table 3.1: Ionic Size of Cations (16)

Ion	Crystallographic Radius (Å)	Hydrated Radius (Å)	Ionization Potential
Li	0.68	10.0	1.3
Na	0.98	7.9	1.0
K	1.33	5.3	0.75
NH_4	1.43	5.37	--
Rb	1.49	5.09	0.67
Cs	1.65	5.05	0.61
Mg	0.89	10.8	2.6
Ca	1.17	9.6	1.9
Sr	1.34	9.6	1.6
Ba	1.49	8.8	1.4

As Figure 3.2 (17) shows, as the valence of the exchange ion increases and as the hydrated radius of the exchange ion decreases, the selectivity or affinity of the ion exchange resin for that ion increases. As should be apparent, if the resin is in an ionic form "A" which has a lower replacing power than an ion "B" in solution, the resin will be converted at equilibrium to the ionic form "B". The selectivity of resins in the hydrogen ion or the hydroxide ion form, however, depends on the strength of the acid or base formed between the functional group and the hydrogen or hydroxide ion. The stronger the acid or base formed, the lower is the selectivity coefficient. It should be noted that these series are not followed in non-aqueous solutions, at high solute concentrations or at high temperatures.

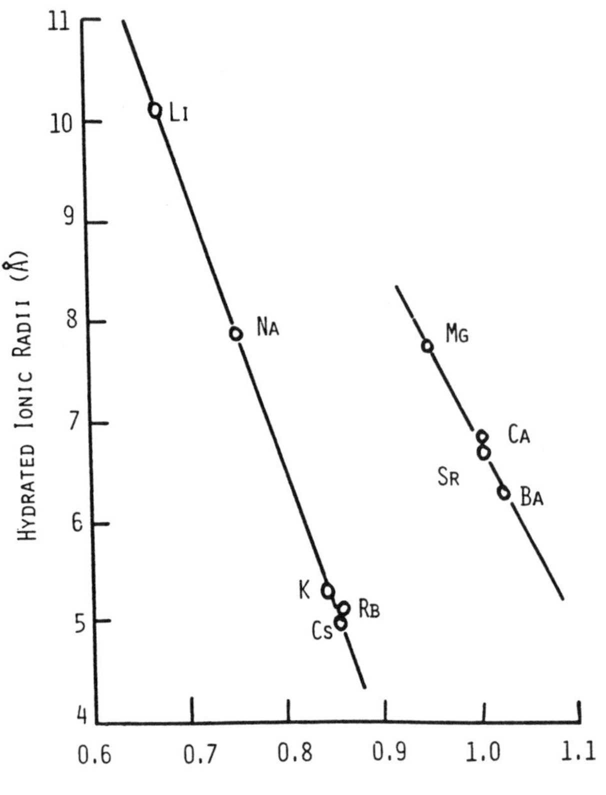

Figure 3.2. Effect of ionic radius on ion exchange in a carbonaceous zeolite (Reference 17).

The dependence of selectivity on the ionic strength of the solution was related through the mean activity coefficient to be inversely proportional to the Debye-Huckel parameter a° (18):

$$\log \gamma_{\pm} = \frac{-A\sqrt{\mu}}{1 + Ba^{o}\sqrt{\mu}} \qquad (3.6)$$

where $\gamma_{\pm}$ is the mean activity coefficient, A and B are constants, and μ is the ionic strength of the solution. The mean activity coefficient in this instance represents the standard free energy of formation ($-\Delta F°$) for the salt formed by the ion exchange resin and the exchanged ion. The Debye-Huckel parameter is a measure of the distance of closest approach which is also related to the ionic hydration size. Figure 3.3 (19) shows this dependence as the ionic concentration of the solution is changed. As the ionic concentration of the solution increases, the differences in the selectivity of the resin for ions of different valence decreases, and in some cases, the affinity may be greater for the lower valence ion.

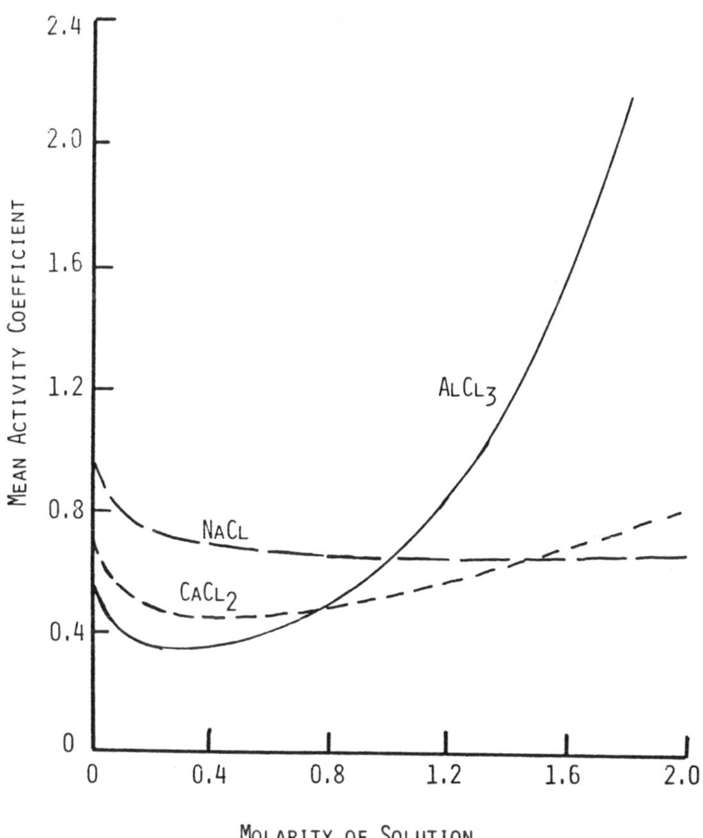

Figure 3.3. Dependence of the activity coefficient on the ionic concentration of aqueous solution (Reference 19).

The selectivity of an ion exchange resin will also depend on its crosslinking. The polymer structure of the ion exchange resin can be thought of as collections of coiled springs which can swell or contract during the exchange of ions (20). The crosslinking of the polymer limits the extent to which the resin may swell: the higher the degree of crosslinking, the lower the extent to which the resin can be hydrated. This limit on the extent to which the resin can be hydrated determines the relative equivalent volumes of hydrated ions which the crosslinked polymer network can accommodate. This is shown in Table 3.2 (21) and in Figure 3.4 (22). As the resin's degree of crosslinking or its fixed ion concentration is lowered, the selectivity of the resin decreases.

Table 3.2: Selectivity and Hydration of Cation Resins with Different Degrees of Crosslinking (21)

	4% DVB		8% DVB		16% DVB	
Cation	K	H	K	H	K	H
Li	1.00	418	1.00	211	1.00	130
H	1.30	431	1.26	200	1.45	136
Na	1.49	372	1.88	183	2.23	113
NH_4	1.75	360	2.22	172	3.07	106
K	2.09	341	2.63	163	4.15	106
Cs	2.37	342	2.91	159	4.15	102
Ag	4.00	289	7.36	163	19.4	102
Tl	5.20	229	9.66	113	22.2	85

K = Selectivity compared to Li

H = Hydration (g H_2O/eq resin)

The degree of crosslinking can affect the equilibrium level obtained, particularly as the molecular weight of the organic ion becomes large. With highly crosslinked resins and large organic ions, the concentration of the organic ion in the outer layers of the resin particles is much higher than in the center of the particle. This is shown in Figure 3.5 for the distribution of methylene blue in carboxylic cation exchange resins (23).

The selectivity of the resin for a given ion is also influenced by the dissociation constants of the functional group covalently attached to the resin (the fixed ion) and of the counterions in solution. Since the charge per unit volume within the resin particle is high, a significant percentage of the functional groups may be un-ionized. This

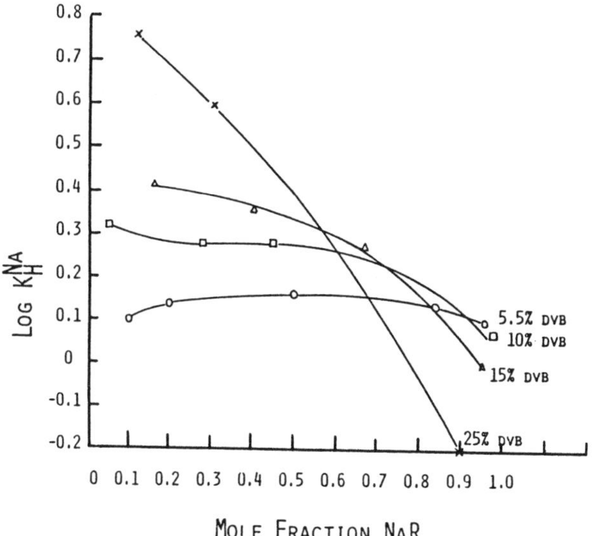

Figure 3.4. Dependence of the selectivity coefficient for the Na-H ionic exchange as a function of resin ionic composition and resin crosslinking (Reference 22).

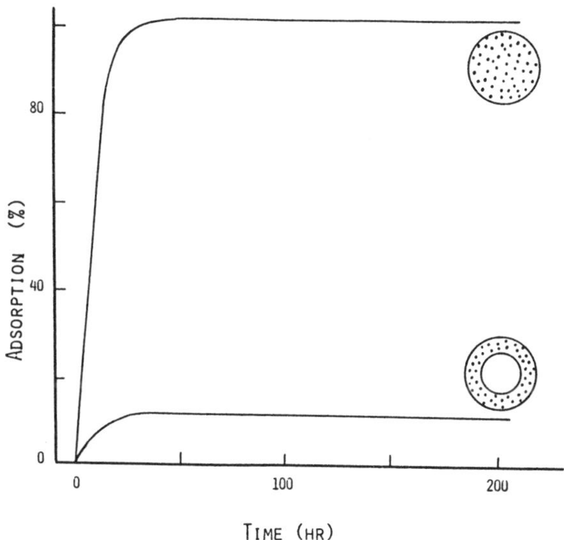

Figure 3.5. Sorption of methylene blue by carboxylic cation exchange in the Na form. The resin for the top curve has 2% crosslinking while that for the lower curve has 10% crosslinking (Reference 23).

is particularly true if the functional group is a weak acid or base. For cation exchange, the degree of dissociation of the functional group increases as the pH is increased. However, the degree of dissociation for the ions in solution decreases with increasing pH. Therefore, if a cation resin had weak acid functionality, it would exhibit little affinity at any pH for a weak base solute. Similarly, an anion resin with weak base functionality exhibits little affinity at any pH for a weak acid solute.

The influence of pH on the dissociation constants for resin with a given functionality can be obtained by titration in the presence of an electrolyte. Typical titration curves are shown in Figure 3.6 for cation resins and in Figure 3.7 for anion resins (24). For sulfonic acid functional groups, the hydrogen ion is a very weak replacing ion and is similar to the lithium ion in its replacing power. However, for resin with carboxylic acid functionality, the hydrogen ion exhibits the highest exchanging power. Table 3.3 (25,26) summarizes the effect different anion exchange resin functionalities have on the equilibrium exchange constants for a wide series of organic and inorganic anions.

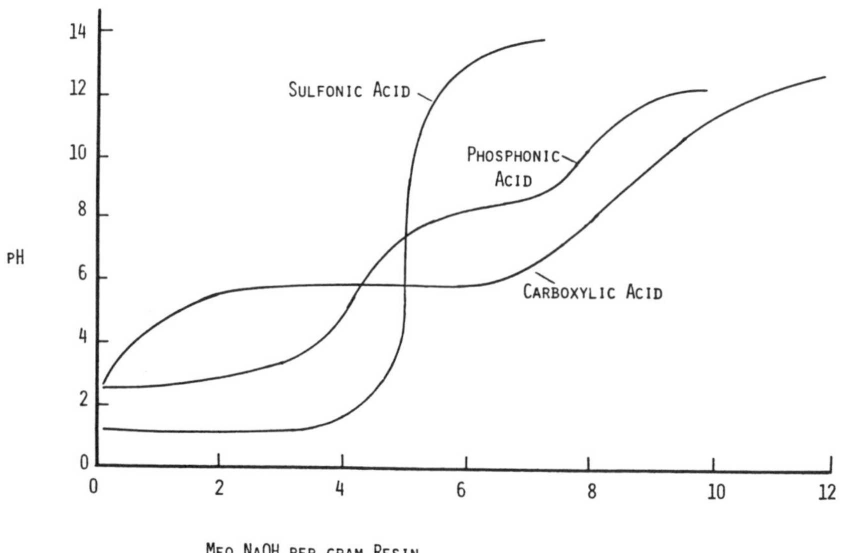

Figure 3.6. Titration curves of typical cation exchange resins (Reference 24).

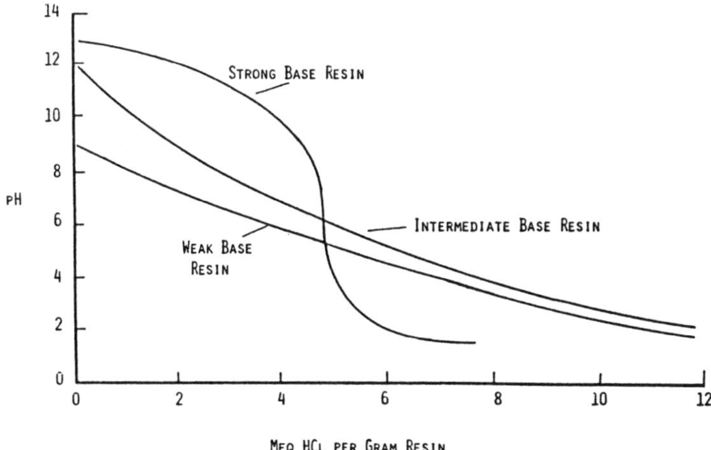

Figure 3.7. Titration curves of typical anion exchange resins (Reference 24).

Table 3.3: Selectivity Coefficients for Strongly Basic Anion Resin (25, 26)

Type I Anion		Type II Anion	
Anion	K^x_{Cl}	Anion	K^x_{Cl}
Salicylate	32.2	Salicylate	28
I^-	8.7	$C_6H_5O^-$	8.7
$C_6H_5O^-$	5.2	I^-	7.3
HSO_4^-	4.1	HSO_4^-	6.1
NO_3^-	3.8	NO_3^-	3.3
Br^-	2.8	Br^-	2.3
CN^-	1.6	CN^-	1.3
HSO_3^-	1.3	HSO_3^-	1.3
NO_2^-	1.2	NO_2^-	1.3
Cl^-	1.00	Cl^-	1.00
HCO_3^-	0.32	OH^-	0.65
H_2PO_4	0.25	HCO_3^-	0.53
$HCOO^-$	0.22	H_2PO_4	0.34
CH_3COO^-	0.17	$HCOO^-$	0.22
$H_2NCH_2COO^-$	0.10	CH_3COO^-	0.18
OH^-	0.09	F^-	0.13
F^-	0.09	$H_2NCH_2COO^-$	0.10

The selectivity is also influenced by the non-exchanging ions (co-ions) in solution even though these ions are not directly involved in the exchange reaction. An example of this influence would be the exchange of calcium ascorbate with an anion exchange resin in the citrate form. Although calcium does not take part in the exchange reaction, sequestering of citrate will provide an additional driving force for the exchange. This effect, of course, would have been diminished had a portion of the ascorbate been added as the sodium ascorbate instead of the calcium ascorbate.

For non-polar organic solutes, association into aggregates, perhaps even micelles, may depress solution activity. These associations may be influenced by the co-ions present.

The selectivity of synthetic ion exchange resins for organic hydrophilic substances is usually low. The selectivity constants for carboxylic cation resins (Na^+ form) for antibiotics of the group kanamycin, gentamycin, monomycin and sisomycin are in the range 0.5-1.0 (27). A procedure was developed (23) to prepare concentrated solutions of these antibiotics. In a batch reactor, instead of merely allowing the cation resin and the antibiotic fermentation broth filtrate to come to equilibrium, the resin is first washed with a complex forming agent before adding the antibiotic filtrate. Such washing results in an additional sorption of the antibiotic and higher elimination of both the other organic constituents and inorganic ions from the ion exchange resin as shown in Figure 3.8.

Many fermentation and biotechnology products are ampholytes or zwitterions which can adsorb on ion exchange resins not only by actual exchange but also by distribution of the zwitterion segments and by the Donnan exclusion of the ampholyte from the resin. A theory of ampholyte interaction with ion exchange resins has been developed (23) which allows a quantitative evaluation of the equilibrium state for these systems. This evaluation can be used to select the optimum conditions for adsorbing and desorbing the ampholyte.

The distribution of a monoaminomonocarboxylic acid

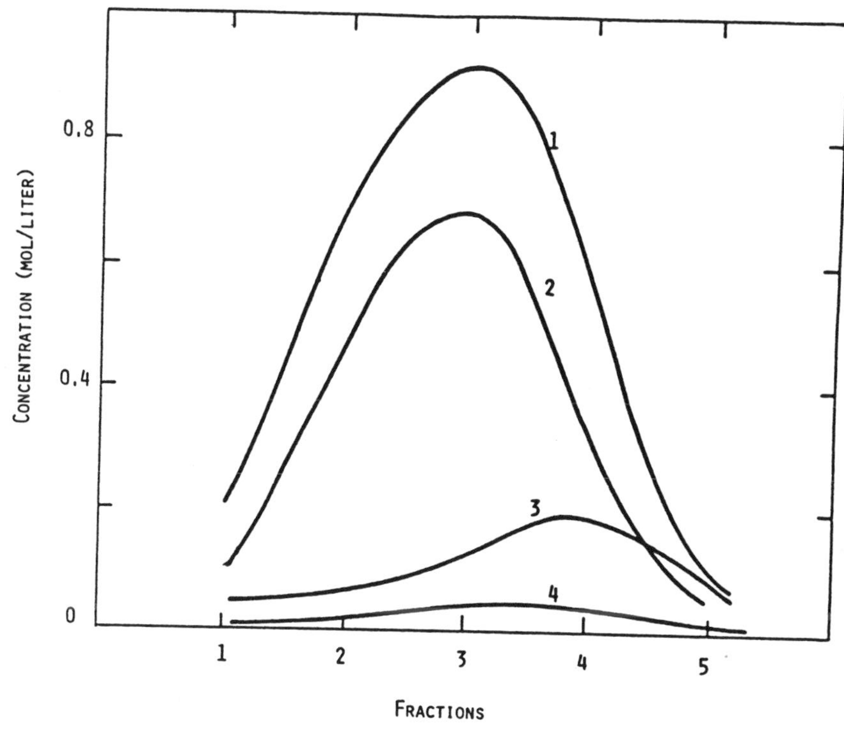

Figure 3.8. Gentamycin desorption with a 0.5 M sulfuric acid. Curves 1 and 4 are for gentamycin and magnesium respectively after washing the cation resin with the complex forming solution. Curves 2 and 3 are for gentamycin and magnesium, respectively, without such washing (Reference 23).

between a strong acid cation resin and the solution may be represented as:

$$\begin{array}{ccccc} \overline{AH}^{+} & \underset{}{\overset{\overline{K}_1}{\rightleftharpoons}} & \overline{A}^{\pm} & \overset{\overline{K}_2}{\rightleftharpoons} & \overline{A}^{-} \\ \Big\updownarrow K_H^{AH} & & \Big\updownarrow K_P & & \Big\updownarrow K_D \\ AH^{+} & \underset{K_1}{\rightleftharpoons} & A^{\pm} & \underset{K_2}{\rightleftharpoons} & A^{-} \end{array} \quad \begin{array}{l} \text{Cation Exchange Resin} \\ \\ \text{Solution} \end{array} \tag{3.7}$$

where AH^+, $A^\pm$, A^-, $\overline{AH}^+$, $\overline{A}^\pm$, $\overline{A}^-$ are cation, zwitterion and ampholyte anion in solution and on the ion exchange resin, respectively.

K_i and $\bar{K}_i$ are the ampholyte ionization constants in solution and in the ion exchange resin, respectively. K_H^{AH} is the constant of ion exchange of ampholyte cation - hydrogen ion, K_P is the distribution coefficient of zwitterions and K_D is the Donnan distribution coefficient of ampholyte as non-exchangeable electrolytes.

The value of these constants for systems containing d-phyenylglycine, 7-aminodesacetoxycephalosporanic acid, ampicillin and cefalexin are shown in Table 3.4 when the ion exchange resins are 4% and 8% crosslinked strong cation resins and the anion is IRA-938. These constants can be combined into "effective distribution coefficients", K_{eff}, which can be plotted as $1/K_{eff}$ versus pH to determine the best adsorption and elution conditions. Examples of these plots are shown in Figure 3.9.

Table 3.4: Electrochemical and Sorption Constants (23)

Constant	Ion Exchange Resin	D-phenyl-glycine	7-aminodesace-toxycephalosporanic acid	ampicillin	cefalexin
pK_1		1.96	2.25	2.38	2.85
pK_2		9.02	4.85	7.07	7.15
$p\bar{K}_1$	KU-2-4	2.4	--	2.7	--
$p\bar{K}_1$	KU-2-8	2.6	2.7	--	--
$p\bar{K}_2$	IRA-938	9.1	4.8	--	7.0
$K_H^{AH^+}$	KU-2-4	3.4	--	6.3	--
K_P	KU-2-4	1.2	--	2.9	--
$K_H^{AH^+}$	KU-2-8	4.6	4.0	--	--
K_P	KU-2-8	1.0	1.4	--	4.6
$K_{cl}^{A^-}$	IRA-938	0.7	1.0	--	1.4
K_p	IRA-938	0.9	1.0	--	1.1

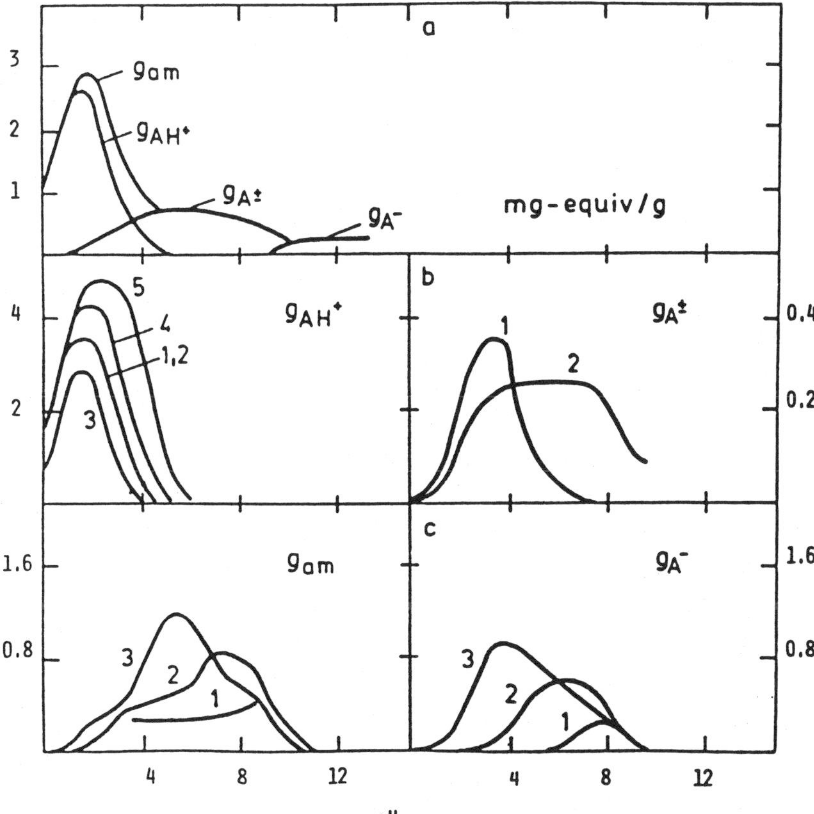

Figure 3.9. Dependence of the amount of the ampholyte and its separate electrochemical forms on the ion exchange resin as a function of pH and mineral ion concentration. (a) d-phenylglycine at 0.05 M, 0.1 M NaCl on an 8% crosslinked cation resin, (b) curve (1) is 0.1 M d-phenylglycine and 0.1 M NaCl; curve (2) is 0.1 M 7-aminodesacetoxycephalosporonic acid and 0.1 M NaCl; curve (2) is 0.1 M 7-aminodesacetoxycephalosporonic acid (7-A D C A) and 0.1 M NaCl; curves (3), (4) and (5) are 0.05 M d-phenylglcyine and 0.1, 0.01 and 0.001 M NaCl, respectively. All are on an 8% crosslinked cation resin. (c) Curves (1), (2) and (3) are 0.1 M d-phenylglycine, cefalexin and 7-A D C A, respectively, and 0.1 M NaCl on Amberlite IRA-938 anion resin (Reference 23).

Selectivity calculations may be used to define the operational limits of an ion exchange process in terms of maximum exhaustion capacity, degree of regeneration or the "leakage" from a column operation. The "leakage" is the appearance of a low concentration of an undesired solute in the column effluent during the operating portion of the cycle. This may be due to the ionic form of the resin at the bottom of the column and to the composition of the solution moving down the column.

Example 3.1

Anderson (28) used selectivity calculations to evaluate the removal of calcium ions from a 40% monosodium glutamate solution. When the feedstream has 0.15% calcium, the resin has a total exchange of 2 eq/L and a selectivity of Ca^{++} over Na^{+} of 3, and the monosodium glutamate has an equivalent weight of 169 g/eq, the relative equivalent concentrations of cations are:

$$\left[Na^{+}\right] = \frac{400 \text{ g/l}}{169 \text{ g/eq}} = 2.37 \text{ eq/l} \tag{3.8}$$

$$\left[Ca^{++}\right] = \frac{1.5 \text{ g/l}}{20 \text{ g/eq}} = 0.075 \text{ eq/l} \tag{3.9}$$

The equivalent fraction of Ca in solution is:

$$X_{Ca^{++}} = \frac{0.075}{2.45} = 0.031 \tag{3.10}$$

The equivalent fraction of Ca in the resin is:

$$\frac{\overline{X}_{Ca^{++}}}{(1 - \overline{X}_{Ca^{++}})} = K^{Ca^{++}}_{Na^{+}} \frac{\overline{c}}{c} \frac{X_{Ca^{++}}}{(1 - X_{Ca^{++}})^2} \tag{3.11}$$

Since $\overline{c}$ = 2 eq/L and c = $[Na^{+}]$ + $[Ca^{++}]$ = 2.45 eq/L,

$$\frac{\overline{X}_{Ca^{++}}}{(1 - \overline{X}_{Ca^{++}})} = 3 \frac{2}{2.45} \frac{0.031}{(0.969)^2} = 0.081 \tag{3.12}$$

So that

$$\overline{X}_{Ca^{++}} = 0.07 \tag{3.13}$$

This means that only 7% of the resin capacity, at best, can be used under these circumstances. Therefore it would be advisable to use other methods to reduce the Ca^{++} concentration in the feedstream.

3.2.2 Kinetics

Theoretical considerations for the selectivity of ion exchange resins are well developed. Unfortunately, this is not the case with the kinetics of ion exchange reactions. The problems posed by systems more complex than the "ideal" case of binary ion exchange of strong electrolytes with either strong acid cation or strong base anion resins continue to be the subject of research papers (29) and conferences (30).

The overall exchange process may be divided into five sequential steps: (1) the diffusion of ions through the solution to the surface of the ion exchange particles, (2) the diffusion of these ions through the ion exchange particle, (3) the exchange of these ions with the ions attached to the functional group, (4) the diffusion of these displaced ions through the particle, and (5) the diffusion of these displaced ions through the solution. Each step of the diffusion, whether in the resin or solution phase, must be accompanied by an ion of the opposite charge to satisfy the law of electroneutrality.

Kinetics of ion exchange is usually considered to be controlled by mass transfer in ion exchange resin particles or in the immediately surrounding liquid phase. The theory used to describe mass transfer in the particle is based on the Nernst-Planck equations developed by Helfferich (31) which accounted for the effect of the electric field generated by ionic diffusion, but excluded convection.

The Nernst equation is (32):

$$\psi_o = k + \frac{RT}{zF} \ln a_p \qquad (3.14)$$

where R is the ideal gas constant, F is the Faraday constant, T is the absolute temperature, z is the charge on the potential determining ion, a_p is the activity of the potential determining ion in the bulk solution and k is a constant.

From this, the generalized Nernst-Planck equation can be written as:

$$\underset{\text{flux}}{J_i} = \underset{\text{diffusion}}{-D_i\,(\text{grad } c_i} + \underset{\text{electric transference}}{z_i\, c_i \frac{F}{RT} \text{grad } \phi)} \qquad (3.15)$$

Whereas the kinetics of adsorption is described using Fick's Law models, the Nernst-Planck equations, which take into account the principal coupling effect of ionic fluxes, provide a significant improvement over Fick's Law in describing ion exchange kinetics with only a minor increase in complexity.

The important distinction between the kinetics of ion exchange compared to the kinetics of adsorption is that the forward and reverse exchange of the same two ions of different mobilities will occur at different rates and with different behavior of the concentration profiles. This is shown in Figure 3.10 (33).

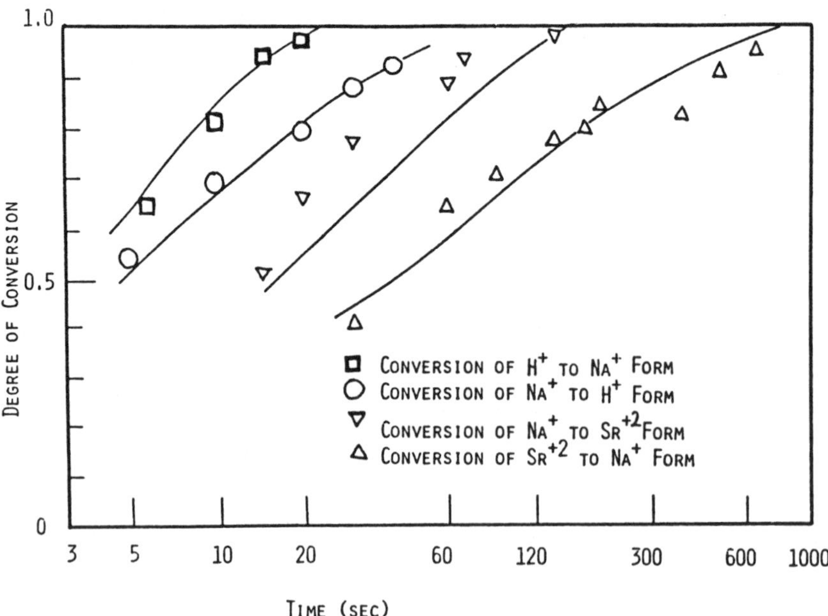

Figure 3.10. Comparison of forward and reverse ion exchange rates (Reference 33).

The Nernst-Planck equations for ion exchange kinetics are still the dominant theory in use today (34). It is recognized that the theory fails to take into account the effect of swelling and particle size changes which accompany ion exchange or to take into account the slow relaxation of the resin network which causes the diffusion coefficient to vary with time. However, the approximations which the Nernst-Planck equations provide are a reasonable starting point and will most likely be found to be sufficient for the biotechnology engineer. Any further refinements would rapidly lead to diminishing returns. Therefore, the mass transfer in the liquid phase is usually described according to the Nernst film concept using a version (35) of the Nernst-Planck equation or Glueckauf's (36) simpler linear driving force approximation.

The earliest theories of ion exchange kinetics assumed a mechanism which was controlled by a chemical exchange reaction coupled with diffusion. Helfferich (37) developed a detailed mathematical theory of kinetics of ion exchange which is accompanied with fast ionic reactions, such as dissociation equilibrium of weak acid groups. In these cases, the rate is controlled by diffusion as the limiting process, but it is dramatically affected by reaction equilibrium, which controls the concentration of diffusing species.

There are five models (38) which can be used to represent the kinetics in ion exchange systems which involve liquid phase mass transfer, solid phase mass transfer and chemical reaction at the exchange group.

In Model 1, the liquid phase mass transfer with a linear driving force is the controlling element. This model assumes that there are no concentration gradients in the particle, that there is a quasi-stationary state of liquid phase mass transfer, that there is a linear driving force and that there is a constant separation factor at a given solution concentration.

In Model 2, the rate-controlling step is diffusion within the ion exchange particles. This model assumes that there are no concentration gradients in the liquid phase and that there is no convection either through solvent uptake or release in the solid phase.

Model 3 is controlled by the exchange reaction at the fixed ionic groups. This model assumes that the slowness of the exchange reaction allows sufficient time for mass transfer to establish and maintain equilibrium so that no concentration gradient exists in either the ion exchange particles or in the liquid phase.

Model 4 is a variation of Model 3 in which the counterions from the solution do not permeate beyond the portion of the particle which has been converted to the exchanging ionic form. The boundary of the unreacted core reduces with time, such that this is called the shrinking core model. It is this sharp boundary between the reacted and unreacted portion of the particles that distinguishes Model 4 from Model 3.

In Model 5, the rate controlling step is the diffusion of the counterion across the converted portion of the particle. Since the exchange groups undergo a fast and essentially irreversible reaction with the counterions, their type of reaction affects the rate of reaction and the geometry of the diffusing zone.

Table 3.5 (38) summarizes the effect of operating parameters (particle size, solution concentration, separation factor, stirring rate, resin exchange capacity, and temperature) on ion exchange kinetics described by these different models in batch reactors.

For the cases of interest, the rate of ion exchange is usually controlled by diffusion, either through a hydrostatic boundary layer, called "film diffusion" control or through the pores of the resin matrix, called "particle diffusion" control.

In the case of film diffusion control, the rate of ion exchange is determined by the effective thickness of the film and by the diffusivity of ions through the film. When resin particle size is small, the feedstream is dilute or when a batch system has mild stirring, the kinetics of exchange are controlled by film diffusion.

In the case of particle diffusion control, the rate of ion exchange depends on the charge, spacing and size of the diffusing ion and on the micropore environment. When the resin particle size is large, the feedstream is concentrated or

Table 3.5: Dependence of Ion Exchange Rates on Experimental Conditions (38)

Factor	Model 1	Model 2	Model 3	Model 4	Model 5 (a)
Particle size (r)	$\alpha\,1/r$	$\alpha\,1/r^2$	independent	$\alpha\,1/r$	$\alpha\,1/r^2$
Solution concentration (c)	$\alpha\,c$	independent	$\alpha\,c$	$\alpha\,c$	$\alpha\,c$ (b)
Separation factor (α)	independent up to a specific time when ≥ 1; $\alpha\,\alpha$ when $\alpha \ll 1$.	independent (c)	independent	independent	independent (c)
Stirring rate (rpm)	sensitive	independent	independent	independent	independent
Resin exchange capacity ($\bar{c}$)	$\alpha\,1/\bar{c}$	independent	independent	independent	$\alpha\,1/\bar{c}$
Temperature (T)	~4%/°K	~6%/°K	function of E_{Act}	function of E_{Act}	~6%/°K

(a) Applicable to forward exchange only.

(b) Provided partition coefficient is independent of solution concentration.

(c) For complete conversion and constant solution composition.

when a batch system has vigorous stirring, the kinetics are controlled by particle diffusion.

The limits at which one or the other type of diffusion is controlling have been determined by Tsai (39). When $Kk^2\delta > 50$, the rate is controlled by film diffusion. When $Kk^2\delta < 0.005$, the rate is controlled by particle diffusion. In these relationships, K is the distribution coefficient, k^2 is the diffusivity ratio (D_p/D_f), δ is the relative film thickness ((b-a)/a), D_p and D_f are the pore diffusion and film diffusion coefficients, respectively, and (b - a) is the film thickness on a resin particle with a radius a. Between these two limits, the kinetic description of ion exchange processes must include both phenomena.

The characteristic Nernst parameter "δ", the thickness of the film around the ion exchange particle, may be converted to the mass transfer coefficient and dimensionless numbers (Reynolds, Schmidt and Sherwood) that engineers normally employ (40).

In terms of the solute concentration in the liquid, between 0.1 and 0.01 mol/L, the rate limiting factor is the transport to the ion exchange bead. Above this concentration, the rate limiting factor is the transport inside the resin beads (41). During the loading phase of the operating cycle, the solute concentration is in the low range. During regeneration, however, in which the equilibrium is forced back by addition of a large excess of regenerant ions, the solute is above the 0.1 mol/L limit.

One of the important factors in the kinetic modeling of organic ions is their slow diffusion into the ion exchange resin. The mean diffusion time is listed in Table 3.6 as a function of resin particle size for different size classifications of substances (42). With the larger organic ions, the contact time for the feed solution and the resin must be increased to have the ion exchange take place as a well defined process such as occurs with the rapidly diffusing ions of mineral salts.

Table 3.6: Characteristic Diffusion into Spherical Resin Particles for Various Substances (42)

Coefficient of Diffusion (Order of Magnitude) (cm^2/sec)	Type of Sorbed Substance (Ion)	Particle Radius (cm)	Mean Time of Intraparticle Diffusion
10^{-6}	Ions of mineral salts bearing a single charge	0.5	3 min
		0.01	7 sec
		0.005	1.8 sec
10^{-7}	Ions of mineral salts bearing several charges, amino acids	0.05	30 min
		0.01	1.2 min
		0.005	18 sec
10^{-8}	Tetraalkylammonium ions, antibiotic ions on macroporous resins	0.05	5 hr
		0.01	12 min
		0.005	3 min
10^{-9}	Dyes, alkaloids, antibiotics in standard ion exchange resins	0.05	over 2 days
		0.01	2 hr
		0.005	0.5 hr
10^{-10}	Some dyes, polypeptides and proteins	0.05	over 20 days
		0.01	over 20 hr
		0.005	over 5 hr

A mathematical model has been developed for the kinetics of ion exchange in batch systems (43). The fractional attainment of equilibrium at a specific time was developed in terms of the parameters ξ, α, K, k and δ.

The parameter ξ represents the ratio of the forward chemical reaction rate to the intraparticle diffusion rate. As ξ approaches infinity, the surface chemical reaction becomes instantaneous.

The parameter α represents the ratio of the volume of the liquid solution to the volume of the solid ion exchange resins. As would be expected, the exchange rate increases as the proportion of the resin increases.

K represents the distribution coefficient of the resin since it is the ratio of the rate constants of the forward and reverse ion exchange reactions, each multiplied by the solute concentration in the particle and in the solution, respectively. The value of K decreases as the solute concentration in the liquid phase increases. Likewise, the reaction rate increases as K decreases. This is shown in Figure 3.11.

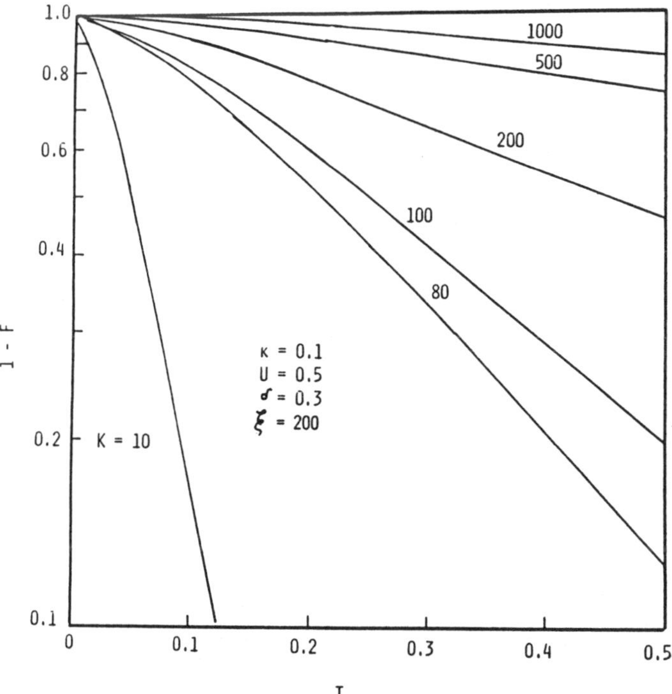

Figure 3.11. Plots of log (1-F) vs T for various values of K (Reference 43).

The parameter k compares the intraparticle diffusion rate to the film diffusion rate. For small value of k, the exchange rate is controlled by a combination of surface chemical reaction and intraparticle diffusion. For large values of k, the rate is controlled by a combination of surface chemical reaction and film diffusion. This is shown in Figure 3.12.

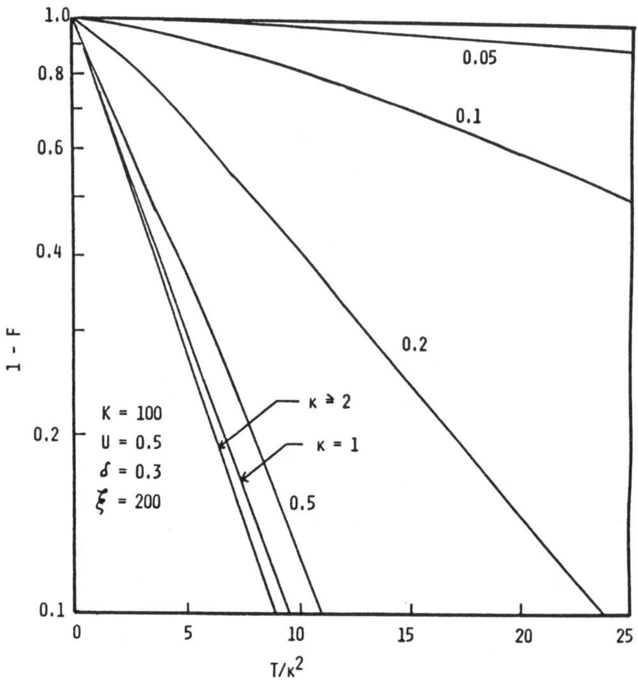

Figure 3.12. Plots of log (1-F) vs T/k for various value of k (Reference 43).

The parameter δ measures the ratio of the film thickness to the radius of the ion exchange particle. When δ is small, the film diffusion resistance to mass transfer is negligible so that the exchange rate is controlled by a combination of surface chemical reaction and intraparticle diffusion. As δ increases, the effect of film diffusion becomes important in controlling the exchange rate.

Example 3.2

With a particle diffusion of $1 \times 10^{-8} cm^2/sec$ for an amino acid such as lysine, the parameter k changes from

0.0383 to 0.0436 to 0.0470 as the ion exchange bead diameter increases from 420 to 710 to 840 microns for a 0.05M solution stirred at 300 rpm. The corresponding values of $Kk^2\delta$ calculated range from 0.173 to 0.012 to 0.003, respectively. In the first two cases, the value is between 0.005 and 50 so that the exchange rate should be expected to be controlled by both film and intraparticle diffusion. In the last case, the bead size has become sufficiently large that the exchange rate should be controlled by particle diffusion.

It should be apparent that the kinetics and mass transfer equations describing the ion exchange resin column operation are in many respects similar to what was seen with column operations in the chapter on adsorption.

The set of equations to be solved to model the ion exchange process in fixed bed operations are:

$$D \frac{\partial^2 c}{\partial z^2} = \frac{\partial c}{\partial t} + u \frac{\partial c}{\partial z} + \frac{\rho}{\varepsilon} \frac{\partial q}{\partial t} \tag{3.16}$$

$$\frac{\partial q}{\partial t} = K_a (c - c^*) \tag{3.17}$$

$$c^* = b(q)^a \tag{3.18}$$

$$K_a = A\, e^{-B(q/q_\infty)} + D\, e^{-E(q/q_\infty)} \tag{3.19}$$

As with mass transfer in adsorption, these equations have been solved using numerical analysis and analytical and graphical solutions (44-47). The solution at short times is given by (44):

$$c = c^o\, e^{\lambda}\, \mathrm{erfc}\, \frac{\lambda}{2\sqrt{\tau}} \tag{3.20}$$

where dimensionless time, τ, is equal to:

$$\tau = \left(t - \frac{\alpha x}{v}\right) \frac{\bar{D}}{R^2} \tag{3.21}$$

and the dimensionless column length, λ, equals:

$$\lambda = 3\,(1 - \alpha)\,\frac{K_d\,\bar{D}\,x}{R^2\,v} \tag{3.22}$$

For long times, the solution can be written in terms of modified Bessel functions to give:

$$c = c^o\,e^{-2\lambda}\left[1 + \int_0^{2\pi\sqrt{2\lambda(\tau - 0.13\lambda)}} e^{-\xi^2/8\lambda}\,I_1(\xi)\,d\xi\right] \tag{3.23}$$

These two solutions can be combined to describe the mass transfer in an ion exchange column at any value of τ. Breakthrough curves using these equations are shown in Figure 3.13.

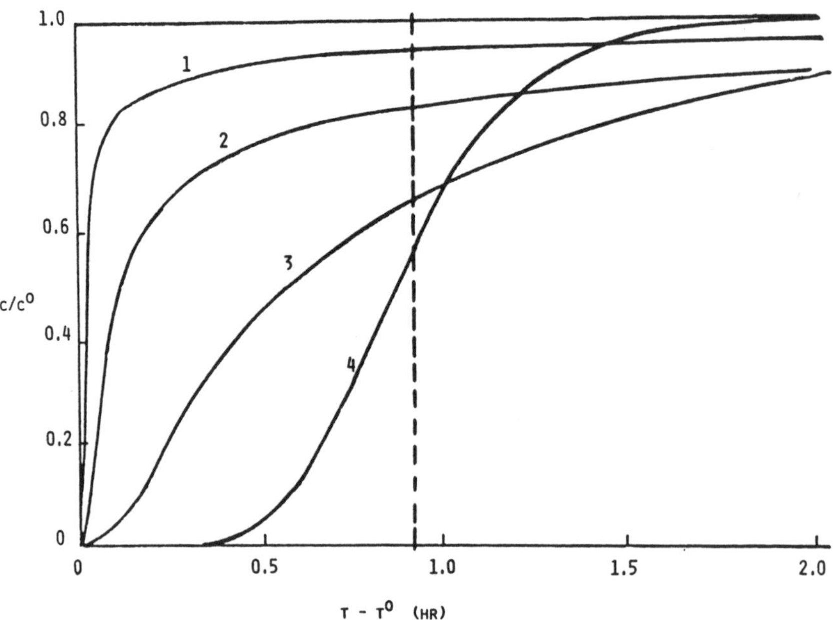

Figure 3.13. Breakthrough curves (x = 26.6 cm, ν = 100 ml-cm^{-2} hr^{-1}, K_a = 5, α = 0.3 and R = 0.005 cm). Irregular process: Curve (1) D = 10^{-11} cm^2/sec, λ = 0.004; Curve (2) D = 10^{-10} cm^2/sec, λ = 0.04; Curve (3) D = 10^{-9} cm^2/sec, λ = 0.4; Regular Process: Curve (4), D = 10^{-8} cm^2/sec, λ = 4.0 (Reference 42).

Example 3.3

Breakthrough curves have been calculated using these

equations for the exchange of morphocycline onto a strong acid cation resin. They are compared in Figure 3.14 with experimental data. While the initial portion of the curves are calculated to be greater than the experimental values, the agreement becomes better as the value of τ is increased.

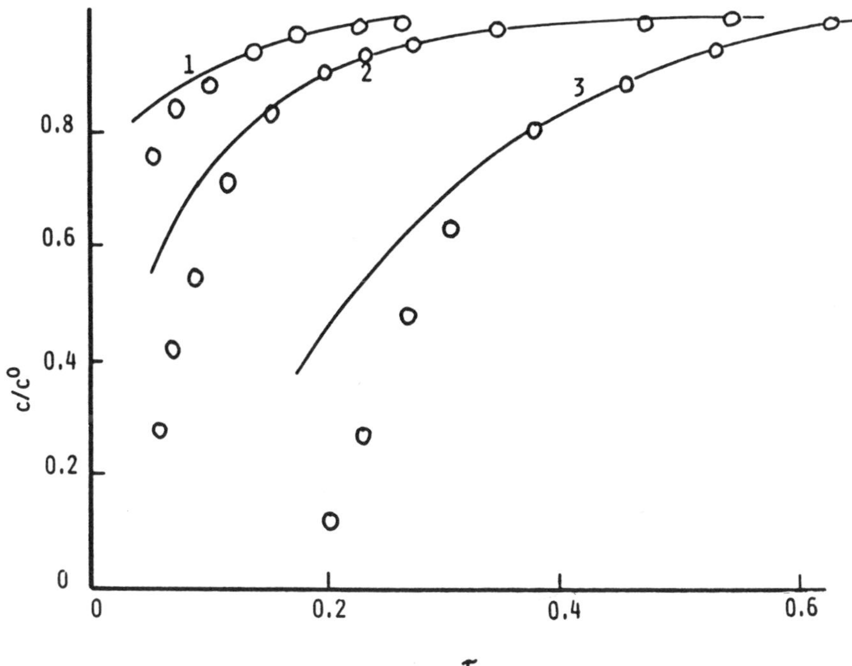

Figure 3.14. Comparison of calculated and experimental breakthrough curves for morphocycline sorbed onto a 4% crosslinked cation resin at λ values of 0.1, 0.25, and 0.8 for curves (1), (2) and (3), respectively (Reference 42).

The shape of the breakthrough curve is largely dependent on the type of exchange isotherm obtained under equilibrium conditions (48). Three different types of isotherms, linear, convex and concave, are possible as model situations (Figure 3.15). Actual isotherms may have inflection points due to variations in selectivity with resin composition.

The simplest type, a linear isotherm, is only valid when the exchange is between ions of the same valence or when a second ion is present in great excess in the feedstream. A

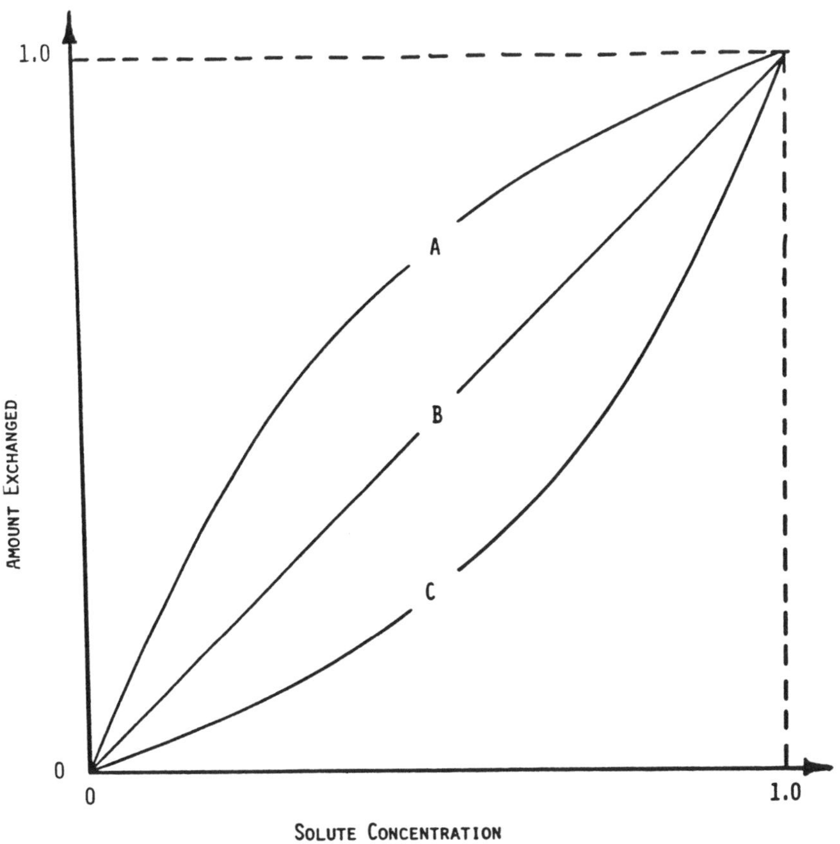

Figure 3.15. Types of exchange isotherms: (A) convex, (B) linear and (C) concave.

linear isotherm was used in calculating the values in Example 3.3. A closer fit may be obtained using an alternate isotherm as shown in Figure 3.16. The more typical isotherm, the convex type, occurs when the ion to be taken up has a higher affinity for the resin than the original ion in the resin. The concave isotherm represents the situation where the ion to be taken up is attracted less strongly to the resin than the original ion.

In principle, fluidized ion exchange beds are similar to stirred tank chemical reactors. The general equations of kinetics and mass transfer can be applied to the individual fluidized units in an identical manner to those for chemical reactors. The primary difference lies in accounting for the behavior of suspended particles in the turbulent fluid (49).

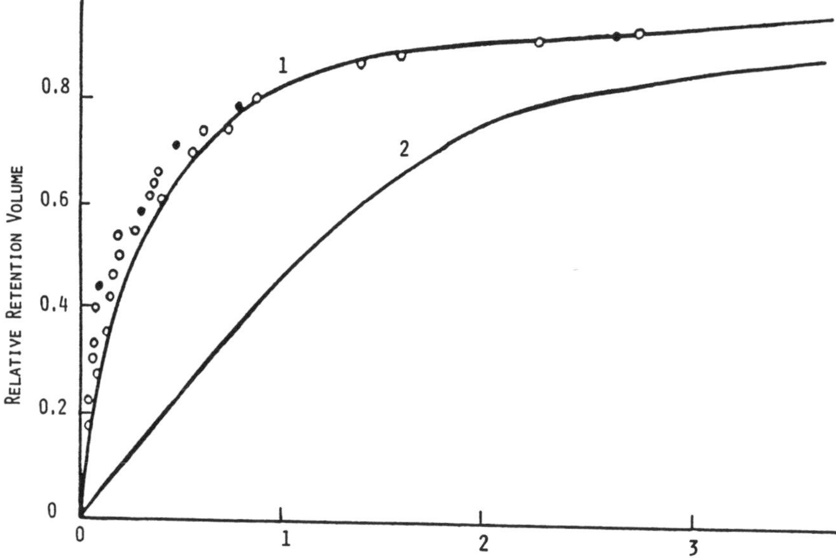

Figure 3.16. Relative retention volumes based on the infection points of breakthrough curves for oxyglucocycline (∘) and morphocycline (•) plotted as a function of the dimensionless parameter λ. The solid lines are calculated using (1) rectangular and (2) linear exchange isotherms (Reference 42).

The operation of these fluidized ion exchange beds is identical to that of fixed beds, with the exception that the resin of each stage is confined by perforated plates and maintained in a fluidized suspension using liquid flow or impellers.

The critical design parameter for fluidized beds is the loss or leakage of the solute through a given stage. The design equation for a single stage bed is given by (50):

$$\frac{c}{c_o} = 1 - \left[\frac{K\,a}{U + K\,a}\right] \exp\left(- \frac{U}{mV_r}\int_0^t \frac{K\,a\,dt}{U + K\,a}\right) \tag{3.24}$$

where c is the concentration of the solute leaving the bed at time t; c_o is the concentration of the solute in the feedstream; K is the overall mass transfer coefficient; m is the solute distribution coefficient; U is the solution flow rate; a is the transfer surface area per stage; and V_r is the settled volume of resin in the stage. Once the flow rate, feed concentration and desired final effluent concentration are specified, this equation can be used to determine the combination of m, V_r, K and t which can be used.

When multiple stage operations are employed, it is necessary to determine the loss from each prior stage to determine the loss from stage "n". The general equation is:

$$\frac{c_n}{c_o} = (1 - \beta)\left(\frac{c_{n-1}}{c_o}\right) + \frac{\beta^2}{m} e^{-\beta N/m} \int_0^N e^{\beta N/m} \left[\frac{c_{n-1}}{c_o}\right] dN \qquad (3.25)$$

where

$$\beta = \frac{K a}{U + N a} \qquad (3.26)$$

and N, the number of bed volumes of solution, is equal to:

$$N = \frac{t Q}{n V_r} \qquad (3.27)$$

For the second and third stage, this equation becomes:

$$\frac{c_2}{c_o} = 1 - \beta \left[e^{-\beta N/m}\right]\left(2 - \beta + \frac{\beta^2 N}{m}\right) \qquad (3.28)$$

and

$$\frac{c_3}{c_o} = 1 - \beta e^{-\beta N/m}\left(3 - 3\beta + \beta^2 + \frac{3\beta^2 N}{m} - 2\beta^2 N + \frac{\beta^4 N^2}{2m^2}\right) \qquad (3.29)$$

Equation 3.25 can then be used along with Equations 3.17 - 3.19 to model the ion exchange process in fluidized bed operations.

Example 3.4

Experimental data for a three stage system using Dowex 1x4 resin to adsorb thorium nitrate are shown in Figure 3.17 for different experimental conditions. The three curves are calculated using (I) a single stage bed with 900 cc of resin, (II) a two stage bed with 450 cc of resin each and (III) a three stage bed with 300 cc of resin each. The data should follow curve III if there is no intercompartment mixing of resin. It is obvious from the data that stacked, multicompartment beds should be fluidized to the minimum extent possible to allow passage of suspended particles while minimizing interstage mixing of the resin.

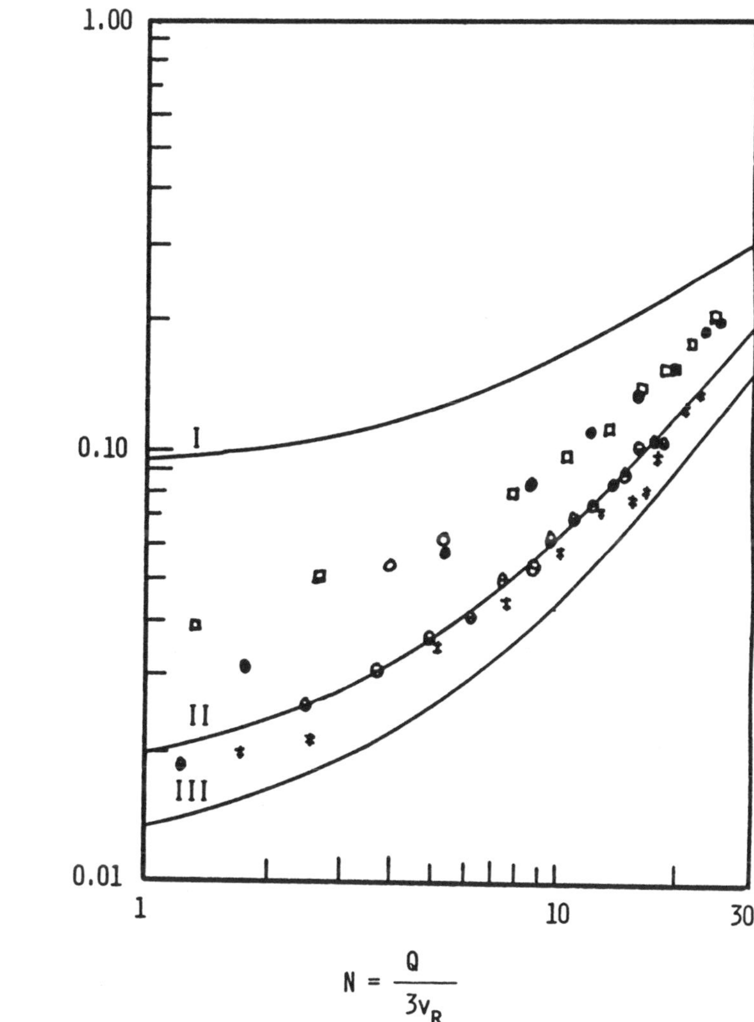

Figure 3.17. Experimental data and calculated curves for three stage multicomponent beds (Reference 50).

3.3 ION EXCHANGE MATERIALS AND THEIR PROPERTIES

Ion exchange materials are a special class of polyelectrolytes. The chemical and physical properties of an ion exchange material play a more important role in determining its suitability for a biochemical application than for other types of applications. The chemical properties to be considered are the matrix and the ionic functionality attached to the matrix. The important physical properties are the pore size, the pore volume, the surface area, the density and the particle size.

3.3.1 Ion Exchange Matrix

These materials can be broadly categorized into those which are totally inorganic in nature and those that are synthetic organic resins.

Inorganic ion exchangers (51) include both naturally occurring materials such as mineral zeolites (sodalite and clinoptilolite), the greensands, and clays (the montmorillonite group) and synthetic materials such as gel zeolites, the hydrous oxides of polyvalent metals (hydrated zirconium oxide) and the insoluble salts of polybasic acids with polyvalent metals (zirconium phosphate).

A history of the development of synthetic ion exchange resins and a review of current methods of synthesizing polymeric ion exchange resins has been presented by Millar (52).

The synthetic organic resins consist of a crosslinked polymer matrix which is functionalized to provide their ion exchange capacity. The matrix usually must undergo additional reactions to provide the strong acid cation, strong base anion, weak acid cation or weak base anion functionality.

The most commonly used materials as the matrix for synthetic ion exchangers are polymers based on the copolymerization of styrene crosslinked with divinylbenzene (Figure 3.18). These copolymers can be used to form strong acid cation, strong base anion (Type I and Type II), and weak base anion resins.

Weak acid cation exchangers (Figure 3.19) may be formed by the direct polymerization of acrylic acid or methacrylic acid with divinylbenzene as the crosslinking agent. Similarly, the acrylate- or methacrylate-divinylbenzene copolymer may be formed which can be further reacted to have strong base anion or weak base anion functionality.

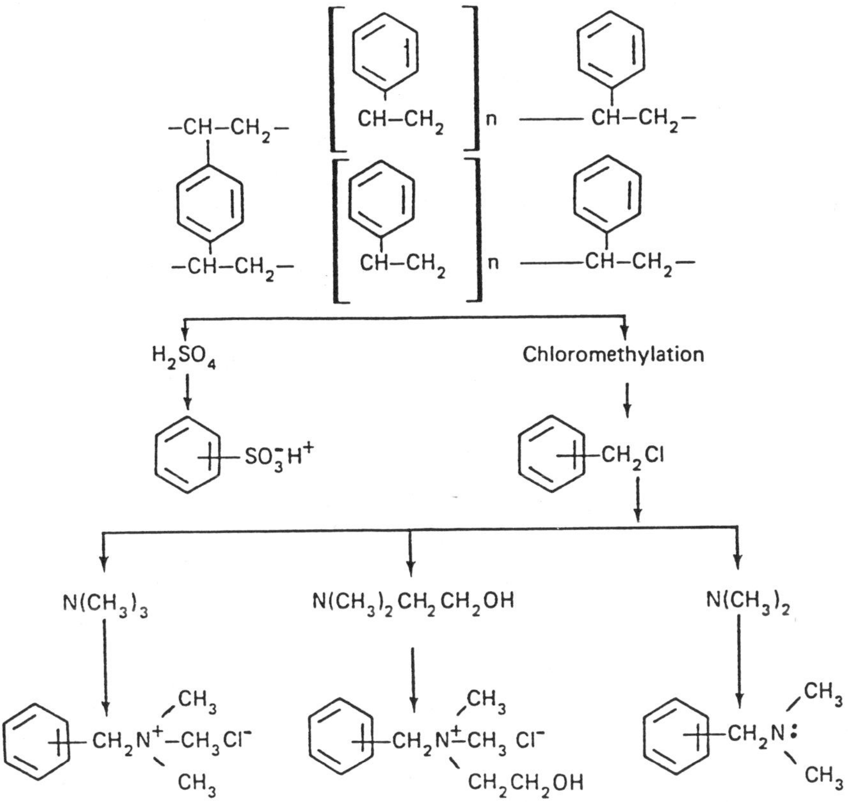

Figure 3.18. Styrene-divinylbenzene resins functionalized with strong acid, strong base I, strong base II and weak base groups.

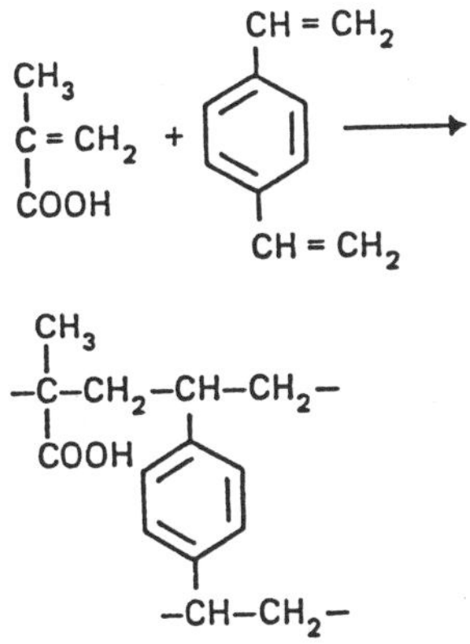

Figure 3.19. Methacrylic acid-divinylbenzene resins forming weak cation resins.

Epoxy-polyamine resins (Figure 3.20) are formed by the direct polymerization of a chloroepoxide with ammonia or an amine. The molecular weight of this type of polymer is comparatively low with respect to the acid-adsorbing nitrogen sites. This results in a high capacity acid-adsorbing weak base resin.

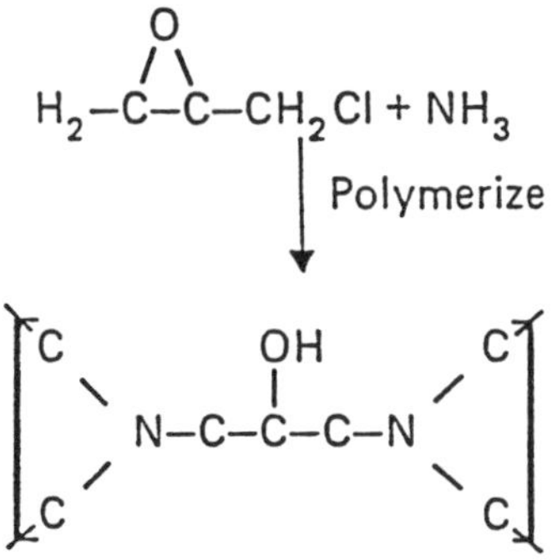

$$H_2C{-}CH{-}CH_2Cl + NH_3 \xrightarrow{\text{Polymerize}} \left[(C)_2N{-}C{-}C(OH){-}C{-}N(C)_2 \right]$$

Figure 3.20. Epoxy-polyamine resin formed from chloroepoxide and ammonia.

In the past, ion exchange matrices have been formed by the reaction of phenol and formaldehyde (Figure 3.21). These resins have, to a large extent, been replaced by other resins with their inherently higher capacity, higher thermal stability and longer life.

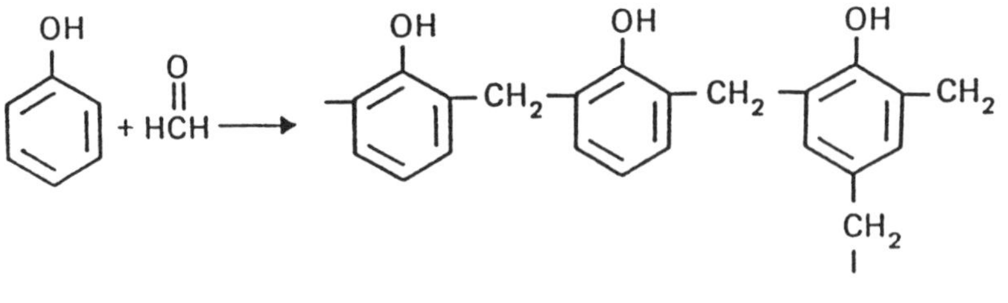

Figure 3.21. Phenol-formaldehyde resin matrix.

These four synthetic organic resins are the most commonly used in industrial applications and have been used in biochemical applications and even in protein purifications and enzyme immobilizations. However, the hydrophobic matrices have the disadvantages that they might denature the desired biological material or that the high charge density may give such strong binding that only a fraction of the adsorbed material might be recovered.

Resins with cellulosic matrices (Figure 3.22) are much more hydrophilic and these do not tend to denature proteins. Cellulosic resins have been used extensively in the laboratory analysis of biological material, enzyme immobilization and small scale preparations. The low capacity and poor flow characteristics limited the usefulness of these matrices for larger applications.

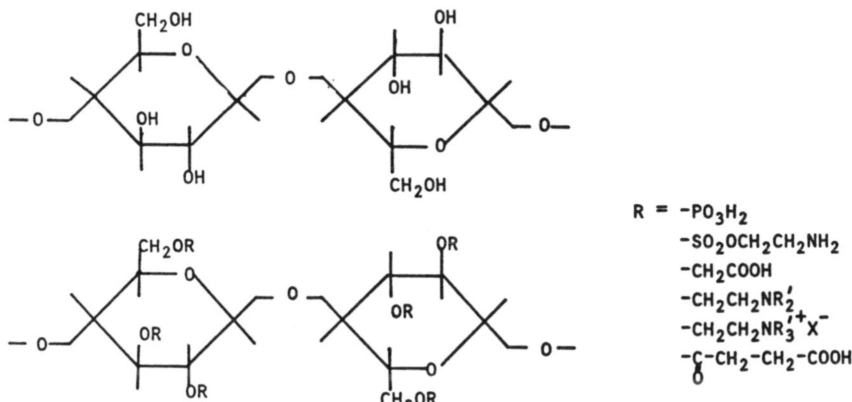

Figure 3.22. Cellulosic resins.

Recently, diethylaminoethyl(DEAE)-silica gel was shown (53) to be an improvement over typical cellulosic matrix resins for the separation of acidic and neutral lipids from complex ganglioside mixtures. The specific advantages claimed were:

(1) An increase in flow rate was possible through the DEAE-silica gel.

(2) The DEAE-silica gel was able to be equilibrated much more rapidly with the starter buffer.

(3) The DEAE-silica gel was more easily regenerated.

(4) The DEAE-silica gel was less susceptible to microbial attack.

(5) The preparation of DEAE-silica gel from inexpensive silica gel was described as a simple method that could be carried out in any laboratory.

The kinetic and electrochemical properties of ion exchange resins depend on the nature of the constituents used in their synthesis. The changes in the initial diffusion coefficients of organic ions inside the resin particles is shown in Figure 3.23 for cation resins synthesized from various monomers (23). The use of maleic acid provides a significant increase in the exchange kinetics. However, such cation resins are capable of dissociation at higher pH compared to resins synthesized with acrylic acid. The pK = 7.0 and 5.5 for resins synthesized from maleic acid and acrylic acid, respectively.

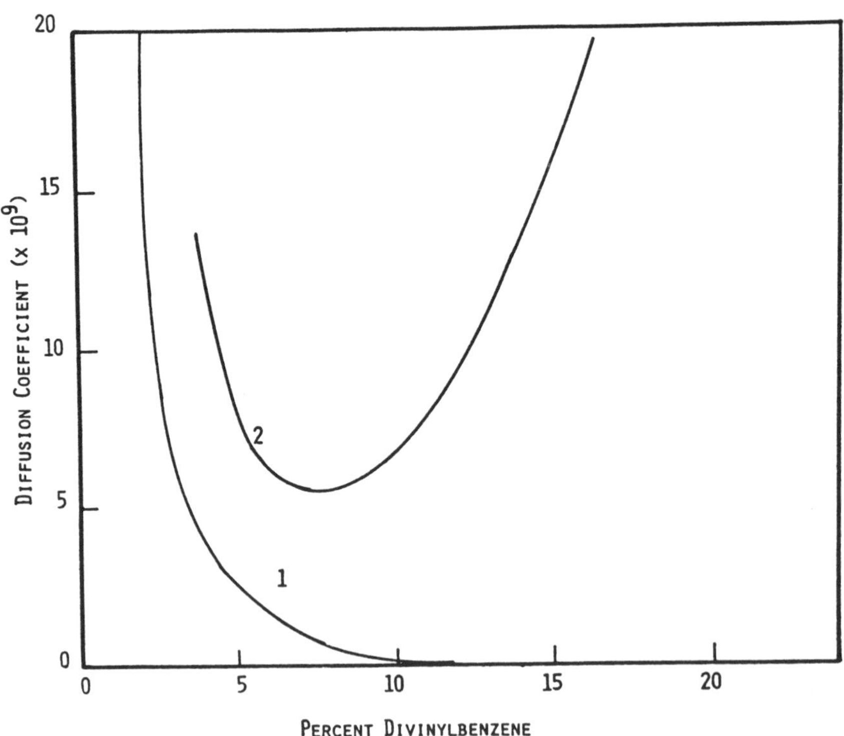

Figure 3.23. Diffusion coefficients (cm^2/sec) of streptomycin in the antibiotic sorption by (1) gel and (2) modified carboxylic cation exchangers with various crosslinking (Reference 23).

3.3.2 Functional Groups

The strong acid cation exchange resins are made by the sulfonation of the matrix copolymer. The resulting resin for styrene-divinylbenzene copolymer has the polymer structure shown in Figure 3.18 with the SO_3^- attached. Associated with that functional group is the exchangeable ion, in this case, the hydrogen ion. Strong acid cation resins are characterized by their ability to exchange cations or split neutral salts. They will function throughout the entire pH range.

The synthesis of weak acid cation resins has been described above. The ability of this type of resin to split neutral salts is very limited. This resin has the greatest affinity for alkaline earth metal ions in the presence of alkalinity. Only limited capacities for the alkali metals are obtained when alkalinity other than hydroxide is present. Effective use is limited to solutions above pH 4.

The anion exchange resins require the synthesis of an active intermediate. This is usually performed in the process called chloromethylation. The subsequent intermediate is reactive with a wide variety of amines which form different functional groups. Figure 3.18 illustrates the structure of the Type I strong base anion resin, the Type II strong base anion resin, and a tertiary amine weak base anion exchange resin.

The Type I resin is a quaternized amine resin resulting from the reaction of trimethylamine with the chloromethylated copolymer. This functionalized resin has the most strongly basic functional group available and has the greatest affinity for weak acids. However, the efficiency of regenerating the resin to the hydroxide form is somewhat lower than Type II resins, particularly when the resin is exhausted with monovalent anions.

The Type II resin results when dimethylethanolamine is reacted with the chloromethylated copolymer. This quaternary amine has lower basicity than that of the Type I resin, yet it is high enough to remove the anions of weak acids in most applications. While the caustic regeneration efficiency is significantly greater with Type II resins, their thermal and chemical stability is not as good as Type I resins.

Weak base resins may be formed by reacting primary or secondary amines or ammonia with the chloromethylated copolymer. Dimethylamine is commonly used. The ability of the weak base resins to adsorb acids depends on the basicity of the resin and the pK of the acid involved. These resins are capable of adsorbing strong acids in good capacity, but are limited by kinetics. The kinetics may be improved by incorporating about 10% strong base capacity. While strong base anion resins function throughout the entire pH range, weak base resins are limited to solutions below pH 7.

The desired functionality on the selected matrix will be determined by the nature of the biochemical solute which is to be removed from solution, its isoelectric point, the pH restrictions on the separation and the ease of eventually eluting the adsorbed species from the resin.

Some resins have been developed with functional groups specifically to adsorb certain types of ions. The resins shown in Table 3.7 are commercially available.

Table 3.7: Commercial Resins with Special Functional Groups

Functionality	Structure
Iminodiacetate	$R\text{-}CH_2N(CH_2COOH)_2$
Polyethylene Polyamine	$R\text{-}(NC_2H_4)_mH$
Thiol	R-SH
Aminophosphate	$R\text{-}CH_2NHCH_2PO_3H_2$
Amidoxime	$R\text{-}C(NH_2){=}N\text{-}OH$
Phosphate	$R\text{-}PO_3H$

The selectivity of these resins depends more on the complex that is formed rather than on the size or charge of the ions. Generally they are effective in polar and non-polar solvents. However, the capacity for various ions is pH sensitive so that adsorption and elution can be accomplished by pH changes in the solution.

These chelating resins have found most of their use in metal ion recovery processes in the chemical and waste recovery industries. They may find use in fermentation applications where the cultured organisms requires the use of metal ion cofactors. Specific ion exchange resins have also been used in laboratory applications that may find eventual use in biotechnology product recovery applications (54).

A review of selective ion exchange resins has been compiled by Warshawsky (55). A diaminotetraacetic polymer developed by Mitsubishi (56) was developed for the purification of amino acid feed solutions. The conversion of chloromethylpolystyrene into thiolated derivatives for peptide synthesis has been described by Warshawsky and coworkers (57).

3.3.3 Porosity and Surface Area

The porosity of an ion exchange resin determines the size of the molecules or ions that may enter an ion exchange particles and determines their rate of diffusion and exchange. Porosity is inversely related to the crosslinking of the resin. However, for gel-type or microporous resins, the ion exchange particles has no appreciable porosity until it is swollen in a solvating medium such as water.

The pore size for microporous resins is determined by the distances between polymer chains or crosslinking subunits. If it is assumed that the crosslinking is uniform throughout each ion exchange particle, the average pore diameter of these resins can be approximated from the water contained in the fully swollen resin. The moisture content of cation resins as a function of the degree of crosslinking is shown in Figure 3.24 and of anion resins in Figure 3.25. The calculated average pore size for sulfonic cation resins ranges from 16 to 20 Å as the resin crosslinking is decreased from 20 to 2% divinylbenzene. The calculated average pore size of the anion resins ranges from 18 to 14 Å as the crosslinking is decreased from 12 to 2%. Even at low crosslinking and full hydration, microporous resins have average pore diameters of less than 20 Å. The dependence of pore size on the percent of crosslinking is shown in Table 3.8 for swollen microporous resins of styrene-divinylbenzene (58).

Sulfonic acid and carboxylic acid resins have also been equilibrated with a series of quaternary ammonium ions of different molecular weights to measure the average pore size when these resins have increasing degrees of crosslinking. These results are shown in Figures 3.26 and 3.27.

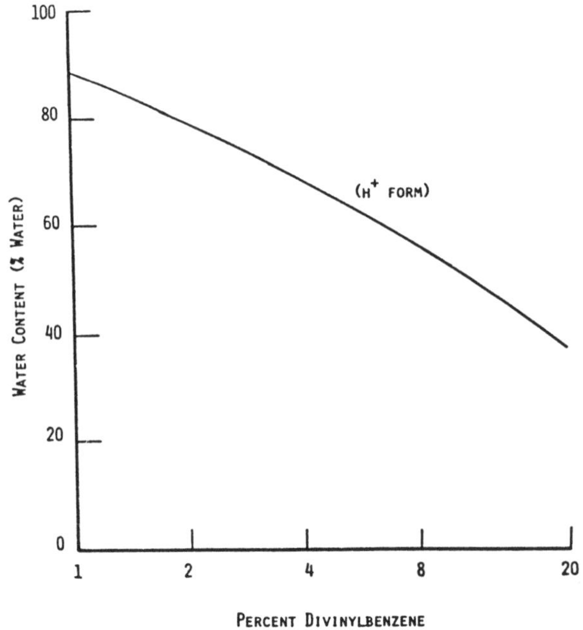

Figure 3.24. Moisture content of strong acid cation resins as a function of divinylbenzene content.

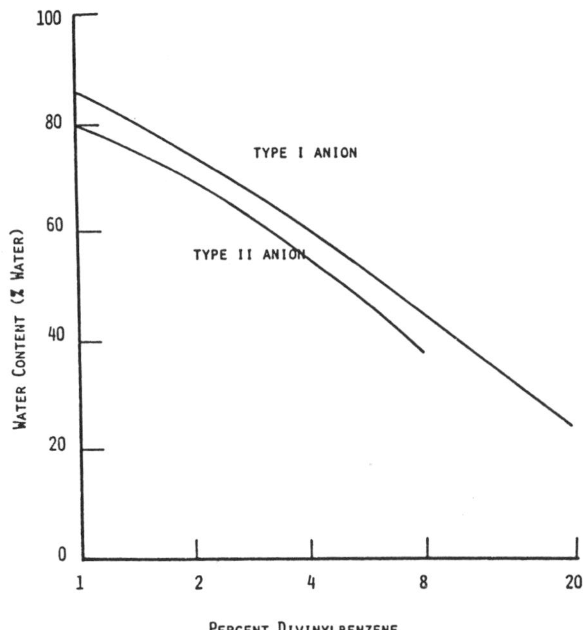

Figure 3.25. Moisture content of strong base anion resins as a function of divinylbenzene content.

Table 3.8: Average Swollen Diameter of Crosslinked Polystyrene Beads in Tetrahydrofuran (58)

Divinylbenzene Concentration (Crosslinking) (%)	Swollen Pore Diameter (%)
1	77
2	54
4	37
8	14
16	13

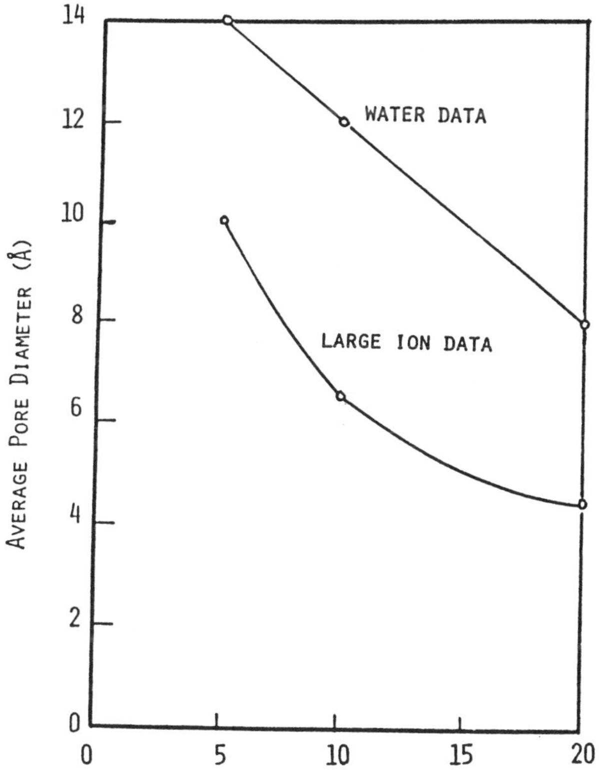

Figure 3.26. Average pore diameter of sulfonic acid cation exchange resin as a function of degree of crosslinking (Reference 15, page 46).

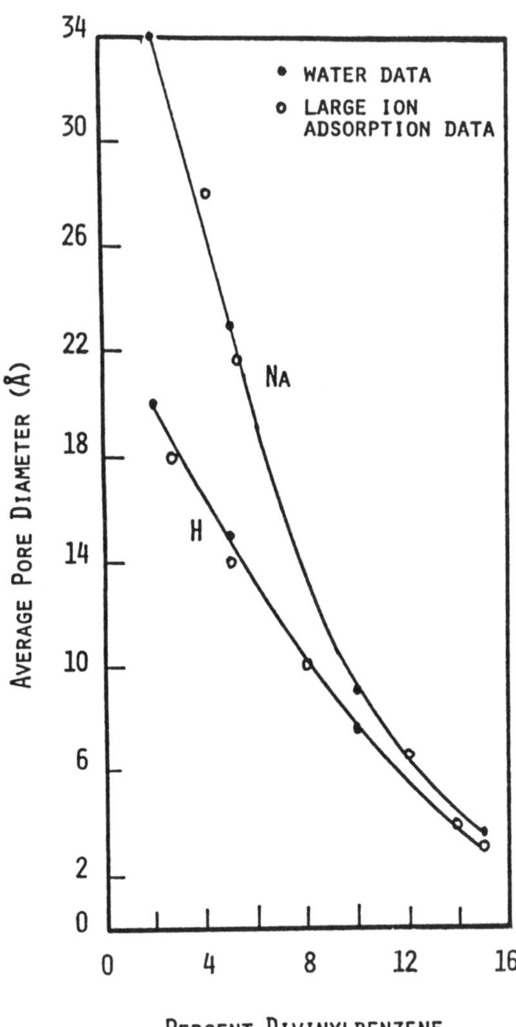

Figure 3.27. Average pore diameter of carboxylic acid cation exchange resin as a function of degree of crosslinking (Reference 15, page 47).

Figure 3.28 shows the change in the ionic diffusion coefficients of tetraalkylammonium ions in strong acid cation resins as a function of mean effective pore diameter of the resins (59). As the pore diameter is increased, the penetrability of the resins with respect to the large ions also increased.

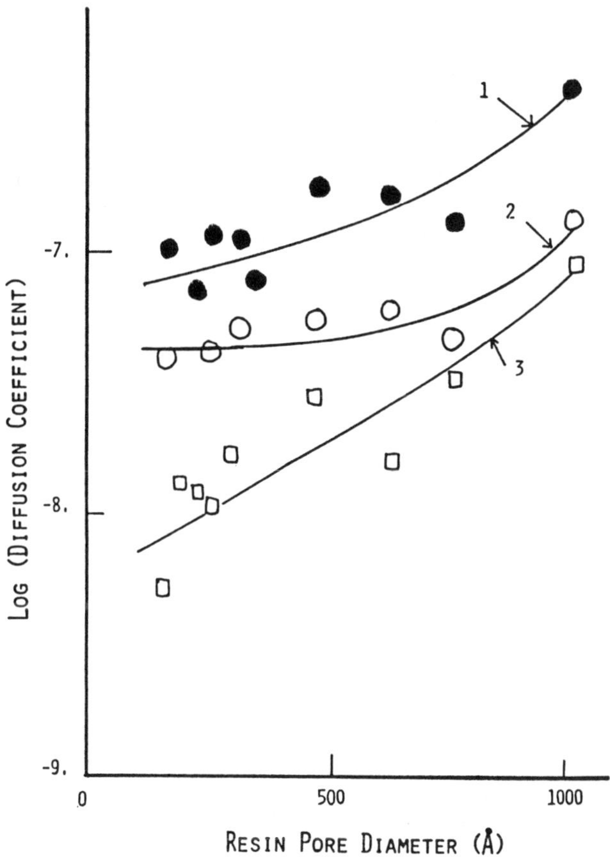

Figure 3.28. Ion ((1) tetramethyl; (2) tetraethyl; (3) tetrabutyl ammionium ions) diffusion coefficients in macroporous sulfonated cation resins (Reference 59).

If an inert diluent (porogen) is incorporated into the monomer mixture before the copolymer is formed, it is possible to form a structure containing varying degrees of true porosity or void volume (60,61). Variations in the amount of divinylbenzene crosslinking and diluent allow for a range of particle strengths and porosity to be made. Subsequent reactions with the appropriate chemicals result in the introduction of the same functional groups as discussed above. These are called macroporous or macroreticular resins.

A list of the porogens can be compiled from the Ashahi

patents (62-64) for specific monomer and comonomer systems (Table 3.9). These substances can be classified according to their solubility coefficient, δ, their polarity and according to their solvent capability for homopolymers. The solvent capability is divided into three categories (65):

(1) where it is a solvent for all homomonomers of the monomer used

(2) where it is non-solvent for all homopolymers of the monomers used

(3) where it is a solvent for some and a non-solvent for other homopolymers of the polymers used.

Table 3.9: Classification of Porogens from Asahi Patents (65)

Solvent	Polarity	Solvent Capability Category[a] Examples 1	2	3	4	5	Solubility Coefficient, δ
hexanes	--	2	2	2	2	2	~7
heptanes	--	?	2	2	?	2	7.5
octanes	--	2	2	2	2	2	7 - 7.5
decanes	--	2	?	2	?	2	7.5 - 8
cyclohexane	--	2	2	2		2	8
diisopropyl ketone	+					3	8
ethyl benzoate	+	3	1	3			8.2
ethyl propionate	+	3	1			3	8.4
methyl isobutyl ketone	+	3		[3]	3		8.4
benzonitrile	(+)	1	1				8.4
butyl acetates	+	3				3	8.3 - 8.5
n-butyl propionate	+	3		3		3	8.8
xylenes	--	3	1	3	1		8.8
ethylbenzenes	--	3		3	1	3	8.8
toluene	--	3	1	3	1	3	8.9
ethyl acetate	+	3		3	3	3	9.1
chloroform	(+)				1		9.3
methyl ethyl ketone	+	3	1	[3]	3		9.3
2-nitropropane	(+)	1	1	[3]			9.9
cyclohexanone	+	1	1	[3]			9.9
o-dichlorobenzene	(+)	3	1		1		10
amyl alcohols	+	3	3	2			10 - 11
octanols	+	3	3	2	2		~10.5
methyl benzoate	+	3	1	3			10.5
butanols	+	3	3				10.5 - 11.5
hexanols	+	3	3	2	2		~10.7
cyclohexanol	+	3	3	2	2		11.4

[a] Examples 1-5 refer to monomers/comonomers in the patents
( ) = poor hydrogen bonding
[ ] = in acrylonitrile copolymers only

The simplest system is merely the use of a porogen from category 3. When a porogen from category 2 is added to a porogen from category 3, the pore diameter of the final resin increases. Conversely, when a porogen from category 1 is added to a category 3 porogen, the pore diameter decreases. A mixture of porogens from categories 1 and 2 results in a wide variation in pore size distribution as the ratio of the porogens is changed. Fine control on the pore size distribution can be obtained by using mixtures of porogens from category 3.

Since each bead of a given external diameter that is made by the inert diluent process will contain some void volume, there is actually less polymer available per unit volume for the introduction of functional groups. Therefore, these macroporous resins are inherently of lower total exchange capacity than gel-type resins of the same composition.

Macroporous resins are most useful when extremely rigorous osmotic shock conditions are encountered, when the very high porosity is desirable from the standpoint of the molecular weight of the material being treated or when non-polar media are involved. The drawbacks of using macroporous resins are poorer regeneration efficiencies, lower total exchange capacities and higher regeneration costs.

Until the advent of macroporous resins, the synthetic organic ion exchange resins were of such low porosity that large proteins and other macromolecules would be adsorbed or interact only with the exterior exchange sites on the resins. Therefore, although the microporous resins may have higher total exchange capacities than macroporous resins, the effective capacity of macroporous resins for protein or macromolecule absorption may often times be greater than that of microporous resins.

Typical macroporous ion exchange resins may have average pore diameters ranging from 100 Å to 4000 Å. Table 3.10 shows the pore sizes of several resins of different matrices that have been used in enzyme immobilization (66). Pore volumes for macroporous resins may range from 0.1 to 2.0 cc/g. The relationship between pore diameter and porosimeter pressure is shown in Table 3.11 for mercury intrusion porosimetry based upon the Kelvin equation.

Table 3.10: Physical Properties and Capacities for Ion Exchange Resins (66)

Resin Matrix	Functionality	Pore Size	Surface Area	Resin Capacity	Adsorption Capacity for Enzyme
phenolic	3° polyethylene polyamine	250 Å	68.1 m^2/g	4.38 meq/g	3.78 meq/g
phenolic	partially 3° polyethylene polyamine	290 Å	95.3 m^2/g	4.24 meq/g	3.57 meq/g
polystyrene	polyethylene polyamine	330 Å	4.6 m^2/g	4.20 meq/g	3.92 meq/g
polystyrene	polyethylene polyamine	560 Å	5.1 m^2/g	4.75 meq/g	4.32 meq/g
polyvinyl chloride	polyethylene polyamine	1400 Å	15.1 m^2/g	4.12 meq/g	3.72 meq/g

Table 3.11: Relationship Between Mercury Intrusion Pressure and Pore Diameter for Polystyrene Beads

Hg Porosimeter Pressure (Bar)	Pore Diameter (Å)
28	5000
70	2000
125	1000
275	500
690	200
1240	100
2760	50
6900	20
12400	10

Normally, as the mean pore diameter increases, the surface area of the resin decreases. These surface areas can be as low as 2 m^2/g to as high as 300 m^2/g. Table 3.10 also points out that the total exchange capacity is not utilized in these biochemical fluid processes. Whereas in water treatment applications, one can expect to utilize 95% of the total exchange capacity, in biotechnology applications it is often possible to use only 20 to 30% of the total exchange capacity of gel resins. Macroporous resins have increased the utilization to close to 90% for the immobilization of enzymes, but biochemical fluid processing applications where the fluid flows through an ion exchange resin bed still are limited to about 60% utilization even with macroporous resins.

Table 3.12 shows the molecular size of some biological macromolecules for comparison to the mean pore size of the resins. When selecting the pore size of a resin for the recovery or immobilization of a specific protein, a general rule is that the optimum resin pore diameter should be about 4 to 5 times the length of the major axis of the protein. Increasing the pore size of the resin beyond that point will result in decreases in the amount of protein adsorbed because the surface area available for adsorption is being decreased as the pore size is increased. An example of this optimum adsorption of glucose oxidase, as defined by enzyme activity, is shown in Figure 3.29 (67). Enzyme activity is a measure of the amount of enzyme adsorbed and accessible to substrate.

3.3.4 Particle Density

The typical resin densities may range from 0.6 g/cc to 1.3 g/cc for organic polymers. Silicate materials may be more dense, up to 6 g/cc. Since the fermentation broth or other biochemical fluid may be denser than water, the slow flow rates that are usually involved may require resins that have a greater density than water. A minimum flow rate may be necessary to maintain a packed bed when a fluid denser than water is being processed by a medium density resin. If this is not possible, an upflow operation or batch process may be necessary. This will be discussed in more detail in section 3.6.3.

Table 3.12: Molecular Size of Biopolymers

Biopolymer	Molecular Weight	Maximum Length of Biopolymer
Catalase	250,000	183 Å
Glucose Isomerase	250,000 - 100,000	75 - 100 Å
Glucose Oxidase	15,000	84 Å
Lysozyme	14,000	40 Å
Papain	21,000	42 Å

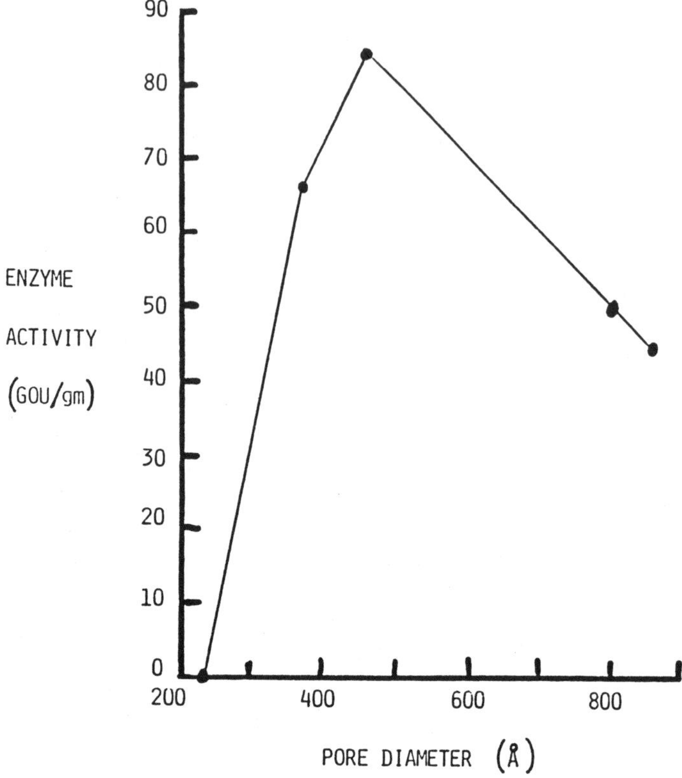

Figure 3.29. Effect of a resin pore diameter on the enzyme activity of glucose oxidase (Reference 67).

The lower-density resins are usually associated with a highly porous structure which has less mechanical strength than the typical gel or macroporous resins. When the mean pore diameter of a resin is greater than 2000 Å, the resin would be subject to attrition in a stirred tank or to collapse in a tall column.

3.3.5 Particle Size

Many of the resins used in early biochemical separations were much smaller (75-300 microns). With the development of macroporous resins, enzyme immobilizations and protein purifications were performed with resins of the 400-1000 micron size since the macroporous structure allowed sufficient surface area for adsorption almost independent of particle size.

3.4 LABORATORY EVALUATION RESIN

The total exchange capacity, the porosity, the operating capacity and the efficiency of regeneration need to be evaluated in the laboratory when comparing resins for a given application.

The total exchange capacity is usually determined by titrating the resin with a solution of acid or base to a specific endpoint. This type of information is readily available from the manufacturers of commercial ion exchange resins.

The pore size of a microporous resin can be determined using water soluble standards, such as those listed in Table 3.13 (68). A typical pore size distribution is shown in Figure 3.30 for a strong cation-resin.

Table 3.13: Water Soluble Standard Samples for Pore Measurements (68)

Sample	Mean Pore Diameter (Å)
D_2O	3.5
Ribose	8
Xylose	9
Lactose	10.5
Raffinose	15
Stachyose	19
T-4 [a]	51
T-10	140
T-40	270
T-70	415
T-500	830
T-2000	1500

[a] The T-Standards are Dextrans from Pharmacia

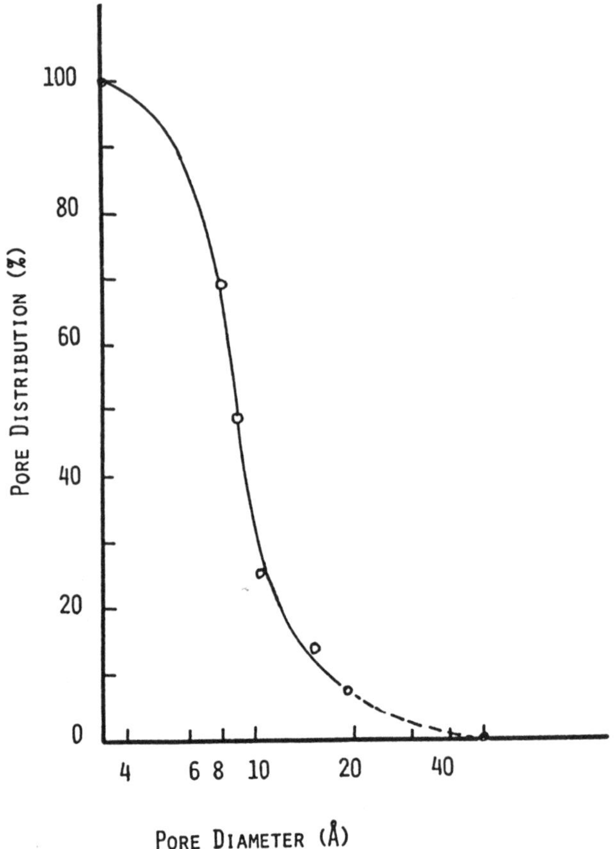

Figure 3.30. Pore size distribution of the cation exchange resin Hamilton HC-X7 (Na).

If the resin is made with an inert, extractable diluent to generate the macroporous structure, it is easier to determine the mean pore size and pore size distribution. Care must be taken so that the pores are not collapsed during the removal of the water from the resin. Martinola and Meyer (69) have devised a method of preparing a macroporous resin for BET surface analysis or pore size analysis by mercury porosimetry.

(1) Convert the ion exchange resin to be desired ionic form.

(2) Add 500 ml of water-moist resin to a round bottom flask with an aspirator. Add one liter of anhydrous isopropyl

alcohol and boil under reflux at atmospheric pressure for one hour. Then suck the liquid off. Repeat the isopropyl addition, boiling and aspirating four times. After this procedure the resin will contain less than 0.1% water.

(3) After drying to constant weight at 10^{-3} torr and 50°C, the resin sample is ready for pore size analysis.

The design of an ion exchange unit requires knowledge of the capacity of the resin bed and the efficiency of the exchange process. The "theoretical" capacity of a resin is the number of ionic groups (equivalent number of exchangeable ions) contained per unit weight or unit volume of resin. This capacity may be expressed as milliequivalents (meq) per milliliter (ml) or per gram.

The amount of dissolved impurities in a feedstream is usually expressed in terms of equivalent calcium carbonate, abbreviated to "as $CaCO_3$". Calcium carbonate provides a good reference because it has a molecular weight of 100, which facilitates calculations. An alternative method of expressing this is in parts per million of the ions themselves. The interrelationships are given in Table 3.14. If the impurities are expressed in terms of parts per million as equivalents of the ion (meq/L), the amounts in terms of ppm must be divided by the equivalent weight of the ions. The equivalent weight of an ion is its molecular weight divided by its valence.

When deciding which resin to use for a given operation, batch testing in a small beaker or flask will allow resin selection and an approximation of its loading capability. A useful procedure is to measure out 1, 3, 10 and 30 milliliter volumes of resin and add them to a specific volume of the feedstream. These volumes were chosen to have even spacing on a subsequent log-log plot of the data.

After the resin bed feed solution has been mixed for at least one-half hour, the resin is separated from the liquid phase. The solute concentration remaining in the solution is then determined. The residual concentration is subtracted from the original concentration and the difference is divided by the volume of the resin. These numbers and the residual

Table 3.14: Conversion Factors to Express as Parts per Million as $CaCO_3$

Ions	Ionic Weight	Equivalent Weight	Factor
Cations			
Aluminum	27.0	9.0	5.56
Ammonium	18.0	18.0	2.78
Barium	137.4	68.7	0.78
Calcium	40.1	20.0	2.49
Copper	63.6	31.8	1.57
Hydrogen	1.0	1.0	50.0
Ferrous	55.8	27.9	1.80
Ferric	55.8	18.6	2.69
Magnesium	24.3	12.2	4.10
Manganese	54.9	27.5	1.82
Potassium	39.1	39.1	1.28
Sodium	23.0	23.0	2.18
Anions			
Bicarbonate	61.0	61.0	0.82
Bisulfate	97.1	97.1	0.51
Bisulfite	81.1	81.1	0.61
Carbonate	60.0	30.0	1.67
Chloride	35.5	35.5	1.41
Fluoride	19.0	19.0	2.63
Hydroxide	17.0	17.0	2.94
Nitrate	62.0	62.0	0.81
Phosphate (primary)	97.0	97.0	0.51
Phosphate (secondary)	96.0	48.0	1.04
Phosphate (tertiary)	95.0	31.7	1.58
Sulfate	96.1	48.0	1.04
Sulfide	32.1	16.0	3.12
Sulfite	80.1	40.0	1.25

concentration are plotted on log-log paper and frequently give a straight line.

A vertical line drawn at the feed concentration intersects at a point extrapolated from the data points to give an estimate of the loading of the solute on the resin.

Example 3.5

Figure 3.31 shows such a plot for glutamic acid

adsorbed on an 8% crosslinked strong acid cation microporous resin in a fermentation broth with 11 mg/ml of amino acid. The resin was placed in a beaker with 250 ml of broth. The extrapolation of the line for the 1, 3, 10 and 30 ml resin adsorption data indicates that the loading of glutamic acid on this resin is expected to be 60 g/liter resin.

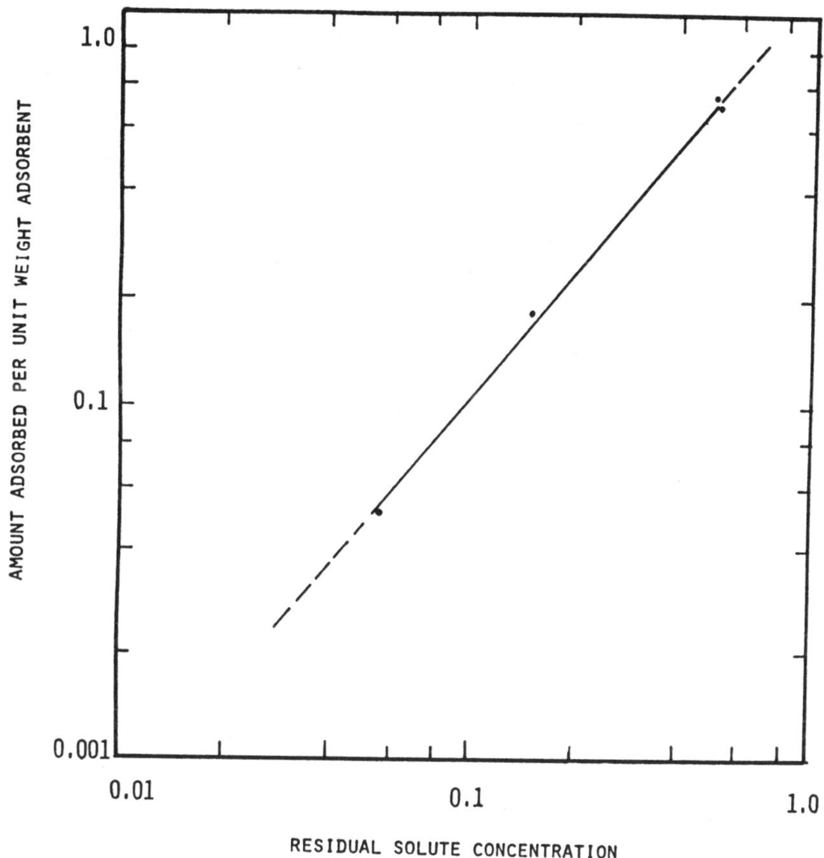

Figure 3.31. Plot of glutamic acid adsorbed on an 8% crosslinked strong cation resin.

After several resins have been tested in this manner, the resin is selected for column evaluation which has a high loading per ml of resin or a low residual with larger resin quantities.

In practice, the ion exchange resin is generally operated at a level considerably below its theoretical capacity. Since the ion exchange reactions are equilibrium reactions, an impractically large quantity of regenerant would be required to drive the reaction to completion. The "operating" capacity of a resin is the number of ionic groups actually utilized per unit weight or volume of resin under a given set of operating conditions.

The operating capacity of a resin is not directly proportional to the amount of regenerant used. "Efficiency" is the concept used to designate the degree of utilization of the regenerant. Column efficiency is the ratio of the operating exchange capacity of a unit to the exchange that theoretically could be derived from a specific weight of applied regenerant.

It is recommended that operating capacity and column efficiency be run initially on a small, laboratory scale to determine if the reaction desired can be made to proceed in the desired direction and manner. The column should be at least 2.5 cm in diameter to minimize wall effects. The preferential flow in a resin column is along the wall of the column. The percentage of the total flow along the wall of the column decreases as the column diameter increases and as the resin particle size decreases.

The bed depth should be at least 0.5 m and the flow rate should be about 0.5 bed volumes per hour for the initial trial. These conditions are good starting points since it is desirable that the transition zone not exceed the length of the column. Using much larger columns would require quantities of the feedstream which are larger than may be readily available. A suggested operating procedure is outlined below.

(1) Soak the resin before adding it to the column to allow it to reach its hydrated volume.

(2) After the resin has been added to the column, backwash the resin with distilled water and allow the resin to settle.

(3) Rinse the column of resin with distilled water for ten minutes at a flow rate of 50 mL/min.

(4) Start the treatment cycle. Monitor the effluent to develop a breakthrough curve, such as shown in Figure 3.32, until the ion concentration in the effluent reaches the concentration in the feed solution.

(5) Backwash the resin with distilled water to 50-100% bed **expansion** for 5 to 10 minutes.

(6) Regenerate the resin at a flow rate that allows at least forty-five minutes of contact time. Measure the ion concentration of the spent regenerant to determine the elution curve (Figure 3.33) and the amount of regenerant actually used.

(7) Rinse column with distilled water until the effluent has reached pH 7.

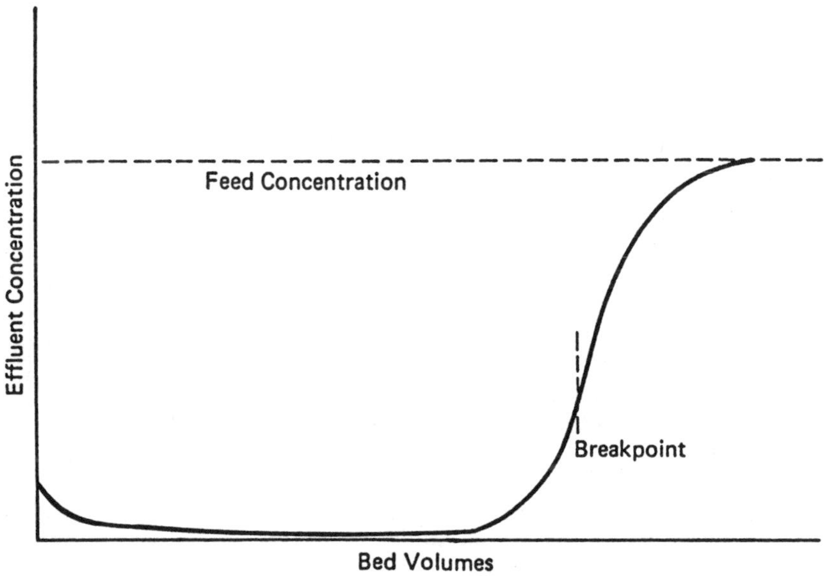

Figure 3.32. Concentration of adsorbed species in column effluent during column loading.

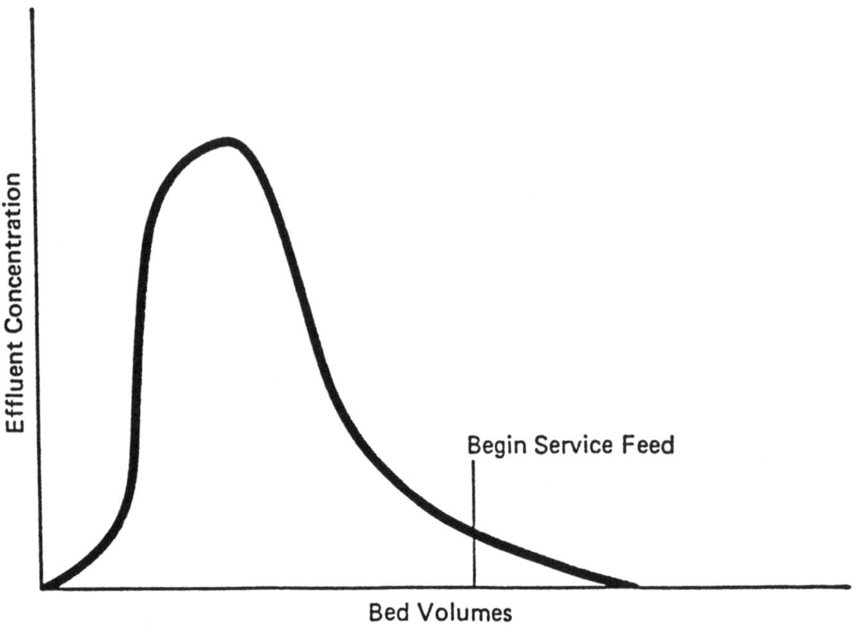

Figure 3.33. Elution curve showing concentration of adsorbed species eluted during resin regeneration.

Feed concentrations, flow rates, and regenerant dosages may be varied to develop the relationship between resin utilization and regenerant efficiency so that the optimum operating conditions can be selected for the system.

The first portion of the breakthrough curve in Figure 3.32 shows the quality of product that can be obtained under the processing conditions. An integration of the area up to the breakthrough point provides an estimate for commercial column capacities for the space velocities used in the experiment. The velocity at which the mass transfer zone is moving through the column is given by dividing the length of the column by the time it takes to detect the solute in the column effluent. The difference between that time and the time at which the selected breakthrough concentration appears in the effluent, when multiplied by the velocity of the mass transfer zone, results in an approximation of the mass transfer zone.

During the column test, the starting volume and the final volume of the resin should be measured. If there is a change of more than 5%, progressive volume changes as the resin is operated through several cycles should be recorded. These changes may be significant enough to affect the placement of laterals or distributors in the design of commercial equipment. For instance, carboxylic resins may expand by 90% when going from the hydrogen form to the sodium form. This type of volume change may dictate how the resin must be regenerated to prevent the breakage of glass columns due to the pressure from the swelling resin.

Gassing, the formation of air pockets, within the resin bed is to be avoided. Gassing may occur because of heat released during the exchange reaction. It will also occur if a cold solution is placed in a warm bed or if the liquid level falls below the resin level. Keeping the feed solution 5°C warmer than the column temperature should prevent the gassing due to thermal differences.

It is necessary to configure the experimental apparatus to insure that the feedsteam moves through the column at a steady rate to maintain a well-defined mass transfer zone. Possible methods of maintaining constant flow are shown in Figure 3.34.

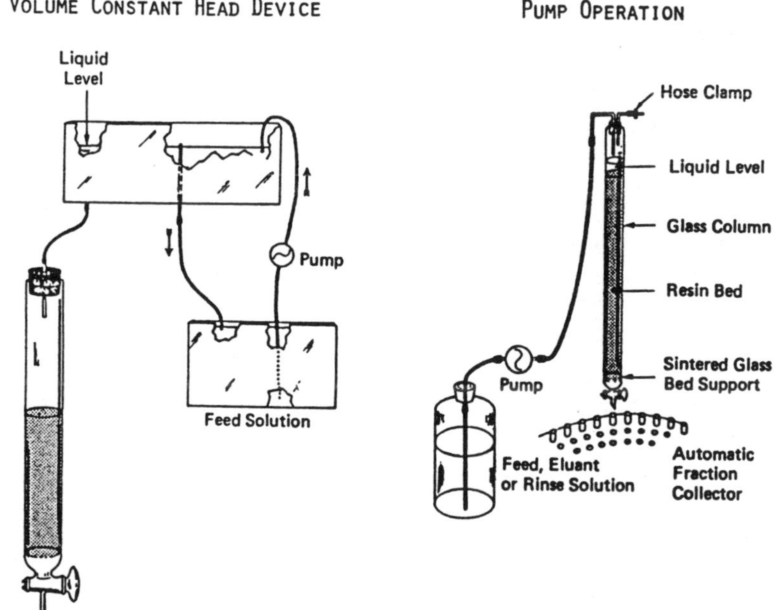

Figure 3.34. Equipment for laboratory evaluation of ion exchange resins.

Once it is determined that the reaction will proceed as desired, subsequent optimization of the system in the laboratory calls for setting up a packed resin column of approximately the bed depth to be used in the final equipment, typically one to three meters.

3.5 ION EXCHANGE PROCESSES

3.5.1 Process Categories

Processes involving ion exchange resins usually make use of ion interchange with the resin. Examples of these processes are demineralization, conversion, purification and concentration. Chromatographic processes with ion exchange resins, which merely make use of the ionic environment that the resins provide in separating solutes, will be discussed in Chapter 4.

3.5.1.1 Demineralization: Demineralization is the process in which the salts in the feed stream are removed by passing the stream through a cation exchange column in the hydrogen ion form, followed by an anion exchange column in the hydroxide or "free base" form. Water is the most common feed stream in demineralization. It may also be necessary to remove the salts from a feed stream before fermentation.

High metallic ion concentrations and high total salt content in the carbohydrate feed have been found to decrease the yield in citric acid fermentation (70). These ions can be removed by passing the carbohydrate solution through cation and anion exchange resin beds. The salts required for optimum microorganism activity can be added in the desired concentration prior to fermentation.

$$\text{DEXTROSE} + \text{NaCl} + \text{R-H} \rightleftharpoons \text{R-Na} + \text{HCl} + \text{DEXTROSE} \tag{3.30}$$

$$\text{DEXTROSE} + \text{HCl} + \text{R-OH} \rightleftharpoons \text{R-Cl} + \text{H}_2\text{O} + \text{DEXTROSE} \tag{3.31}$$

3.5.1.2 Conversion: Conversion or metathesis is the process in which salts of acids are converted to the corresponding free acids by reaction with the hydrogen form of a strong acid cation resin. One such example would be the conversion of calcium citrate to citric acid.

$$\text{Ca Citrate} + 2\ \text{R-H} \rightleftharpoons \text{R}_2\text{-Ca} + 2\ \text{Citric Acid} \tag{3.32}$$

The term may also be used to describe a process in which the acid salt is converted to a different salt of that acid by interaction with an ion exchange resin regenerated to the desired ionic form.

3.5.1.3 Purification: Many fermentation products may be purified by adsorbing them on ion exchange resins to separate them from the rest of the fermentation broth. Once the resin is loaded, the product is eluted from the column for further purification or crystallization.

Adsorbing lysine on an ion exchange resin is probably the most widely used industrial method of purifying lysine. The fermented broth is adjusted to pH 2 with hydrochloric acid and then passed through a column of strong acid cation resin in the NH_4^+ form. Dilute aqueous ammonia may be used to elute the lysine from the resin (71).

Gordienko (72) has reported that treating the resin with a citrate buffer solution of pH 3.2 and rinsing with distilled water before elution results in an 83-90% yield of lysine, with a purity of 93-96%.

3.5.1.4 Concentration: Ion exchange can be used to concentrate valuable or toxic products of fermentation reactions in a manner similar to purification. The difference between the two processes is in the lower concentration of the desired product in the feed solution of concentration processes.

Shirato (73) reported the concentration process for the antibiotic tubercidin produced from fermented rice grain using the microorganism, *Streptomyces tubercidicus.* Macroporous strong acid cation resin was used to concentrate the antibiotic from 700 μg/ml in the fermentation broth to 13M μg/ml when eluted with 0.25N HCl. The yield of the antibiotic was about 83%.

3.5.2 Purification Procedures

A general procedure has been outlined (74) for the use of ion exchange resins to purify mixtures containing organic acids or bases.

3.5.2.1 Purification of Strong Bases:

(A) No Stability Limitations - Selectivity and

removal of weaker bases can be obtained by carrying out the adsorption at high pH (either strong or weak acid resin can be used) or by carrying out the adsorption at low pH and washing the resin adsorbate with high pH buffers. Weaker bases can be preferentially removed from the resin. Either procedure yields a resin adsorbate containing only strong bases. These can be removed from the resin by eluting with an appropriate salt for a strong acid resin or with a strong acid for a weak acid resin.

(B) Unstable at Alkaline pH; Stable at Neutral and Acidic pH - Under these conditions separation from bases of intermediate pK may be impossible. The adsorption should be carried out using a weak acid resin (carboxylic) at the highest pH compatible with the stability limitation. The elution is by acidification with a strong acid.

(C) Unstable at Acidic pH; Stable at Neutral and Alkaline pH - Adsorption conditions would be the same as (A) with no limitations on resin type. Elution should be carried out with a salt using a cation preferred by the resin.

(D) Stable Only at Neutral pH - Adsorption should be like (B) while elution must be performed with a salt.

(E) Stable Only at Acidic pH - Separation from weak bases is not possible. Adsorption on a strong acid resin followed by salt elution yields a mixture of all types of bases.

(F) Stable Only at Alkaline pH - Adsorb at high pH and elute with salt at high pH.

3.5.2.2 Purification of Strong Acids:

(A) No Stability Limitations - Adsorb on either strong or weak base resin at low pH. Elute from a strong resin with a salt or from a weak base resin with alkali.

(B) Unstable at Alkaline pH; Stable at Neutral and Acid pH - Carry out adsorption on a strong base resin at the minimum pH compatible with stability. Elute with an appropriate salt. If the pH range is satisfactory, use a weak base resin and elute with alkali.

(C) Unstable at Acidic pH; Stable at Neutral and Alkaline pH - Adsorb at low pH. Elute with an appropriate salt.

(D) Stable Only at Neutral pH - A purification based on pK is not possible but a mixture of acids can be obtained by adsorption on a strong base resin and elution with a salt.

(E) Stable Only at Acidic pH - If there is no lower pH limit on stability, the adsorption can be carried out at low pH on either strong or weak base resins. The elution is carried out with strong acid or salt.

(F) Stable Only at Alkaline pH - A separation based on pK is not possible. A mixture of all acid solutes can be isolated by adsorption on a strong base resin followed by elution with a salt. Elution from an intermediate strength resin can be accomplished with alkali.

3.5.2.3 Purification of Weak Bases:

(A) No Stability Limitation - Adsorb on a strong acid resin; elute with a weak base. The strong base compounds will remain on the resin.

(B) Unstable at Alkaline pH; Stable at Neutral and Acidic pH - Unless the compound is stable at about 2 pH units below its pK_B, ion exchange separation is unlikely to succeed. If the adsorption step at acidic pH can be carried out, the weak base can be isolated by elution with a weak base.

(C) Unstable at Acidic pH; Stable at Neutral and Alkaline pH - Separation from strong bases may not be possible. May be adsorbed at acidic pH and eluted with a salt.

(D) Stable Only at Neutral pH - Separation from neutral compounds may not be possible. Strong bases are removed by adsorption on strong acid resin. Acidic compounds are removed by adsorption on strong base resins. Elution is with salt.

(E) Stable Only at Acidic pH - Separations from strong bases is not possible. Bases are separated by adsorption on a strong acid resin and elution with a salt.

(F) Stable Only at Alkaline pH - Acids and strong bases are removed by adsorption with appropriate resins. Neutral compounds and weak bases remain.

3.5.2.4 Purification of Weak Acids:

(A) No Stability Limitations - Adsorb on a strong base resin at alkaline pH; elute with a weak acid. Strong acids remain on the resin.

(B) Unstable at Alkaline pH; Stable at Neutral and Acidic pH. Separation from strong acids may not be possible. Adsorb acids at alkaline pH and elute with a salt.

(C) Unstable at Acid pH; Stable at Neutral and Alkaline pH - Separation from neutral compounds may not be possible. Strong acids and all bases are removed by appropriate ion exchange treatment.

(D) Stable Only at Neutral pH - Strong acids and bases removed by appropriate ion exchange treatment. Weak bases and neutral compounds may not be separated.

(E) Stable Only at Acidic pH - Weak acids and neutral compounds remain after ion exchange removal of strong acids and all bases.

(F) Stable Only at Alkaline pH - Separation from strong acids may not be possible. Acids are adsorbed onto a strong base resin and eluted with a salt.

Example 3.6

The fermentation of valine using *Arthrobacter paraffineus* will also result in the production of small amounts of glutamic acid. The glutamic acid can be removed from the fermentation broth by passing the fluid through a strong base anion resin at a pH of 7.0. Then the eluant may be adjusted to a pH of 2 and passed through a strong acid cation resin to concentrate and purify the valine. Elution of the valine from the resin occurs with 1N ammonium hydroxide.

3.5.3 Biopolymer-Resin Interactions

As the biopolymer becomes more complex, the mechanisms by which it adsorbs to and desorbs from resins may involve mechanisms in addition to simple ionic exchange. The attachment of biochemicals to the surface of the ion exchange resin may involve a balance between London-van der Waals forces in addition to the electrostatic forces of electrical double layers (75).

Table 3.15 shows the types of biochemical interactions that are possible with ion exchange resins and insoluble polymer matrices. The degree of difficulty in forming the specific interaction - without denaturing a protein or biopolymer - is related to the strength of the bond which is formed. The stronger the interaction, the more difficult it is to be formed without adversely affecting the biopolymer. With ionic bonds and physical adsorption, simply bringing the resin and the biopolymer into close proximity will cause the interaction. Likewise, changing the pH or ionic strength of the solution will cause the interaction's reversal or regeneration of the resin.

Specifically for interactions with ion exchange resins, the typical biopolymer can be thought of as a macroscopic

Table 3.15: Biopolymer Interactions with Ion Exchange Resins and Polymer Matrices

	Covalent Bond	Ionic Bond	Physical Adsorption	Entrapment
Formation	Complicated	Simple	Simple	Complicated
Possibility of Biopolymer Denaturing	High	Low	Low to High	High
Bonding Strength	Strong	Weak	Weak	Strong
Regeneration	No	Possible	Possible	No

ion having a number of electrostatically charged surface sites. The ionic interactions can be represented as:

$$R{-}N^{+}(CH_3)_3Cl^{-} \;+\; {}^{-}OOC{-}\underset{H}{\overset{R'}{\underset{|}{\overset{|}{C}}}}{-}NH_2 \;\rightleftharpoons\; R{-}N^{+}(CH_3)_3^{-}OOC{-}\underset{H}{\overset{R'}{\underset{|}{\overset{|}{C}}}}{-}NH_2 \;+\; Cl^{-} \qquad (3.33)$$

for a positively charged (anion) resin and a negatively charged biopolymer. Similarly, for a negatively charged (cation) resin and a positively charged biopolymer, the interaction is represented as:

$$R{-}SO_3^{-}H^{+} \;+\; H_3N^{+}{-}\underset{COOH}{\overset{R'}{\underset{|}{\overset{|}{C}}}}{-}H \;\rightleftharpoons\; R{-}SO_3^{-}H_3N^{+}{-}\underset{COOH}{\overset{R'}{\underset{|}{\overset{|}{C}}}}{-}H \;+\; H^{+} \qquad (3.34)$$

There are two other ionic interactions that are possible with the assistance of multivalent ions or polyelectrolytes in the feed stream. Thus, the presence of a multivalent cation (P^{+2}) in the solution would allow a cation resin to adsorb a negatively charged biopolymer, represented by:

$$R{-}SO_3^{-}H^{+} \;+\; P^{+2} \;+\; {}^{-}OOC{-}\underset{H}{\overset{R'}{\underset{|}{\overset{|}{C}}}}{-}NH_2 \;\rightleftharpoons\; R{-}SO_3^{-+}P^{+-}OOC{-}\underset{H}{\overset{R'}{\underset{|}{\overset{|}{C}}}}{-}NH_2 \;+\; H^{+} \qquad (3.35)$$

Likewise, multivalent anions (N^{-2}) allow an anion resin to adsorb a positively charged biopolymer according to:

$$R{-}N^{+}(CH_3)_3Cl^{-} \;+\; N^{-2} \;+\; H_3N^{+}{-}\underset{COOH}{\overset{R'}{\underset{|}{\overset{|}{C}}}}{-}H \;\rightleftharpoons\; R{-}N^{+}(CH_3)_3^{-}N^{-}H_3N^{+}{-}\underset{COOH}{\overset{R'}{\underset{|}{\overset{|}{C}}}}{-}H \;+\; Cl^{-} \qquad (3.36)$$

A single protein can, at the same time, participate in specific ion exchange reactions with the resin, form bonds with water-soluble polyelectrolytes and attach to other adsorbent surfaces. As has been seen, the mathematical representation for these complex interactions combines the kinetics of diffusional processes, ion exchange reactions and adsorption isotherms. Figure 3.35 (75) shows the adsorption of bacterial cells on ion exchange resins and Figure 3.36 shows the desorption. It should be noted that there is usually some irreversible adsorption of biopolymer on ion exchange resins, as Figure 3.36 shows. The extent of this irreversible bonding will define the limitations of using a particular resin for purifying a biochemical solution. This irreversible adsorption property can be utilized when an enzyme is to be immobilized on an ion exchange resin.

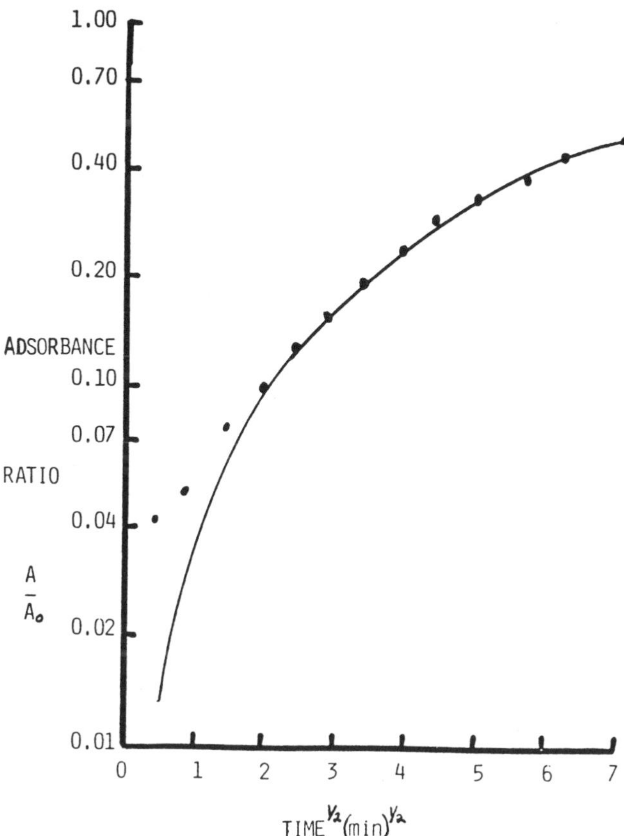

Figure 3.35. Adsorption of bacterial cells onto an anion exchange resin (Dowex 1 X 8, Chloride form), A = 0.598, pH = 3.53 (Reference 75).

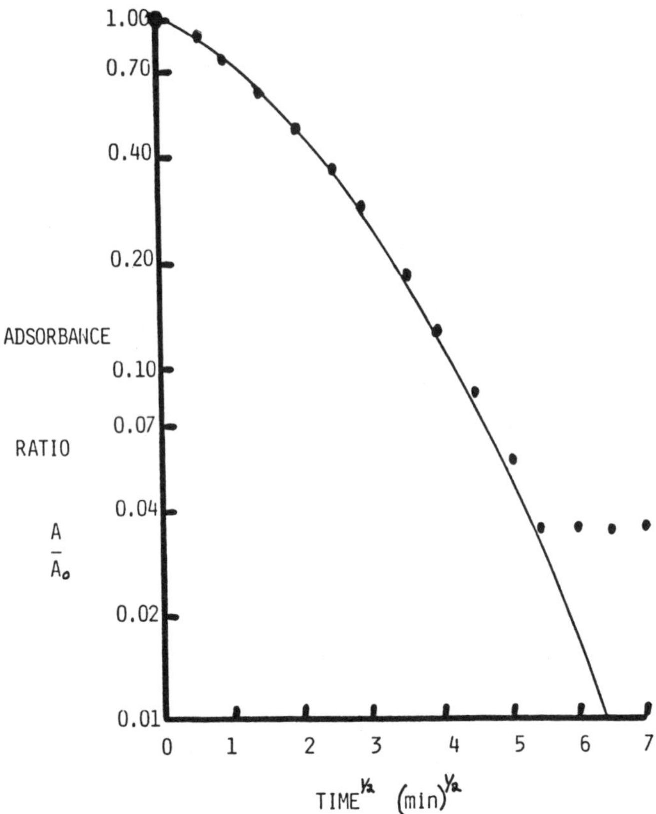

Figure 3.36. Desorption of bacterial cells from the anion resin in Figure 3.35, pH = 4.12, 1 M KCl (Reference 75).

3.6 ION EXCHANGE OPERATIONS

The typical cycle of operations involving ion exchange resins include pretreatment of the resin and possibly the feed solution, loading the resin with the solutes to be adsorbed by contacting the resin with the feed solution, and elution of the desired material from the resin. The scale of operation has ranged from analytical applications with a few milliliters of resin and microgram quantities of material to commercial production units containing several cubic meters of resin and producing metric tons of material. The loading may be applied batchwise, to a semi-continuous batch slurry, or to a column filled with resin which may be operated in a semi-continuous or continuous manner.

3.6.1 Pretreatment

The pretreatment of the feed solution described in Section 2.5.1 for adsorption applications should also be followed for ion exchange applications. It may be necessary to pretreat the ion exchange resin.

Should the resin be used to purify food feedstreams, such as sugar liquids or corn syrup, it is necessary to pretreat the resin to assure that the extractable level of the resin in use complies with Food Additive Regulation 21 CRF 173.25 of the Federal Food, Drug and Cosmetic Act. The pretreatment recommended for a column of resin in the backwashed, settled and drained condition is:

(1) Add three bed volumes of 4% NaOH at a rate sufficient to allow 45 minutes of contact time.

(2) Rinse with five bed volumes of potable water at the same flow rate.

(3) Add three bed volumes of 10% H_2SO_4 or 5% HCl at a flow rate that allows 45 minutes of contact time.

(4) Rinse with five bed volumes of potable water.

(5) Convert the resin to the ionic form desired for use by applying the regenerant that will be used in subsequent cycles.

If the column equipment has not been designed to handle acid solutions, a 0.5% $CaCl_2$ solution or tap water may be used in step 3 in place of H_2SO_4 or HCl for cation resins. Similarly, for anion resins, a 10% NaCl solution could be used in such equipment in place of acids.

3.6.2 Batch Operations

The batch contactor is essentially a single stage stirred reactor with a strainer or filter for separating the resin from the reaction mass once the reaction is complete. This type of contactor has an advantage in some fermentation operations because of its ability to handle slurries. Additional advantages include the low capital cost and simplicity of design.

In a batch operation, an ion exchange resin in the desired ionic form is placed into a stirred reaction vessel containing the solution to be treated. The mixture is stirred until equilibrium with the exchangeable ion is reached (about 0.5 to 3 hours). Then the resin is separated from the liquid phase by filtration. The adsorbate is removed from the resin by rinsing with the eluting solution. An additional step may be required to reconvert the resin to the regenerated form if this is not done by the eluting solvent. The cycle may then be repeated.

The batch system is basically inefficient since the establishment of a single equilibrium will give incomplete removal of the solute in the feed solution. When the affinity of the resin for this solute is very high, it is possible that the removal will be sufficiently complete in one stage. The batch process has the advantage over fixed bed processes that feedstreams containing suspended solids may be treated. In these cases, the resin particles, loaded with the adsorbate and rinsed from the suspended solids, may be placed in a column for recovery of the adsorbate and for regeneration.

The batch contactor is limited to use with reactions that go to completion in a single equilibrium stage or in relatively few stages. Difficulties may also arise with batch contactors if resin regeneration requires a greater number of equilibrium stages than the service portion of the cycle.

3.6.3 Column Operations

3.6.3.1 Fixed Bed Columns: Column contactors allow multiple equilibrium stages to be obtained in a single unit. This contactor provides for reactions to be driven to their desired level of completion in a single pass by adjusting the resin bed depth and the flow conditions. The main components of a column contactor are shown in Figure 3.37. At the end of its useful work cycle, the resin is backwashed, regenerated and rinsed for subsequent repetition of the work cycle. Typically, this non-productive portion of the cycle is a small fraction of the total operating cycle.

Column contactors may be operated in co-current, countercurrent, or fluidized bed mode of operation. The co-current mode means that the regenerant solution flows through the column in the same direction as the feed solution (Figure 3.38). The countercurrent mode has the regenerant flowing in the opposite direction as the feed solution (Figure 3.39).

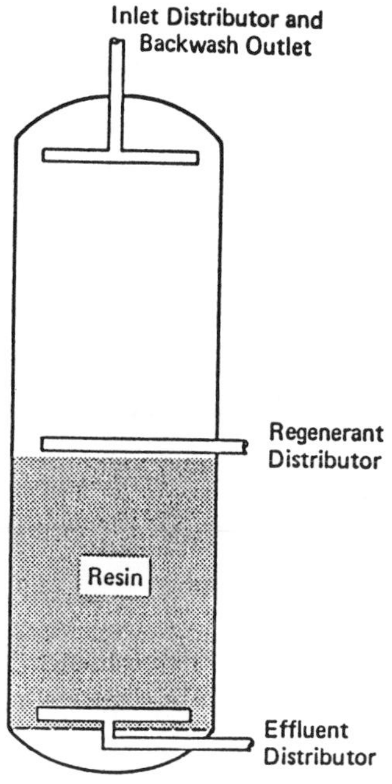

Figure 3.37. Ion exchange column contactor.

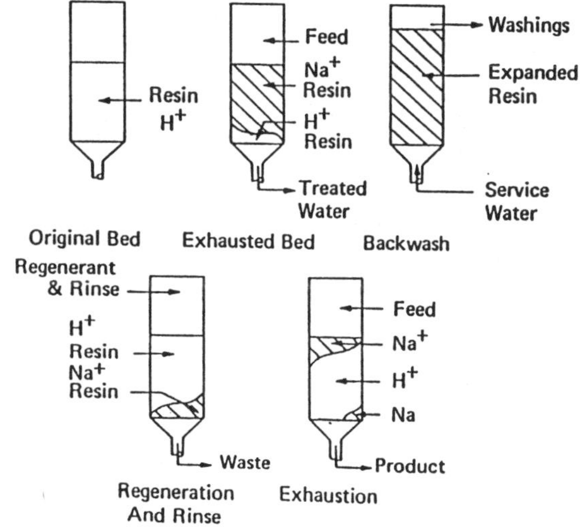

Figure 3.38. Co-current column operation with H^+ form cation resin removing Na^+ ions from feedstream.

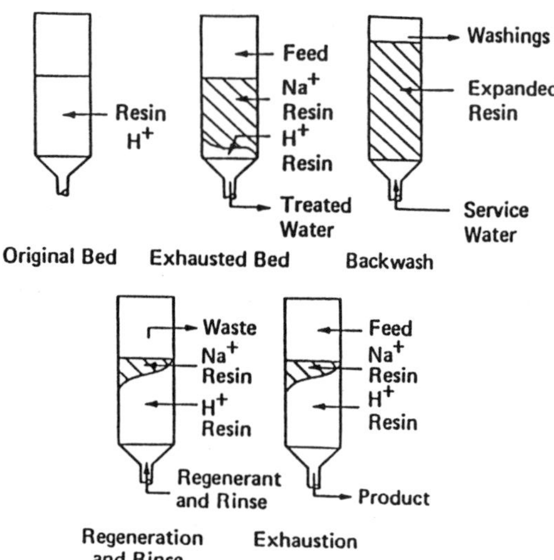

Figure 3.39. Countercurrent column operation with H^+ form cation resin removing Na^+ ions from feedstream.

Countercurrent operation of a column may be preferred to reduce the ion leakage from a column. Ion leakage is defined as the amount of the ion being removed from solution which appears in the column effluent during the course of the subsequent exhaustion run. Comparison of Figure 3.38 and Figure 3.39 shows that the leakage caused by reexchange of nonregenerated ions during the work cycle of cocurrently regenerated resin is substantially reduced with countercurrent regeneration.

The fixed bed column is essentially a simple pressure vessel. Each vessel requires a complexity of ancillary equipment. Each column in a cascade will require several automatic control valves and associated equipment involving process computer controls to sequence the proper flow of different influent streams to the resin bed.

Combinations of column reactors may sometimes be necessary to carry out subsequent exchange processes, such as in the case of demineralization (Figure 3.40). For demineralization a column of cation exchange resin in the hydrogen form is followed by a column of anion exchange resin in the hydroxide form. A mixed bed, such as shown in Figure 3.41, may likewise be used for demineralization.

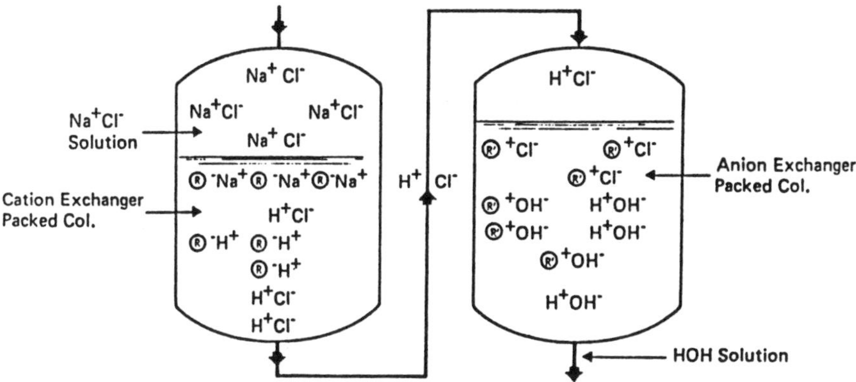

Figure 3.40. Demineralization ion exchange column scheme.

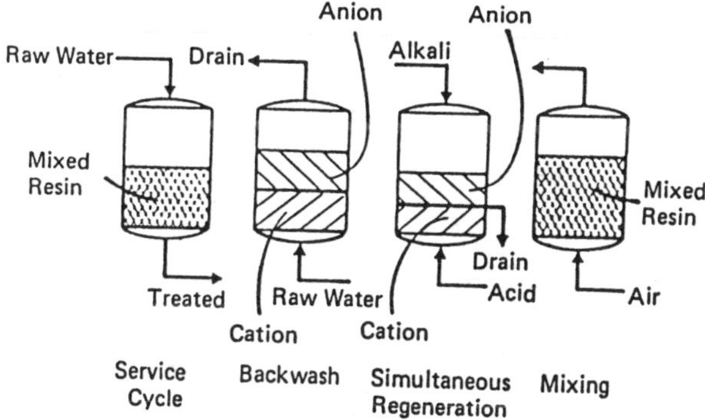

Figure 3.41. Operation of a mixed bed demineralization ion exchange column.

Mixed bed operation has the advantage of producing a significantly higher quality effluent than the concurrently regenerated beds of Figure 3.40, but has the added difficulty of requiring separation of the resin prior to regeneration.

The important requirements for the successful operation of a mixed bed include the careful separation of the strong base anion exchange resin from the strong acid cation exchange resin by backwash fluidization. This is followed by contact of each of the components with their respective regenerant in a manner to minimize the cross contamination of the resins with the alternate regenerant. This requires that the quantity of resin, particularly the cation exchange resin, be precisely maintained so that the anion-cation interface will always be at the effluent distributor level. Typically, matched pairs of resins are used so that an ideal separation can be repeatedly obtained during this process. Recently, inert resins have been marketed which enhance the distance between the anion-cation interface and allow less cross contamination during regeneration (76).

The air mixing of the anion and cation must also be performed in such a manner that complete mixing of the resin and minimum air entrapment are obtained at the end of the mixing cycle.

Ion exchange is usually in a fixed bed process. However, a fixed bed process has the disadvantage that it is cyclic in operation, that at any one instant only a relatively small part of the resin in the bed is doing useful work and that it cannot process fluids with suspended solids. Continuous ion exchange processes and fluidized bed systems have been designed to overcome these shortcomings.

3.6.3.2 Continuous Column Operations: When the ionic load of the feed solution is such that the regeneration/elution portion of the operating cycle is nearly as great, or greater than, the working portion, continuous contactors are used instead of column contactors.

Continuous contactors operate as intermittently moving packed beds as illustrated by the Higgins contactor (77) (Figure 3.42), as fluidized staged (compartmented) columns as demonstrated by the Himsley contactor (78-80) (Figure 3.43).

In the Higgins contactor, the resin is moved hydraulically up through the contacting zone. The movement of resin is intermittent and opposite the direction of solution flow except for the brief period of resin advancement when both flows are cocurrent. This type of operation results in a close approach to steady-state operations within the contactor.

Elegant slide valves are used to separate the adsorption, regeneration and resin backwash stages. The contactor operates in predetermined cycles and is an ideal process for feedstreams with no suspended solids.

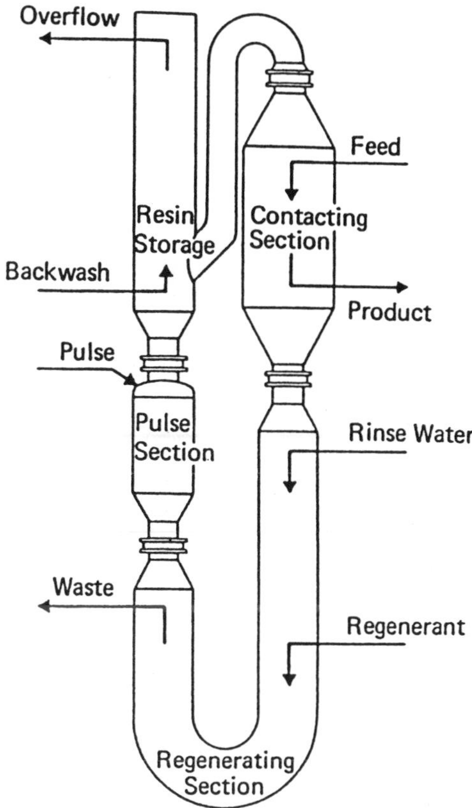

Figure 3.42. The Higgins contactor for continuous operation (Reference 77).

The Higgins type of contactor is able to handle a certain amount of slurry due to the continued introduction of fresh resin material to act as a filter media during the operation. A lower resin inventory should result with continuous contactors than with column reactors handling the same high ionic load feed stream.

The major disadvantage of the Higgins contactor is the lifetime of the resin. Estimates range as high as 30% resin inventory replacement per year due to attrition and breakage of the resin as it passes through the valves (81).

In the Himsley type of contactor, the resin is moved from one compartment or stage to the next, countercurrent to the feed solution flow, on a timed basis that allows for

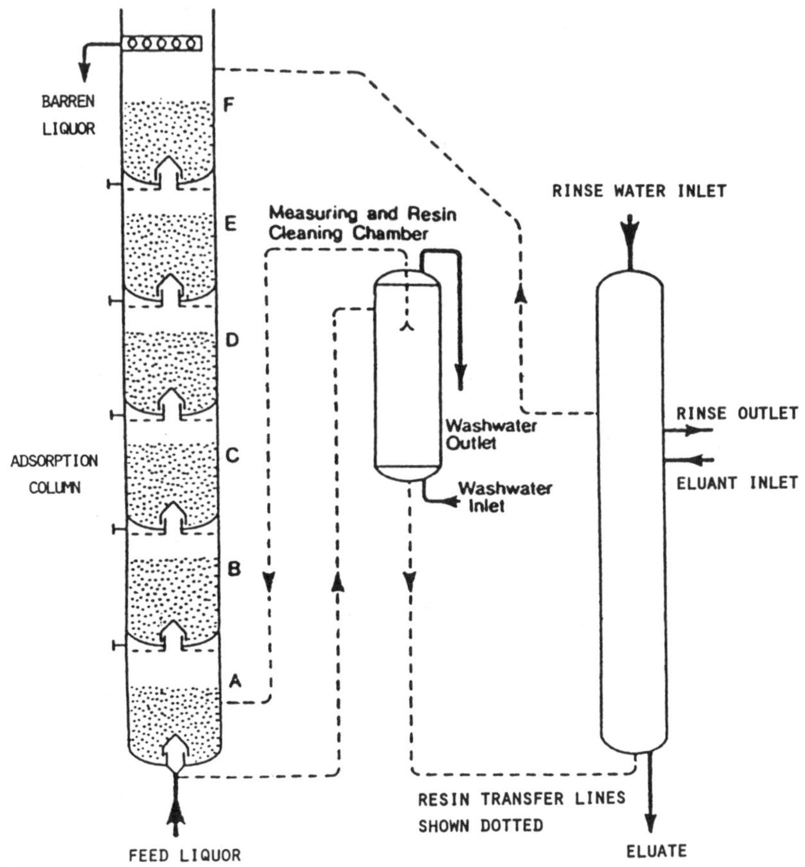

Figure 3.43. The Himsley Continuous Fluid-Solid Contactor (Reference 78).

the rate of equilibrium resin loading in each stage. Thus each compartment or tray is designed to accommodate specific feed compositions and effluent requirements. Equipment from different commercial suppliers differs in the manner of resin transfer.

3.6.3.3 Fluidized Column Operations: Column contactors, when operated with the feedstream in a downflow mode, are poorly suited to handle fermentation slurries because of the excellent filtration characteristics of packed resin beds. For such slurries it is preferable to use a fluidized bed of resin such as is shown in Figure 3.44 (82).

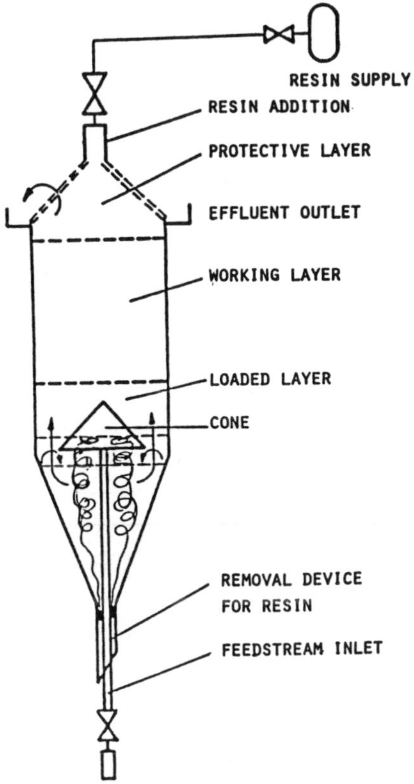

Figure 3.44. Fluid bed ion exchange column (Reference 82).

While most of the commercial applications discussed in the literature pertain to slurries of uranium tailings and paper mill effluent treatment, the equipment may be adapted for use in treating fermentation broths.

The shape of the fluidized bed is important in controlling the position of the resin in the column. The effluent from the column should pass over a vibrating screen, e.g., as SWECO, to retain entrained resin but allow mycelia to pass through.

The Asahi contractor, shown in Figure 3.45 (81), uses conventional pressure vessels as the resin column. These vessels have a resin support grid at the base and a resin screen at the top. The feedstream is fed in upflow through the packed bed. Periodically the liquid contents of the

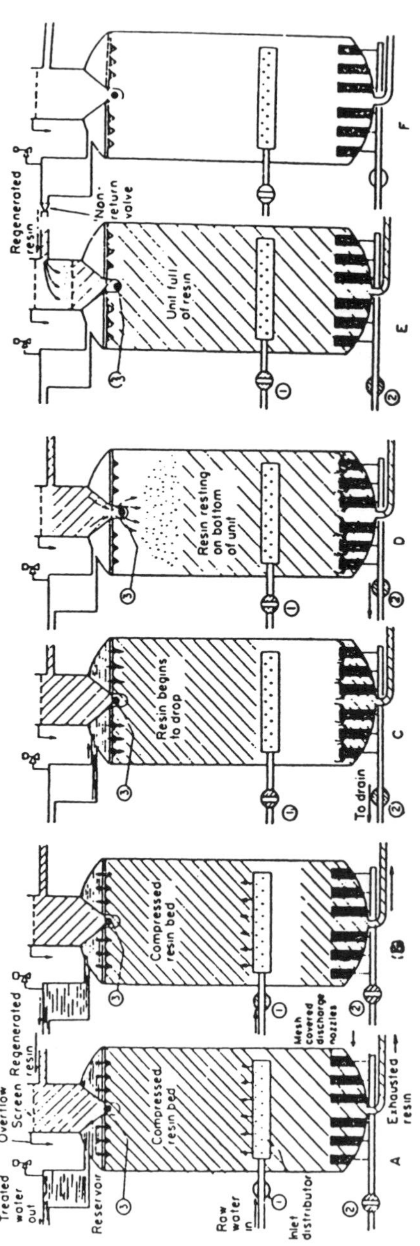

Figure 3.45. Operating sequence of the Asahi Type Countercurrent moving bed contactor (Reference 81).

column are allowed to drain rapidly which causes the resin to flow from the bottom of the bed to a similar vessel for regeneration. At the same time, fresh resin is added to the top of the active column from a resin feed hopper. This hopper contains a ball valve which passes the resin in during downflow operation and seals itself during the upflow portion of the adsorption cycle.

The Cloete-Streat ion exchange equipment is a multistage fluidized bed containing perforated distributor plates (Figure 3.46) (83). The hole size in the plates is greater than the maximum resin particle size. The countercurrent movement of resin occurs due to the controlled cycling of the feedstream. Each cycle the entire amount of resin in one chamber is transferred to the next chamber. Equipment with 4.5 m diameter columns and eight stages for a total height of about 20 m are in commercial operation.

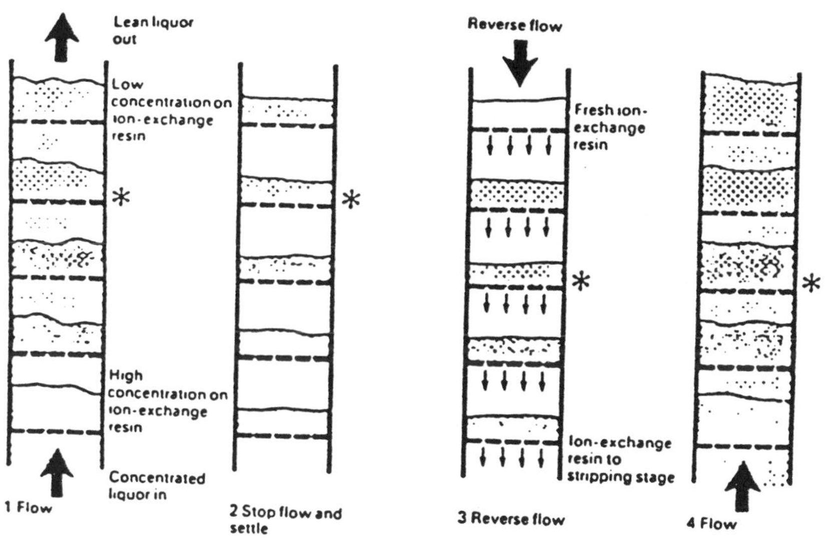

Figure 3.46. Sequence of operation of the Cloete-Streat countercurrent ion exchange process (Reference 83).

The USBM equipment, shown in Figure 3.47 (84), is very similar to the Cloete-Streat system. The differences are in the plate design, the method of transferring solids and in the method of removing the resin. This system and the Cloete-Streat system are able to handle slurry feeds with up to 15% by weight suspended solids.

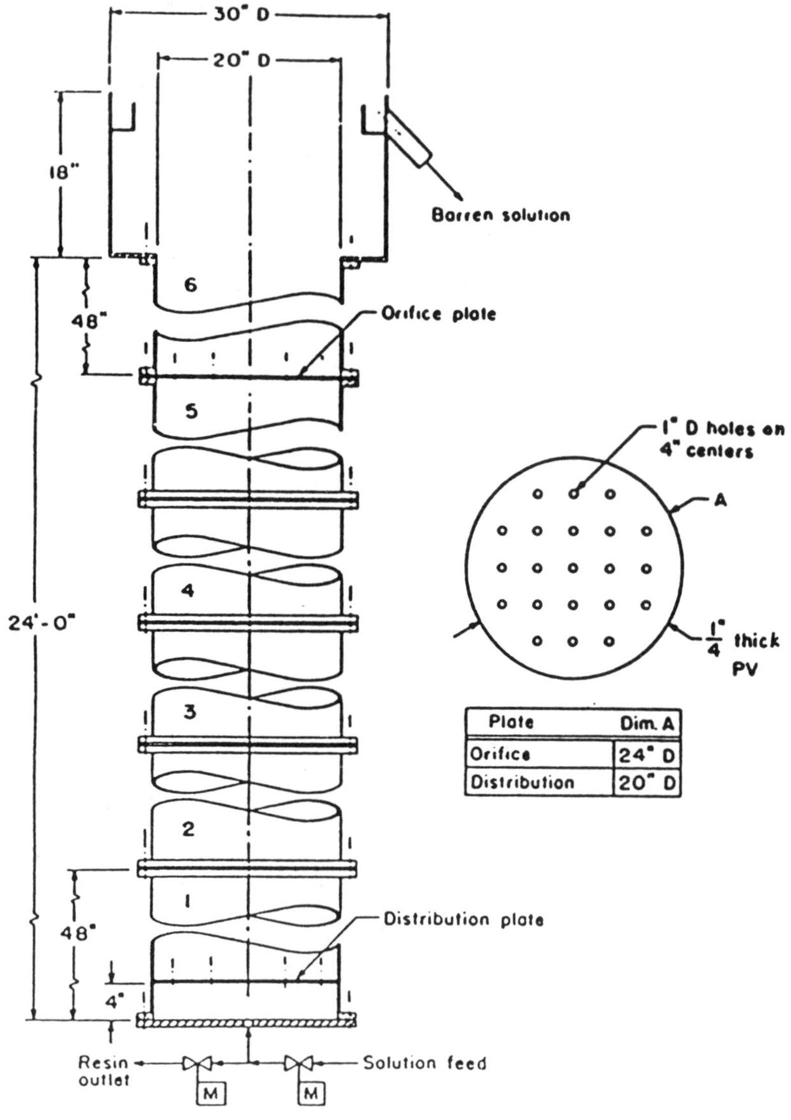

Plate	Dim. A
Orifice	24" D
Distribution	20" D

Figure 3.47. Schematic diagram of the USBM ion exchange column (Reference 84).

The advantage of these fluidized bed columns is the relatively low capital cost, low operating cost, small space requirement, simple instrumentation and control compatability with conventional solvent extraction equipment (81). Only those systems that can accommodate slurries with suspended solids are commercially feasible for biotechnology and fermentation operations. Otherwise the small volume of fermentation feedstreams which need to be processed is not on the scale of operations necessary to make continuous ion exchange processes cost effective.

The principal disadvantage of fluidized bed columns is the mixing of resin in various stages of utilization. This mixing means that breakthrough occurs sooner and the degree of resin capacity utilization is much lower than in packed bed columns. By placing perforated plates in a column (Figure 3.48) (85), the resin beads only mix within a restricted area allowing more complete utilization of the ion exchange resin's capacity.

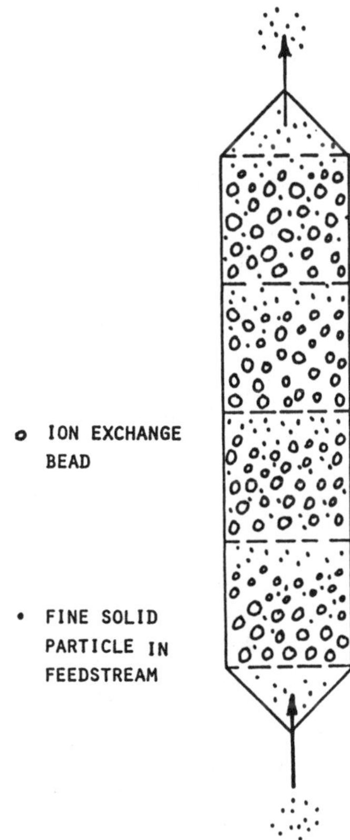

Figure 3.48. The perforated plates in this fluidized bed column allow fine solid particles to flow through the column while the ion exchange beads are, for the most part, confined to individual compartments (Reference 85).

Smaller continuous fluid bed systems, like the one shown in Figure 3.49 (86), have been developed which operate with a high concentration of ion exchange resin and suspended solids. These units are 80% smaller than the conventional resin-in-pulp plants of the type shown in Figure 3.50 which are used in the treatment of uranium ore slurries (87). The pilot plant unit, which would probably be the size needed for processing commercial fermentation broths, had dimensions per contact chamber of 0.82 m x 0.82 m with a 0.82 m fluid bed height and an additional 0.16 m for freeboard. The unit has been successfully operated with 25 to 50% resin and up to 45% suspended solids.

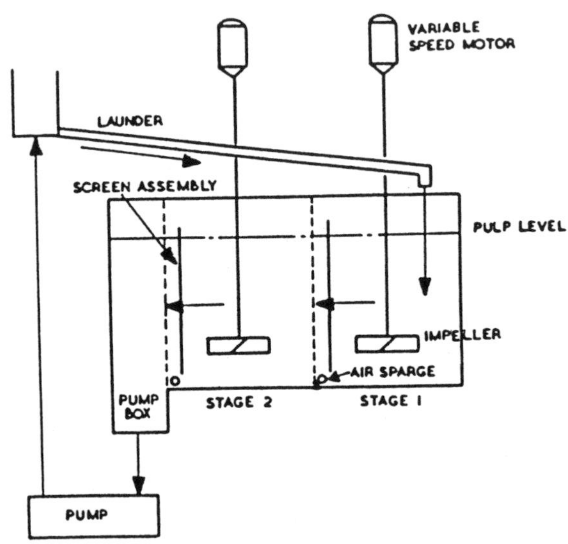

Figure 3.49. Pilot plant for resin-in-pulp contactor unit (Reference 86).

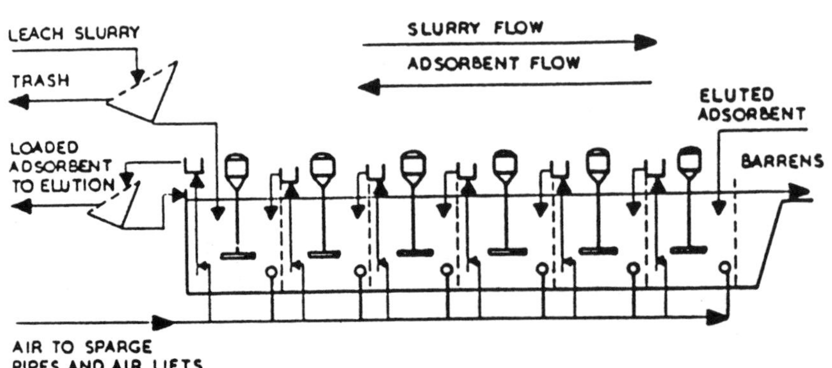

Figure 3.50. Plant design for resin-in-pulp contactor units (Reference 87).

The effect of the degree of regeneration of the resin on the degree of extraction of a solute was measured by Slater (88) using a seven stage unit. The results are shown in Figure 3.51 for the extraction of uranium using a fluidized bed slurry of a strong base resin in a 10% uranium ore leach slurry.

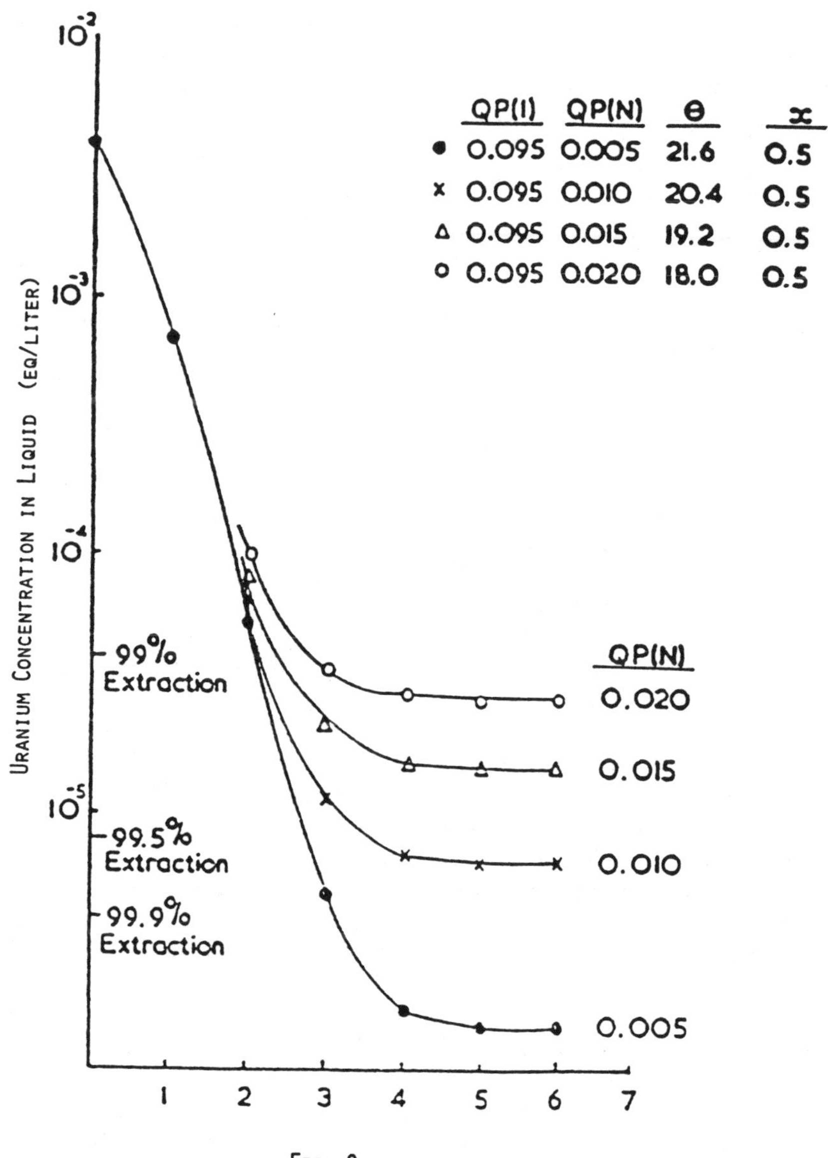

Figure 3.51. Effect of resin feed composition on the efficiency of a staged system (Reference 88).

However, the fluidized bed column, even with perforated plates separating it into as many as 25 compartments, may not be appropriate for applications in which there are strict requirements (<1% of effluent concentration) on the effluent. For readily exchangeable ions, the optimum utilization of this technique occurs when an actual effluent concentration of 10% of the influent is the breakthrough point. At such times, the ion exchange resin capacity would be 70% utilized (85). Should it be acceptable that the breakthrough point occur when the average effluent reaches a concentration of 10% of the influent, the utilization of the ion exchange capacity is 90%. If the ions are not readily exchangeable (low selectivity), the resin utilization would be significantly less and fluidized bed operations should not be used.

3.6.4 Elution/Regeneration

Elution of proteins from ion exchangers can be achieved with buffers containing salts, such as sodium chloride or ammonium acetate, or by an appropriate pH change, provided that the pH change does not result in denaturation of the eluted protein (89). The elution may be performed with a series of stepped changes or with a continuous gradient change in the eluting power of the eluant. With such changes, it is possible to separate different proteins or protein fractions from each other based on their different affinities for the ion exchange resin.

Elution of compounds such as penicillin with either acids or bases will render the penicillin inactive. Although aqueous salt solutions can elute penicillin without inactivating it, the large volumes required make this option impractical. Wolf and coworkers (90) developed an elution solvent combination of organic solvent, water and salt that can elute the penicillin with a minimum volume and no inactivation. The mixture of organic solvent and salt is chosen so that the salt is soluble in the resulting organic solvent-water mixture and the organic substance eluted from the resin is soluble in the elution mixture. Table 3.16 shows the elution volume required to recover the indicated amount of antibiotic when the elution solvent is 70% methanol and 5% or 7.5% ammonium chloride in water.

Regeneration alone is not sufficient to prevent fouling or microbial growth on ion exchange resins. If the resin is left standing in the regenerant during non-operating times, it

Table 3.16: Amount of Antibiotic Recovered with Increasing Volumes of Methanol in Aqueous Ammonium Chloride (90)

Antibiotic	Eluant	Total Volume of Eluant (Bead Volumes)	Amount of Antibiotic Recovered (%)
Dihydronovobiocin	70% MeOH with 5% NH_4Cl	1	50.0
		2	83.3
		3	93.3
		4	96.6
Novobiocin	70% MeOH with 5% NH_4Cl	3.5	99.8
Penicillin	70% MeOH with 7.5% NH_4Cl	0.5	11.0
		1.0	69.0
		1.5	94.0
		2.0	97.1
		2.5	99.7

is possible to suppress the microbial growth (91). The regenerant in this instance was 10 to 20% NaCl. When the initial microbe count was 10 per milliliter, at the end of three weeks in 20% NaCl, the count had risen to just 800/mL compared to 200,000/mL for the resin stored in water. When an alternate regenerant is used (NaOH or HCl), it is preferable to change the storage medium to 20% NaCl since extended time in an acid or base media can adversely affect the resin matrix.

Example 3.7

When a regenerant has a 1% contaminant in a 7% HCl solution, what will be the effect on the sodium leakage in the effluent when a feedstream containing 34 meq/L Ca^{++}, 2 meg/L Mg^{++} and 5 meq/L Na^{+} is treated with a strong cation resin with an exchange capacity of 1.8 meq/ml and a selectivity of $K_{H^+}^{Na^+} = 1.5$?

Selectivity analysis calculations can be used as follows.

The normality of NaCl in the regenerant:

$$\frac{10.0 \text{ g}}{1000 \text{ g}} \times \frac{1,035 \text{ g}}{\text{liter}} \times \frac{1 \text{ eq}}{58.44} = 0.18 \text{ eq/liter} \quad (3.37)$$

The normality of HCl in the regenerant:

$$\frac{70.0 \text{ g}}{1000 \text{ g}} \times \frac{1,035 \text{ g}}{\text{liter}} \times \frac{1 \text{ eq}}{36.46} = 1.99 \text{ eq/liter} \quad (3.38)$$

$$X_{Na^+} = \frac{0.18}{(0.18 + 1.99)} = 0.08 \text{ in the regenerant} \quad (3.39)$$

$$\frac{\bar{X}_{Na^+}}{1 + \bar{X}_{Na^+}} = K_{H^+}^{Na^+} \frac{X_{Na^+}}{1 - X_{Na^+}} = 1.5 \frac{0.08}{1 - 0.08} = 0.13 \quad (3.40)$$

After regeneration, therefore, $\bar{X}_{Na} = 0.115$. During the treatment of the feedstream:

$$\frac{X_{Na^+}}{1 - X_{Na^+}} = \frac{1}{K_{H^+}^{Na^+}} \frac{\bar{X}_{Na^+}}{1 - \bar{X}_{Na^+}} = \frac{1}{1.5} \frac{0.115}{1 - 0.115} = 0.087 \quad (3.41)$$

Therefore, in the effluent from the column, $X_{Na} = 0.080$. This corresponds to 331 ppm Na as $CaCO_3$.

3.6.5 Backwashing

Since most fluids contain some suspended matter, it is necessary to backwash the resin in the fixed bed column on a regular basis to remove any accumulation of these substances. To carry out a backwashing operation, a flow of water is introduced at the base of the column. The flow is increased to a specific rate to classify the resin hydraulically and remove the collection of suspended matter. Figures 3.52 and 3.53 show the types of flow rates which provide certain degrees of expansion of cation and anion resins, respectively. Since the anion resins are significantly less dense than the cation resin shown, it would be necessary to have different amounts of freeboard above the normal resin bed height so that backwashing may be accomplished with only a negligible loss of ion exchange resin. Typically, an anion resin bed may be expanded by 100% during backwashing, while a cation resin bed will only be expanded by 50%.

It is also necessary that the water used for the backwashing be degassed prior to use. Otherwise, resin particles will attach themselves to gas bubbles and be carried out of the top of the column to give an unacceptable increase in resin losses.

When treating fermentation broth filtrates, frequent backwashing of the resin bed is necessary to prevent accumulation of suspended matter. In such cases, the column height should be designed of such a size that the bed is regenerated at least every ten hours. Shorter columns have

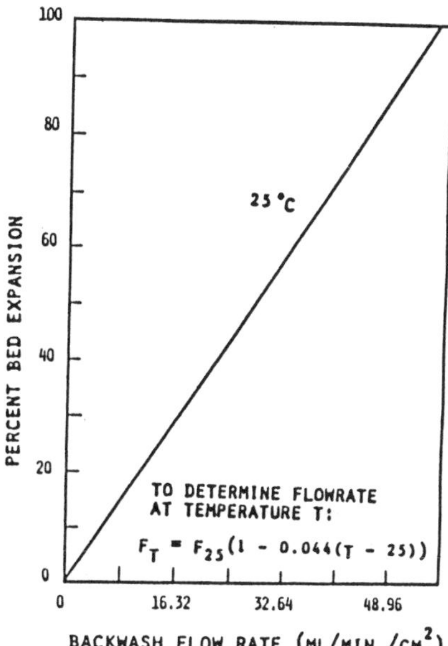

Figure 3.52. Backwash expansion characteristics of a macroporous strong acid cation resin, Dowex 88.

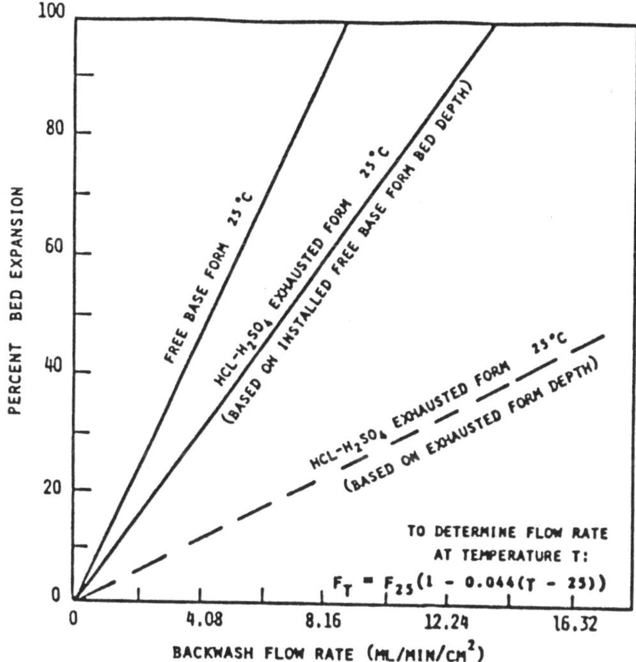

Figure 3.53. Backwash expansion characteristics of regenerated and exhausted macroporous weak base anion resin, Dowex 66.

been designed to be regenerated at least every hour (92). These shorter beds can then use finer resins and achieve a high level of efficiency with lower capital costs. This may be taken to the point of using very fine resins, as with the Powdex^R system (93) which discards the powdered resin after a single use.

After a bed is backwashed, unless it is air-mixed as the level of water is drained down to the surface, the beads or particles classify according to size. The fine beads end up on top and the large beads on the bottom of the column. In co-current operations, the regenerant first contacts the top of the bed. The fast kinetics of the fine particles gives a high regeneration efficiency. However, the large beads on the bottom will regenerate more slowly and may end up only partially regenerated. Thus, when the feedstream is next passed through the resin bed, leakage of undesirable ions may occur from the large beads in the bottom of the column. This may be overcome by using a countercurrent flow arrangement described earlier or by using an air-mixing system during the post-backwash draining (94).

3.7 PROCESS CONSIDERATIONS

3.7.1 Design Factors

The engineer designing an ion exchange column operation usually will prefer to work with the simplest kinetic model and linear driving force approximations. The weakness of this approach is that any driving force law only regards the momentary exchange rate as a function of the solute concentration in the bulk solution and the average concentration in the particle, neglecting the effect of concentration profiles in the particle. Nevertheless, the linear driving force approach provides an approximation that is often sufficiently accurate for the engineer.

3.7.1.1 Scaling-up Fixed Bed Operations: Rodrigues (95) has presented empirical and semi-empirical approaches which may be used to design ion exchange columns when the solute in the feedstream is c_o and the flowrate is u_o. The breakthrough point is usually set at the point when the effluent concentration increases to 5% of c_o. The design equations relate the total equilibrium ion exchange capacity

(Q) to the volume of resin required (V_r) to the time of the breakthrough (t_B).

The empirical approach, the overall mass balance is given by the equation:

$$V_r = c_o \xi t_s/(1 + \xi)Q \tag{3.42}$$

where

$$t_s = \tau(1 + \xi) \tag{3.43}$$

$$\xi = (1 - \varepsilon)Q/\varepsilon Q_o \tag{3.44}$$

$$\tau = \varepsilon V/u_o \tag{3.45}$$

and V is the bed volume with void space ϵ.

It is usually necessary to modify this resin amount by a safety factor (1.2 to 1.5) to adjust for the portion of the total equilibrium capacity that can actually be used at flow rate u and to adjust for any dispersive effects that might occur during operation.

The semi-empirical approach involves the use of the mass transfer zones as was used in the chapter on adsorption. This approach has been described in detail specifically for ion exchange resins by Passino (96). He referred to the method as the operating line and regenerating line process design and used a graphical description to solve the mass transfer problems.

For the removal of Ca^{++} from a feedstream, the mass transfer can be modeled using Figure 3.54. The upper part shows an element of ion exchange column containing a volume v of resin to which is added a volume V_{ex} of the feedstream containing Ca^{++}. It is added at a flow rate (F_L) for an exhaustion time t_{ex}. The concentration of Ca^{++} as it passes through the column element is reduced from x_{ex1} to x_{ex2}. Therefore, the resin, which has an equilibrium ion exchange capacity $\overline{C}$, increases its concentration of Ca^{++} from y_{ex2} to y_{ex1}. In this model, fresh resin elements are continuously available at a flow rate (F_s) = v/t_o, which is another way of saying the mass transfer zone passes down through the column.

The lower part of Figure 3.54 shows the operating lines

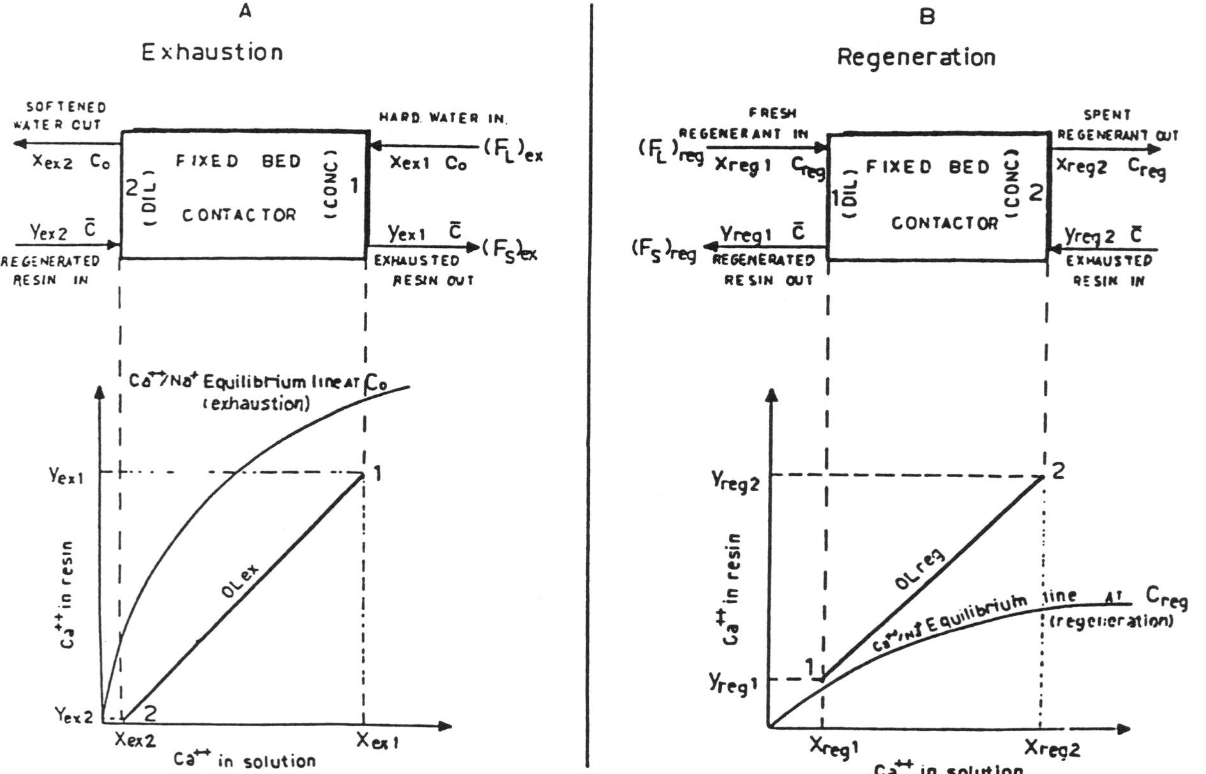

Figure 3.54. Schematic representation of the softening of a hard water feedstream on an ideally continuous bed of strong acid cation resin in the sodium form (Reference 96).

for this process. The ion exchange equilibrium line describes the selectivity in terms of a Freundlich, Langmuir or other appropriate model.

The equations for the points in the lower part are given by:

$$x_{ex1} = x_o \quad (Ca^{++} \text{ in the feedstream}) \tag{3.46}$$

$$y_{ex1} = y_{ex2} + (x_{ex1} - x_{ex2}) \frac{c_o V_{ex}}{\overline{C} v} \quad (Ca^{++} \text{ in exhausted resin}) \tag{3.47}$$

$$x_{ex2} = \frac{\int_0^{V_{ex}} x \, dv}{V_{ex}} \quad (\text{average } Ca^{++} \text{ in the effluent}) \tag{3.48}$$

$$y_{ex2} = 0 \quad (Ca^{++} \text{ in the regenerated resin}) \tag{3.49}$$

and the slope of the operating line:

$$\left(\frac{\Delta y}{\Delta x}\right)_{ex} = \frac{y_{ex1} - y_{ex2}}{x_{ex1} - x_{ex2}} = \frac{c_o V_{ex}}{\overline{C} v} = \frac{c_o (F_L)_{ex}}{\overline{C} F_s} \tag{3.50}$$

The value of c_o, $\overline{C}$ and v are known so that for any V_{ex} value the slope of the operating line can be calculated from Equation 3.50. The specific points: x_o is given, x_{ex2} is obtained by graphic integration from the breakthrough curves. After operation and regeneration, the value of y_{ex2} may not be zero but may be between 0.02 and 0.05 if the regeneration is not complete. The application of this technique has been described in terms of basic design parameters such as number of transfer units, the height of a transfer unit and mass transfer coefficient (97,98).

The data generated with the laboratory column may be scaled-up to commercial size equipment. Using the same flow rate (on a mass basis) as used in the laboratory experiments, the appropriate increase in column size over that used in the laboratory is a direct ratio of the volumes to be treated compared to that treated in the laboratory equipment.

If a reasonable height-to-diameter ratio (approximately 1:1) is obtained in the scale-up using the bed dept involved in the laboratory procedure, then that bed depth is maintained and the cross-sectional area of the column is increased. However, if the sizing is such that the column is much larger in diameter than the bed depth, scale-up should be done to maintain a height-to-diameter ratio of approximately 1. The required resin volume is determined by maintaining the same mass flow conditions (liters of feed solution per minute per cubic meter of installed resin) as was used in the laboratory operation.

Appropriate tank space must be left to accommodate the backwash operation. This is typically 50% of bed depth for cation exchange resins and 100% expansion in the case of anion exchange resins.

Example 3.8

The purification of lysine-HCl from a fermentation broth will be used to illustrate the calculations involved in scaling-up laboratory data.

The laboratory fermentation broth, which is similar to the commercial broth, contained 20.0 g/L lysine, much smaller amounts of Ca^{++}, K^{+} and other amino acids. The broth was passed through 500 ml of strong acid cation resin, DOWEX HCR-S, in the NH_4^+ form. The flow rate was 9 mL/min or 1.77 mL/min per cm^2 of resin. It was determined that the resin capacity averaged 115 g of lysine-HCl per liter of resin. It may be noted that since the equivalent molecular weight of lysine-HCl is 109.6 g and the "theoretical" capacity of DOWEX HCR-S is 2.0 meq/mL, the operating capacity is 52% of theoretical capacity.

The commercial operation must be capable of producing 9M metric tons of lysine (as lysine-dihydrochloride-H_2O) per year. With a 20.0 g/L concentration of lysine in the fermentation broth, the number of liters of broth to be treated each year are:

$$\frac{9000 \text{ mtons}}{\text{year}} \times \frac{146.19 \text{ (MW of lysine)}}{237.12 \text{ (MW of ly-HCl-H}_2\text{O)}} \times \frac{\text{liter}}{20.0 \text{ g}} \times \frac{10^6 \text{ g}}{\text{mton}} = 27.7 \times 10^7 \text{ liter/year} \qquad (3.51)$$

If the plant operates 85% of the time, the flow rate would have to be:

$$27.7 \times 10^7 \text{ l/yr} \times \frac{1}{0.85} \times \frac{1 \text{ year}}{365 \text{ days}} \times \frac{1 \text{ day}}{24 \text{ hours}} = 3.73 \times 10^4 \text{ l/hr} \qquad (3.52)$$

At a resin capacity of 115 gm per liter of resin, the amount of resin that must be available is:

$$\frac{\text{liter (resin)}}{115\ \text{g}} \times \frac{219.12\ \text{(MW of ly-HCl)}}{146.19\ \text{(MW of lysine)}} \times \frac{20.0\ \text{g}}{\text{liter}} \times \frac{3.73 \times 10^{4}\ \text{l}}{\text{hour}} = 9.71 \times 10^{3}\ \text{l(resin)/hr} \quad (3.53)$$

To obtain the maximum utilization of the resin in this operation, series bed operation (Carrousel) operation is recommended. This operation uses three beds of resin in a method having two beds operating in series while the product is being eluted from the third. The freshly regenerated resin is placed in the polishing position when the totally loaded lead bed is removed for regeneration.

The elution/regeneration step, which includes backwashing, eluting and rinsing the resin, might take up to four hours. Therefore, enough resin must be supplied to take up the lysine-HCl presented during that time. The resin requirement for the commercial scale operation would be:

$$\frac{9.71 \times 10^{3}\ \text{l(resin)}}{\text{hour}} \times \frac{4\ \text{hour}}{\text{bed}} = 3.88 \times 10^{4}\ \text{l(resin)/bed} \quad (3.54)$$

Thus, three beds of 39 m^3 resin each is required to produce 9,000 m tons of lysine/year.

3.7.1.2 Comparison of Packed and Fluidized Beds: Belter and coworkers (99) developed a periodic countercurrent process for treating a fermentation broth to recover novobiocin. They found that they were able to scale up the laboratory results to production operations if the two systems have similar mixing patterns and the same distribution of residence times in the respective columns. The mixing patterns are the same when the space velocity (F/V_e) and the volume ratio (V_R/c) are the same. This is shown in Figure 3.55 for the effluent concentration of novobiocin from laboratory and production columns.

The scaling-up of packed beds is subject to the difficulties of maintaining even flow distributions. Removal of solution through screens on side walls is not recommended and the flow of resin from one section into another of much greater area could distort the resin flow profile.

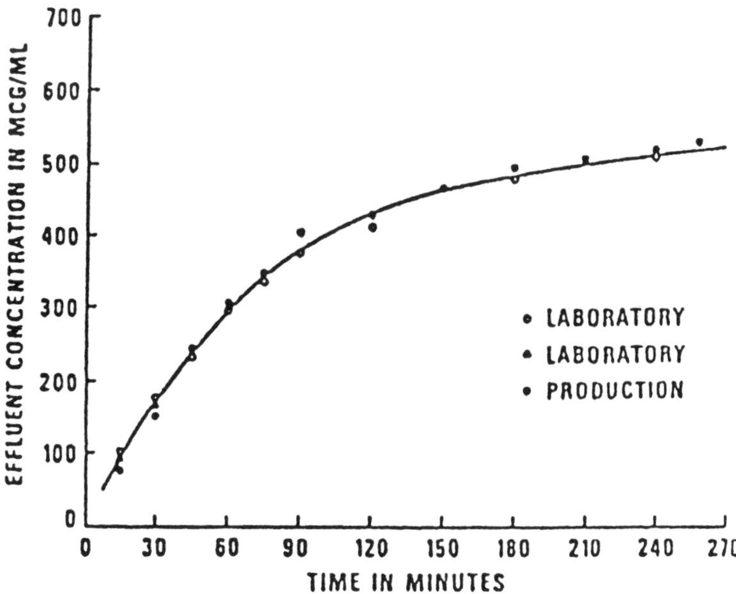

Figure 3.55 Comparison of experimental curves for laboratory and production columns (Reference 99).

The problems of scaling-up fluidized bed operations are more difficult in terms of design calculations, but flow distribution is more easily designed because of the mobility of the resin. The degree of axial mixing of the liquid and the resin has to be taken into account when calculating the changes necessary in the bed diameter and bed height. Figure 3.56 shows the increases in bed height necessary when scaling-up packed and fluidized beds with bed diameter increases (100).

Mass transfer coefficients have been correlated for packed and fluidized beds (101). The mass transfer coefficients for packed beds are 50 to 100% greater than for fluidized beds.

The volume of resin in a packed bed is about half that in a fluidized bed but the packed bed column may be up to eight times smaller. Despite this, a complete fluidized bed operation may still be smaller than a fixed bed operation. Also, fluidized bed columns do not operate at high pressures so they can be constructed more economically.

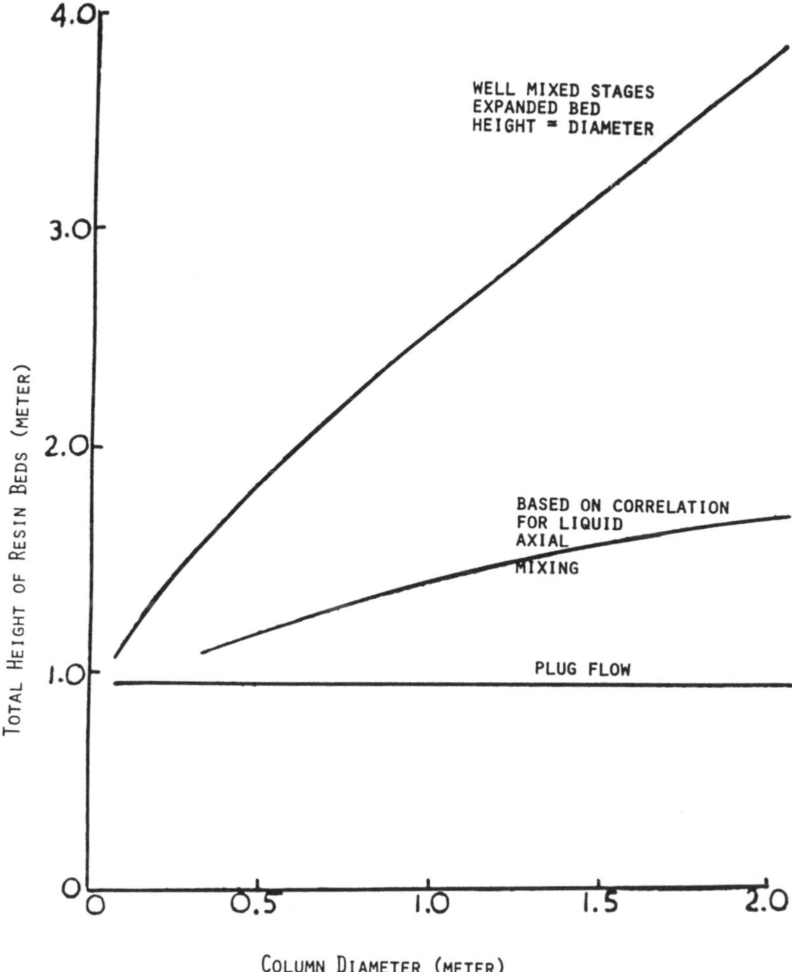

Figure 3.56. Scale-up relationships for fluidized beds (Reference 100).

3.7.1.3 Pressure Drop: The pressure drop across an ion exchange bed has been represented by an equation (102) which depends on the average particle diameter, the void fraction in the bed, an exponent and a friction factor dependent on the Reynolds number, a shape factor, the density and viscosity of the fluid and the flow rate.

While that equation has internally consistent units (English system), the variables are not normally measured in those units. Another disadvantage is that one must check graphs of the exponent and friction factor versus the Reynolds number to use the equation.

For laminar flow with spherical particles, the equation can be simplified to:

$$\frac{\Delta P}{L}(\text{bar/cm}) = \frac{0.0738\ \mu(\text{cp})\ V_o(1/\text{min})\ (1-\varepsilon)^3}{D_p^2(\text{mm}^2)\ \varepsilon^3} \tag{3.55}$$

For most ion exchange resins, the void volume is about 0.38, so that $(1 - \epsilon^3) / \epsilon^3 = 4.34$ and:

$$\frac{\Delta P}{L}(\text{bar/cm}) = \frac{0.32\ \mu(\text{cp})\ V_o(1/\text{min})}{D_p^2(\text{mm}^2)} \tag{3.56}$$

Table 3.17 shows the agreement between results from experiments and those calculated with this equation for several ion exchange resins.

Table 3.17: Pressure Drop for Commercial Ion Exchange Resins

Resin	Flow Rate (ℓ/min)	Mean Bead Diameter (mm)	Pressure Drop (bar/cm)	
			Calculated	Measured
Dowex SBR-P	151.4	.750	5.28	5.08
Dowex WGR-2	22.7	.675	1.00	1.06
Dowex MWA-1	15.1	.675	0.67	0.77
Dowex MSA-1	37.8	.650	1.79	2.13
Dowex MSC-1	30.3	.740	1.08	1.22

Figure 3.57 shows the use of a size factor for resins which is used to develop a pressure factor. This pressure factor can then be used to calculate the pressure drop under conditions of different solution viscosities, flow rates or particle size distributions once the pressure drop is known at one viscosity, flow rate and particle size distribution.

3.7.2 Computer Calculations

Computer methods have been used to predict the shape of the adsorption breakthrough curves for packed bed systems based upon data obtained from small scale batch

WET MESH SIZE	FACTOR
12	0.5
16	1.0
20	2.0
30	4.0
35	6.0
40	8.0
50	16.0
70	32.0
100	64.0

EXAMPLE:

RETAINED ON	%	FACTOR
12 mesh	—	0.5 = 0.0
16 mesh	0.6	1 = 0.6
20 mesh	37.0	2 = 74.0
30 mesh	46.3	4 = 185.2
35 mesh	13.0	6 = 78.0
40 mesh	2.0	8 = 16.0
50 mesh	0.9	16 = 14.4
70 mesh	0.2	32 = 6.4
		374.6

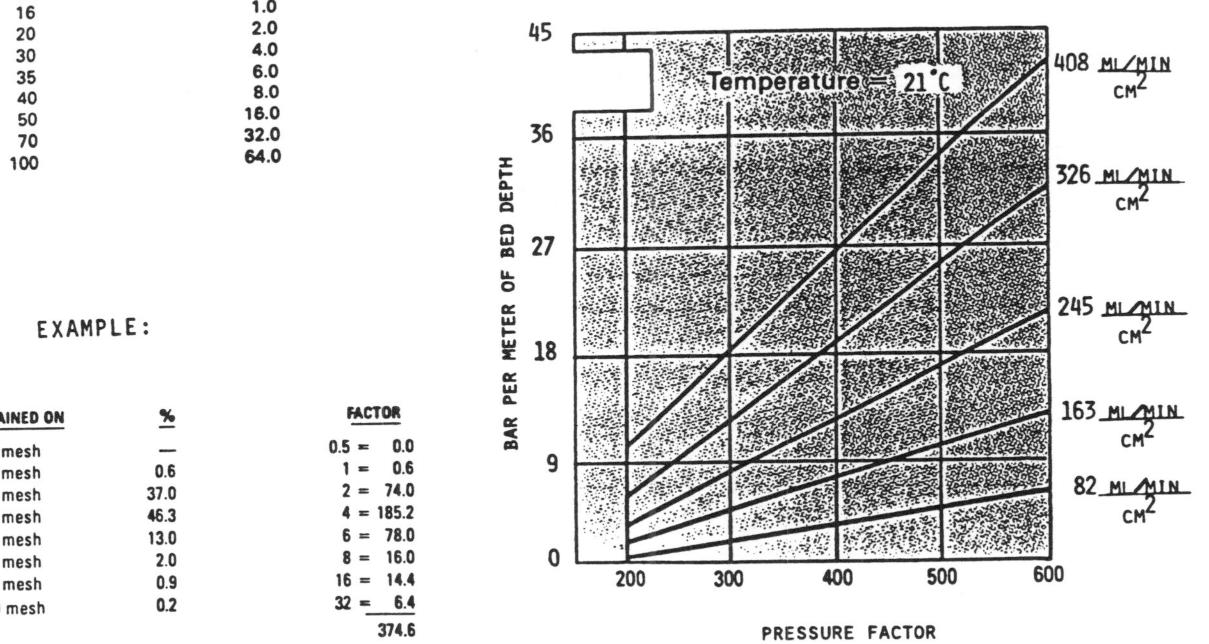

Figure 3.57. Size factor calculations for use in determining estimations for pressure drop in ion exchange columns.

experiments. The computed effects of changes in operating parameters such as flow rate, bed dimensions and inlet solute concentrations have been compared for agreement with experimental values.

The programs developed by Tan (103), which are discussed in Chapter 2, may also be applied to ion exchange fixed bed systems. Liberti and Passino (98) have developed an alternate method for calculating breakthrough curves and a model for regeneration using graphical evaluation. The model is one in which the solid and fluid are viewed as contacting one another countercurrently and continuously at their respective average concentrations. Once the operating and equilibrium lines for a given ion exchange operation have been determined, the graphical method allows fixed bed performances over a wide range of operating conditions to be predicted.

Klein (104) has developed a computer program which computes the equilibrium composition of the fluid phase surrounding an ion exchange resin. The program, shown in Table 3.18 for a TI-59, is based upon an empirically modified mass action law. The memory location of the variables are shown in Table 3.19. The ionic composition and the selectivity of the resin and the valences of the ions are required as input. As many as ten ionic species may be in the fluid phase. The ionic composition for this program can be in terms of the concentration rather than the more usual, and more difficult to measure, activities. The usefulness of this program is as a subroutine for more comprehensive programs used to predict the performance of an ion exchange column treating a multicomponent feedstream.

Klein and coworkers (105) have, in fact, developed such a Fortran program to predict the course of ion exchange in fixed bed columns. The program handles an arbitrary initial composition distribution and feed composition history. The program calculates gradual and step compositional changes along the resin column.

The computer models have allowed predictions to be made on the effects and likely performance when mixtures of two or more solutes with different adsorption parameters are applied simultaneously to an ion exchange column. Effects such as the displacement of a loosely bound protein by a

Table 3.18: Program for Calculating y_i values from Known x_i Values Using a TI-59 Calculator (104)

000	76	LBL
001	44	SUM
002	01	1
003	44	SUM
004	01	01
005	44	SUM
006	02	02
007	44	SUM
008	03	03
009	44	SUM
010	04	04
011	92	RTN
012	76	LBL
013	11	A
014	43	RCL
015	10	10
016	35	1/X
017	42	STO
018	05	05
019	43	RCL
020	40	40
021	55	÷
022	43	RCL
023	30	30
024	95	=
025	42	STO
026	09	09
027	00	0
028	42	STO
029	07	07
030	01	1
031	42.	STO
032	08	08
033	42	STO
034	20	20
035	76	LBL
036	12	B
037	43	RCL
038	51	51
039	42	STO
040	00	00
041	01	1
042	00	0
043	42	STO
044	01	01
045	02	2
046	00	0
047	42	STO
048	02	02
049	03	3
050	00	0
051	42	STO
052	03	03
053	04	4
054	00	0
055	42	STO
056	04	04
057	01	1
058	94	+/-
059	42	STO
060	06	06
061	00	0
062	42	STO
063	52	52
064	76	LBL
065	10	E'
066	22	INV
067	87	IFF
068	01	01
069	14	D
070	53	(
071	53	(
072	43	RCL
073	07	07
074	85	+
075	43	RCL
076	08	08
077	54	)
078	55	÷
079	02	2
080	55	÷
081	43	RCL
082	30	30
083	54	)
084	42	STO
085	09	09
086	76	LBL
087	14	D
088	53	(
089	73	RC*
090	02	02
091	65	×
092	43	RCL
093	09	09
094	45	YX
095	73	RC*
096	01	01
097	54	)
098	45	YX
099	43	RCL
100	05	05
101	65	×
102	73	RC*
103	03	03
104	95	=
105	72	ST*
106	04	04
107	44	SUM
108	06	06
109	65	×
110	73	RC*
111	01	01
112	95	=
113	44	SUM
114	52	52
115	22	INV
116	97	DSZ
117	00	00
118	15	E
119	71	SBR
120	44	SUM
121	61	GTO
122	14	D
123	76	LBL
124	15	E
125	43	RCL
126	50	50
127	32	X:T
128	43	RCL
129	06	06
130	50	I×I
131	77	GE
132	87	IFF
133	92	RTN
134	76	LBL
135	87	IFF
136	22	INV
137	87	IFF
138	01	01
139	18	C'
140	00	0
141	32	X:T
142	43	RCL
143	06	06
144	77	GE
145	17	B'
146	43	RCL
147	40	40
148	42	STO
149	07	07
150	61	GTO
151	12	B
152	76	LBL
153	17	B'
154	43	RCL
155	40	40
156	42	STO
157	08	08
158	61	GTO
159	12	B
160	76	LBL
161	18	C'
162	53	(
163	43	RCL
164	09	09
165	65	×
166	53	(
167	01	1
168	75	-
169	43	RCL
170	06	06
171	65	×
172	43	RCL
173	10	10
174	55	÷
175	43	RCL
176	52	52
177	54	)
178	54	)
179	42	STO
180	09	09
181	65	×
182	43	RCL
183	30	30
184	95	=
185	32	X:T
186	00	0
187	77	GE
188	19	D
189	01	1
190	32	X:T
191	77	GE
192	19	D'
193	61	GTO
194	12	B
195	76	LBL
196	19	D'
197	86	STF
198	01	01
199	61	GTO
200	10	E'
201	76	LBL
202	45	YX
203	93	.
204	05	5
205	42	STO
206	40	40
207	11	A
208	01	1
209	00	0
210	22	INV
211	90	LST
212	91	RTN

Table 3.19: Storage Locations for Problems with Known x's (104)

	0	1	2	3	4	5	6	7	8	9
0	Index Counter	Lowest register number for z_i	K_{i1}	x_i	t_i	$1/z_1$	s	Smallest t_1	Largest t_1	t_1/x_1
1	z_1	z_2	z_3	z_4	z_5	z_6	z_7	z_8	z_9	z_{10}
2	1	K_{21}	K_{31}	K_{41}	K_{51}	K_{61}	K_{71}	K_{81}	K_{91}	$K_{10,1}$
3	x_1	x_2	x_3	x_4	x_5	x_6	x_7	x_8	x_9	x_{10}
4	t_1	t_2	t_3	t_4	t_5	t_6	t_7	t_8	t_9	t_{10}
5	d	n	$\Sigma t_i z_i$	ds/dt_1						

protein with a greater affinity for the resin correspond to what is observed in actual practice.

3.7.3 Economic Analysis

Erskine and Schuliger (106) developed an "operating line" method of relating particular operational performance with relative cost effectiveness for the economic analysis of adsorption and ion exchange systems.

Figure 3.58 shows how the operating line correlates the "superficial liquid retention time" (the net resin volume divided by the superficial liquid flow rate) and the "exhaustion rate" (the loading rate as a function of the degree of regeneration). The data points for this curve are obtained from breakthrough determinations for different bed depths.

The operating lines have the inherent characteristic of exponentially approaching a finite limit at each axis. The limit along the Y-axis, projected to the point on the ordinate representing the minimum exhaustion rate, corresponds to the retention time that provides saturation levels corresponding to equilibrium with the feedstream concentration. The limit along the X-axis, projected to the point on the abscissa representing the minimum retention time, is the contact period required to achieve only the treatment level specified by the desired effluent concentration. Operating points between these two limits are then distributed along a curve whose position depends on the effluent concentration required (Figure 3.59) and type of ion exchange column (Figure 3.60).

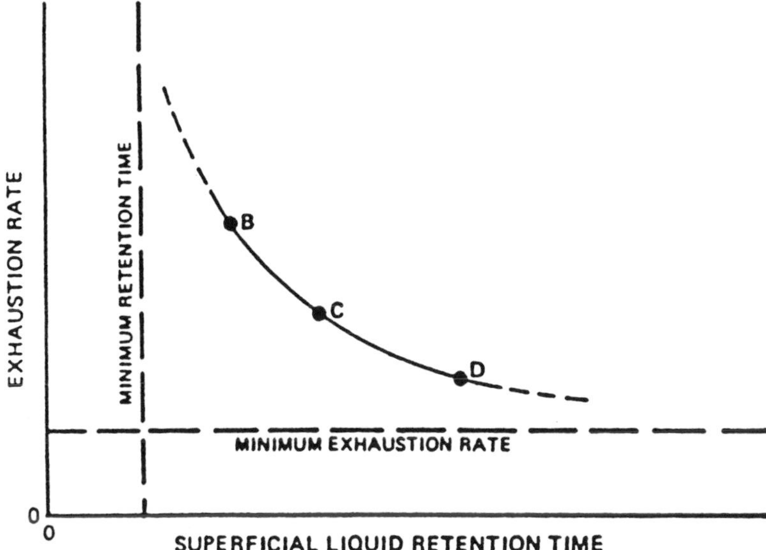

Figure 3.58. Development of the operating line from breakthrough curves at different bed heights (Reference 106).

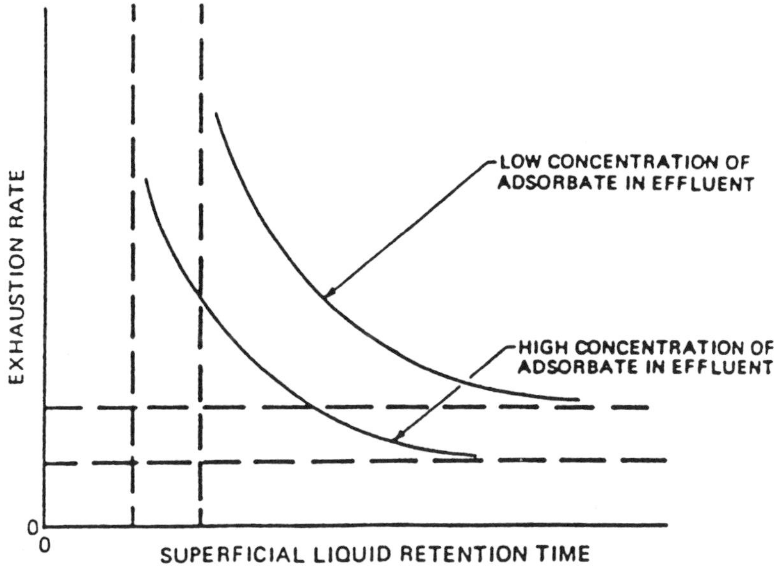

Figure 3.59. Effect of effluent concentrations on the operating line position (Reference 106).

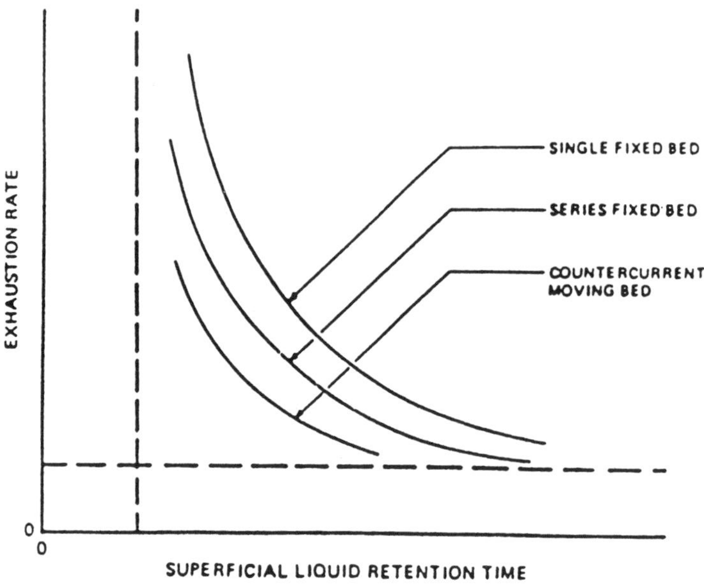

Figure 3.60. Effect of the type of ion exchange column on the operating line position (Reference 106).

Once the operating line is established for a specific set of operating parameters, the economic analysis is carried out as an operating cost optimization where a recovery factor is specified to represent the capital associated for the particular column system. The analysis combines the capitalized costs obtained with the direct operating costs to identify the design and operating parameters that are the optimum for the specific application (107).

3.7.4 Ion Exchange Resin Limitations

When ion exchange resins are used for an extended period of time, the exchange capacities are gradually decreased. Possible causes for these decreases include organic contamination due to the irreversible adsorption of organics dissolved in the feedstream, the oxidative decomposition due to the cleaving of the polymer crosslinks by their contact with oxidants, the thermal decomposition of functional groups due to the use of the resin at high temperature and the inorganic contaminations due to the adsorption of inorganic ions.

When a resin bed has been contaminated with organic

foulants, a procedure is available that can help to restore the resin (108). Three bed volumes of a 10% NaCl - 2% NaOH solution are passed through the resin bed at 50°C. The first bed volume is passed through at a flow rate not greater than 1 L/s-m^3. The second bed volume is introduced at the same rate but is kept in the bed for at least 8 hours, with occasional agitation by air lancing. The last bed volume is then passed through at the same flow rate, followed by a thorough rinsing and two regenerations with the standard regenerate.

If the fouling is due to microbial contamination, the same authors recommend backwashing the bed and then filling the entire vessel with a dilute solution ($\leq$ 0.05%) of organic chlorine. This solution should be circulated through the bed for 8 hours at a warm (50°C) temperature. After this treatment, the resin should be backwashed, regenerated and rinsed before returning it to service. This procedure may cause some oxidation of the polymeric resin, thereby reducing its effective crosslinking and strength. Therefore, treating the resin in this fashion should not be a part of the normal resin maintenance program.

Physical stability of a properly made cation or anion exchange resin is more than adequate for any of the typical conditions of operation. These resins can be made to have a physical crush strength in excess of 300 grams per bead.

Perhaps more important are the limitations inherent in the structure of certain polymers of functional groups due to thermal and chemical degradation. Thermally, styrene-based cation exchange resins can maintain their chemical and physical characteristics at temperatures in excess of 125°C. At temperatures higher than this, the rate of degradation increases. Operating temperatures as high as 150°C might be used, depending on the required life for a particular operation to be economically attractive.

Strong base anion exchange resins, on the other hand, are thermally degraded at the amine functional group. Operating in the chloride form, this is not a severe limitation, with temperatures quite similar to those for cation exchange resins being tolerable. However, most strong base anion exchange resins uses involve either the hydroxide form, the carbonate or bicarbonate form. In these ionic forms, the amine functionality degrades to form lower amines and alcohols. Operating temperatures in excess of 50°C should be avoided for Type I strong base anion exchange resins in the hydroxide form. Type II strong base resins in the hydroxide form are more susceptible to thermal degradation and temperatures in excess of 35°C should be avoided.

The amine functionality of weak base resins is more stable in the free-base form than that of strong base resins. Styrene-divinylbenzene weak base resins may be used at temperatures up to 100°C with no adverse effects.

Chemical attack most frequently involves degradation due to oxidation. This occurs primarily at the crosslinking sites with cation exchange resins and primarily on the amine sites of the anion exchange resins. From an operating standpoint and, more importantly, from a safety standpoint, severe oxidizing conditions are to be avoided in ion exchange columns.

Oxidizing agents, whether peroxide or chlorine, will degrade ion exchange resins (109). On cation resins, it is the tertiary hydrogen attached to a carbon involved in a double bond that is most vulnerable to oxidative degradation. In the presence of oxygen, this tertiary hydrogen is transformed first to the hydroperoxide and then to the ketone, resulting in chain scission. The small chains become soluble and are leached from the resin. This chain scission may also be positioned such that the crosslinking of the resin is decreased, as evidenced by the gradual increase in water retention values.

The degradation of anion resins occurs not only by chain scission but also at the more vulnerable nitrogen on the amine functionality. As an example, the quaternary nitrogen on type I strong base anion resins is progressively transformed to tertiary, secondary, primary nitrogen and finally to a nonbasic product.

Oxidative studies on resins with different polymer backbones and functionalities have been performed as accelerated tests (110). The data is shown in Table 3.20 for polystyrene and polydiallylamine resins. Although the polydiallylamine resins have a higher initial capacity, they are much more susceptible to oxidative degradation. When the polystyrene resin has a mixture of primary and secondary amino groups or when a hydroxy-containing group is attached to the amine of the functional group, the susceptibility to oxidation is enhanced. Thus one can understand the lower thermal limit for Type II anion resins compared to Type I resins.

Table 3.20: Oxidation of Polystyrene and Polydiallylamine Resins in a One-Week Accelerated Test at 80°C (110)

Resin Backbone	Functional Group	Initial Base Capacity (meq/g)	Base Capacity Lost During Test (%)
Polystyrene	$R\text{-}CH_2N(CH_3)_2$	4.4	1
Polystyrene	$R\text{-}CH_2N(CH_2CH_3)_2$	4.5	2
Polystyrene	$R\text{-}CH_2N(C_2H_4OH)_2$	3.6	17
Polydiallylamine	N-H	8.3	37
Polydiallylamine	$N\text{-}CH_2CH_3$	7.3	37
Polydiallylamine	$N\text{-}C_3H_7$	6.7	31

The effect of thermal cycling on strong base anion resins has been studied by Kysela and Brabec (111). The average drop in strong base capacity was 2.1×10^{-4} mmol (OH^-)/mL over the 480 cycles between 20°C and 80°C. Figure 3.61 shows the decrease in total exchange capacity and in strong base (salt splitting) capacity for each of the individual resins included in the study.

As would be expected, the combination of thermal and osmotic shocks has been shown to have a large effect on the osmotic strength of anion resins (112). This is illustrated in Figure 3.62, where the decrease is shown to be over 15% in just 150 cycles. The important processing point to remember from this is that if the regeneration of a resin is performed at a temperature more than 20°C different from the temperature of the operating (loading) process, it is advisable first to adjust the temperature of the resin with condensate until it is at the temperature of the regenerant.

It must be noted that nitric acid and other strong oxidizing agents can cause explosive reactions when mixed with organic materials such as synthetic ion exchange resins.

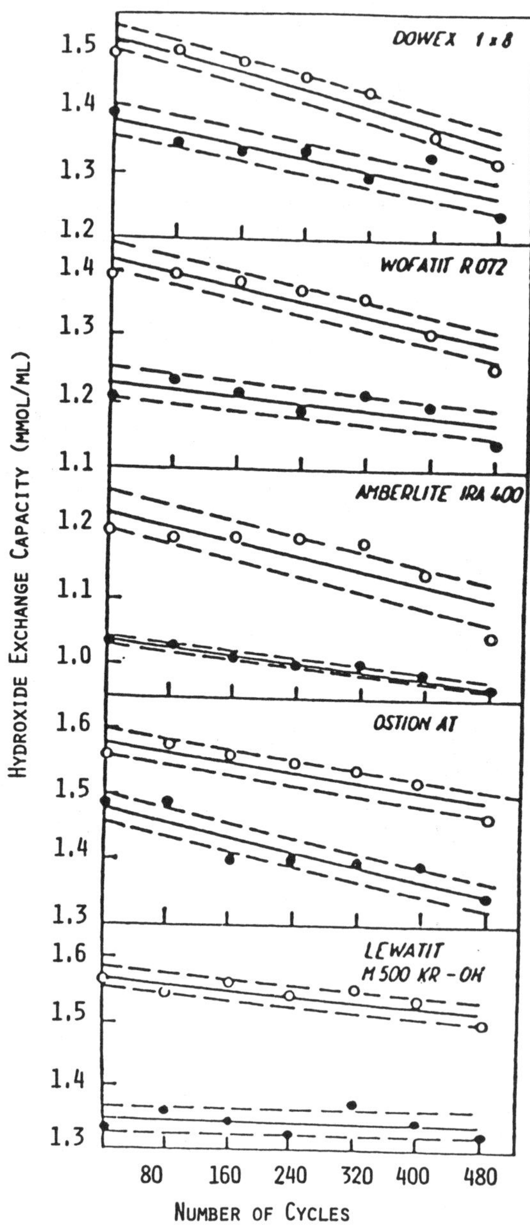

Figure 3.61. Dependence of the strong base capacity on the number of cycles between 20°C and 80°C (Reference 111).

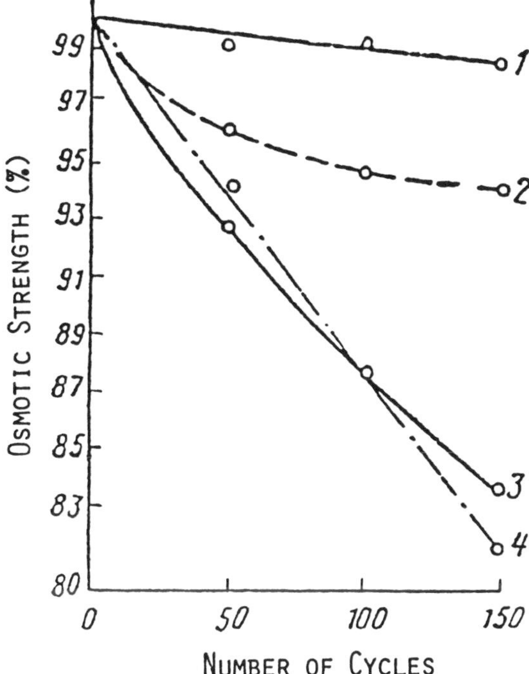

Figure 3.62. The effect of thermal and osmotic cycling on the osmotic strength of a strong base anion resin, AV-17. Curve (1): effect of thermal cycling; Curve (2): effect of osmotic cycling; Curve 3: effect of thermal and osmotic cycling (cold alkali to hot acid); Curve (4): effect of thermal and osmotic cycling (hot alkali to cold acid) (Reference 112).

Nitric acid should NEVER be used as the regenerating acid for cation resins or to place anion resins in the nitrate form.

3.8 BIOTECHNOLOGY APPLICATIONS

It used to be that biotechnology was synonymous with fermentation processes. More recently, the term has been associated primarily with recombinant DNA technology. Actually biotechnology applications range from fermentation, to cell harvesting, to enzyme immobilization, to biological sensors, to microbial manipulation. In essence, they include all applications involving microorganisms, enzymes and biological raw materials. Ion exchange resins have been used, at least in the laboratory, in all of these applications. The outgrowth of the laboratory usage of ion exchange resins has been their use in the commercial recovery and purification of biotechnology products. In fact, most purification schemes for biotechnology operations are merely changes in the scale or size for techniques developed in laboratory biochemical processes.

The industrial purification of biochemicals will usually involve an adsorption or chromatographic operation in conjunction with the strictly ion exchange operation. The applications provided will often give the complete steps required for biochemical purification.

3.8.1 Amino Acid Purification

The amino acids may be grouped into four main types (Figure 3.63):

(1) Basic amino acids - arginine, asparagine, glutamine, histidine, lysine and tryptophan

(2) Dicarboxylic amino acids - glutamic acid and aspartic acid

(3) Neutral amino acids - alanine, cysteine, cystine, glycine, hydroxyproline, isoleucine, leucine, methionine, proline, serine, threonine and valine

(4) Aromatic amino acids - phenylalanine and tyrosine.

Tryptophan could also be categorized with the aromatic amino acids. However, for ion exchange separations, the basic functionality is of more significance.

All of these amino acids are adsorbed on strong acid cation resins in the hydrogen form. This allows their separation from non-ionic impurities, such as may be present in protein hydrolysates. If the feed solution is first adjusted to neutral or basic pH, only the basic amino acids are adsorbed by the strong acid cation resin in the sodium or ammonium form. Basic amino acids are also preferentially adsorbed on weak acid cation exchange resins when the

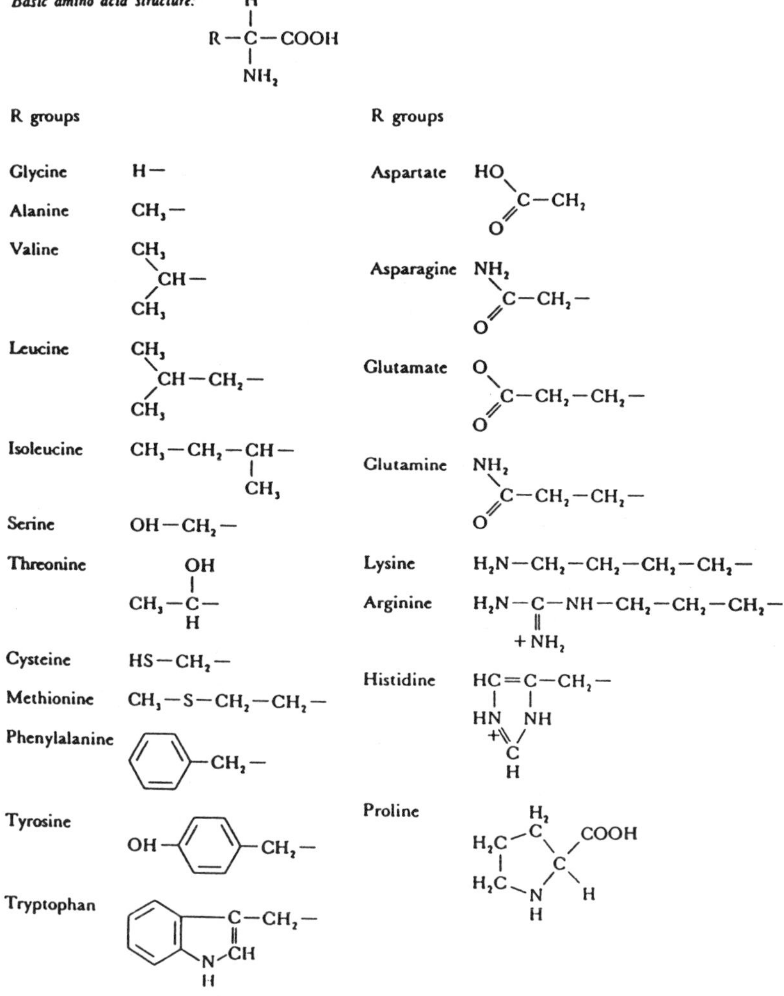

Figure 3.63. Structures of amino acids commonly found in proteins.

feedstream has been adjusted to pH 4.7. The dicarboxylic amino acids can be preferentially adsorbed, compared to the other amino acids, onto weak base anion exchange resins converted to the HCl form. The neutral amino acids can be separated from each other by stepwise elution with changes in the pH of the eluting solution.

Whereas Stein and Moore (113) used a single strong cation resin (Dowex 50 W) with changing elution conditions to separate amino acids, Winters and Kunin (3) used three resins in five columns and different elution conditions (Figure 3.64) to separate amino acids from protein hydrolyzate into their three charge groups (acid, neutral and base) and three individual amino acids (histidine, arginine and lysine). The carboxylic acid resin may offer an advantage for some amino acid purifications because of its higher buffering capacity, compared to strong cation resins, in the critical pH region.

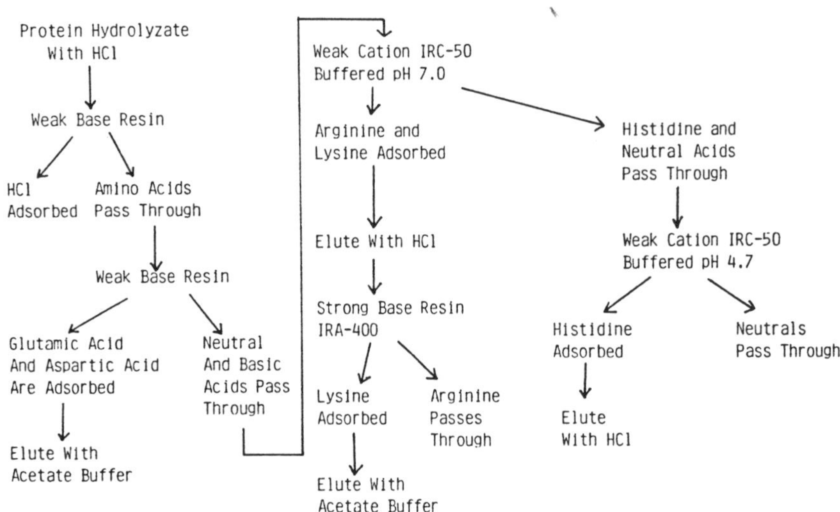

Figure 3.64. Separation of amino acids from a hydrolyzed protein feedstream (Reference 3).

The use of weak base anion exchange resins for recovering glutamic acid and aspartic acid from protein hydrolyzate has been reported (114) for commercial preparations of amino acid solutions of neutral and basic amino acids for intravenous feeding.

Another scheme (Figure 3.65) has been used commercially for separating amino acids from protein hydrolyzate. In these schemes, it can be seen that cystine and cysteine have been removed prior to the ion exchange treatment of the protein hydrolyzate.

According to the scheme in Figure 3.65, the pH of the protein hydrolyzate solution is adjusted with HCl to a level near the isolectric point of glutamic acid to allow the removal of half to two-thirds of the glutamic acid by

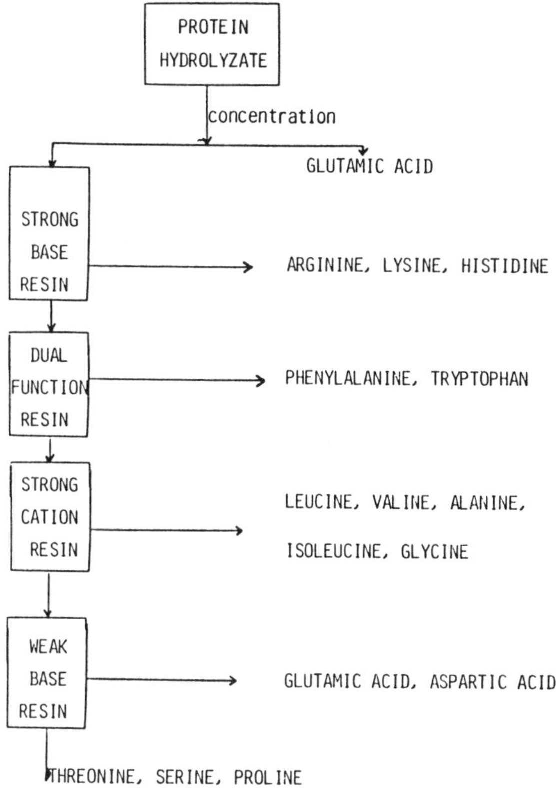

Figure 3.65. Separation of amino acids from a hydrolyzed protein feedstream with controlled elution of individual amino acids.

precipitation. This solution is then passed through a strong base resin which adsorbs the basic amino acids. A duel functionality resin (or a mixture of strong base and weak cation resins) is used next to adsorb the phenylalanine and tryptophan. The majority of the neutral amino acids are adsorbed on a strong cation resin, while a weak base resin adsorbs the acidic amino acids, including any residual glutamic acid. The pH changes of the solution as it passes through the different resins and the pH of the eluting solutions which allow the elution separation of the adsorbed species vary for different amino acid producers.

The recovery of amino acids from these protein hydrolyzates has the limitation that the yield of each amino acid is proportional to its presence in the original proteins.

While protein hydrolyzates will continue to be used as the raw material for some commercial amino acid production, the trend is toward more use of fermentations which produce a specific amino acid.

Amino acid fermentation broths have characteristics which are different from those of protein hydrolyzates. The differences are (115):

(1) In most cases, a significant amount of a single amino acid can be accumulated in the fermentation broth in its free and salt forms and other contaminant amino acids are usually few in number and small in quantity.

(2) Amino acids produced by fermentation are usually the biologically active L-form.

(3) Fermentation broths are aqueous solutions of amino acids whose concentrations vary from 0.1 to 10% (W/V) depending on the individual fermentation process.

(4) Fairly large amounts of microbial cells and soluble protein-like biopolymers are usually contained in fermentation broths.

(5) Some inorganic salts derived from microbial nutrients always exist in fermentation broths.

(6) Other organic impurities derived from microbial nutrients and metabolites, such as sugars and pigments, are also present to some extent.

As an example, L-tryptophan (116) is fermented using *Arthrobacter paraffineus* which requires histidine as a nutrient. The fermentation filtrate is passed through a strong cation resin which selectively adsorbs the L-tryptophan. This is eluted from the resin with 0.5N ammonium hydroxide solution from which it is concentrated and crystallized. The resin used and the elution conditions can be specified with more precision for purification of an amino acid from a fermentation broth since the broad spectrum of amino acids presented in protein hydrolyzates are not in the fermentation broth.

Many other amino acids, a few of which are listed in Table 3.21, have been recovered from fermentation broth using ion exchange resins. After loading the resin to breakthrough with the amino acid feedstream, it is necessary to wash the resin with deionized water before eluting the amino acid so that the glucose, protein (bacterial cell components), salt and color impurities are separated from the amino acid.

Table 3.21: Amino Acids Recovered from Fermentation Broths

Amino Acid	Ion Exchange Resin	Eluant	Reference
N-acetylglutamine	IRA-400 SK-1	0.5N H_2SO_4	117
L-histadine	Strong Acid Cation	Ammonium Hydroxide	118
Lysine	Strong Acid Cation	2N Ammonium Hydroxide	119
Lysine	Weak Acid Cation pH 7.0	0.15N Ammonium Hydroxide	120
L-valine	Strong Acid Cation	1.0N Ammonium Hydroxide	121
L-serine	Strong Acid Cation	Ammonium Hydroxide	122
Glutamic acid	Strong Acid Cation	NaCl	123

The adsorption of amino acids by ion exchange resins is strongly affected by the pH of the solution. Figure 3.66 shows the decrease in the adsorption of lysine by a strong cation resin in the ammonium form as the pH is changed. Other inorganic ions may also compete with the amino acids for exchange sites on the resin, similar to the ion exchange treatment of other mixtures of ions in solution.

In a different use of resins in treating amino acids, a monosodium glutamate mother liquor, containing 4 to 6% monosodium glutamate, has been passed through a weak base resin to remove the color impurities in the solution of the amino acid (124). The flow rate required is approximately 1

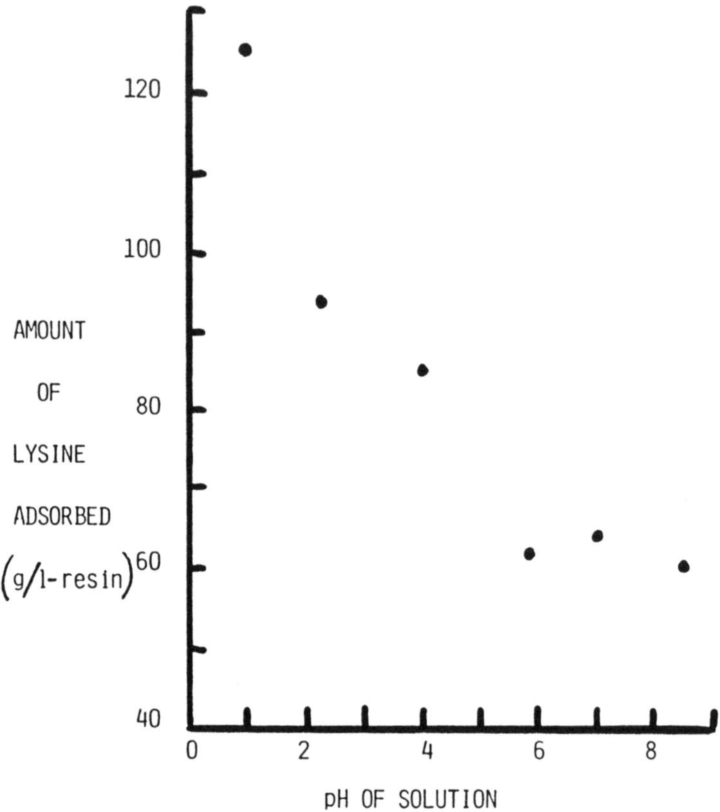

Figure 3.66. The change in the adsorption of lysine on a strong cation resin (Diaion SK1, NH_4^+ form) as a function of the pH of the fermentation broth (Reference 191).

bed volume per hour to reduce the degree of coloration from 0.6 to 0.1. A solution of 4% NaOH is used for removing the colored impurities from the resin. Depending on the color quality of the mother liquor, between 20 and 60 liters of the mother liquor may be treated by one liter of resin.

3.8.2 Antibiotics

The first commercial ion exchange operation for the recovery and purification of biological products occurred in the 1950's when the antibiotic streptomycin was recovered from a fermentation broth using a carboxylic cation exchange resin (125). After the resin, initially in the sodium form, became saturated with streptomycin, dilute mineral acid was used to elute the antibiotic. After elution, the resin was converted back to the sodium form using a dilute sodium hydroxide solution.

Soon after the commercialization of that process for streptomycin, a similar process was developed for the recovery and purification of neomycin (126). In this case, a strong base anion resin was utilized in the chloride form and, after first rinsing the loaded resin with 95% methanol, neomycin was eluted with a solution containing 16% HCl and 84% methanol. The eluate contained 80 to 95% of the antibiotic with a tenfold reduction in fluid volume.

Belter (127) has recently presented an overview of the various types of antibiotics which have been purified or recovered from fermentation broths using ion exchange resins. Currently, antibiotics are classified into four categories: the aminoglycosides, the β-lactams, the macrolides and the tetracyclines. Of these, only the tetracylines do not utilize ion exchange resins in their recovery.

The aminoglycosides are characterized by carbohydrate structures attached to a complex ring structure. Examples of this type of antibiotic are streptomycins, shown in Figure 3.67, neomycins, gentamicins and kanamycins. It is the strongly basic guanidine groups that undergo the adsorption interaction with the ion exchange resin at neutral pH conditions.

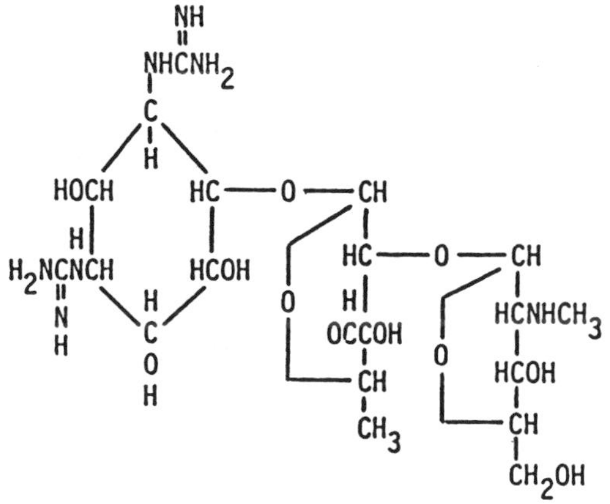

Figure 3.67. Chemical structure of streptomycin.

A fluidized bed can be used to treat the streptomycin-containing fermentation broth directly with the ion exchange resin. The broth flows up through the resin bed, expanding it by about 25%. The height to diameter ratio of these columns is about 3/4. When the resin is completely loaded with the antibiotic, the broth in the fluidized bed is first removed for recycling by rinsing with water and then the antibiotic is eluted down through the packed resin bed. The percentage antibiotic recovery using the fluidized bed technique was about 85% while the recovery from a filtered broth was about 73%.

When streptomycin is adsorbed from a fermentation broth where other impurities, especially ionic ones, are present, the amount adsorbed on the carboxylic resin can be substantially decreased as the ionic impurity concentration is increased. This is shown in Table 3.22 for different amounts of sodium chloride in the streptomycin solution (128). The interference of multivalent ions on the adsorption of streptomycin is even greater. The interference of these ions, however, can be overcome by adding anions to the broth which will cause precipitation or complexation with the multivalent ionic impurities.

Table 3.22: Effect of Sodium Chloride on Streptomycin Adsorption on Weak Acid Cation Resin (128)

Feed Solution Composition (mg/ml)			
Streptomycin Concentration	NaCl Concentration	Relative Adsorption Loading (mg/ml)	Streptomycin "Leakage" During Loading (% of C_0)
0.88	0	1.00	2
0.88	20	.20	2
0.88	30	.15	2
0.88	50	.13	37

The degree of purity obtained after elution is only 70 to 80%. Therefore, the eluant is further purified by passing it through a column of strong base resin, Type I, in the chloride form to remove 90 to 95% of the color; then through a mixed bed of strong acid cation resin and weak base anion resin to remove the calcium and magnesium salts; and finally through a second carboxylic acid resin bed for final purification. The purity of the eluant is raised to over 95%

by these additional steps. It must be added that the cation resin in the mixed bed needs to be highly crosslinked (20%) to avoid adsorption of the streptomycin.

The β-lactams are characterized by a fused four member lactam ring and a five or six member sulfur-containing ring. This category may be subdivided into the penicillins, when the five member ring with sulfur is a thiazolidine and into the cephalosporins when the sulfur containing ring has six members. Figure 3.68 shows the structure of 6-aminopenicillanic acid and Figure 3.69 shows the structure of Cephalosporin C.

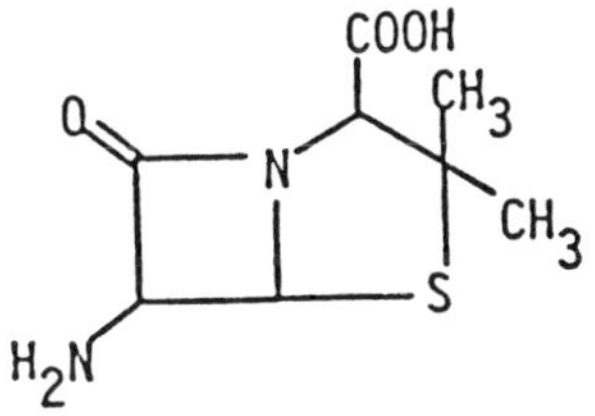

Figure 3.68. Chemical structure of 6-Aminopenicillanic acid.

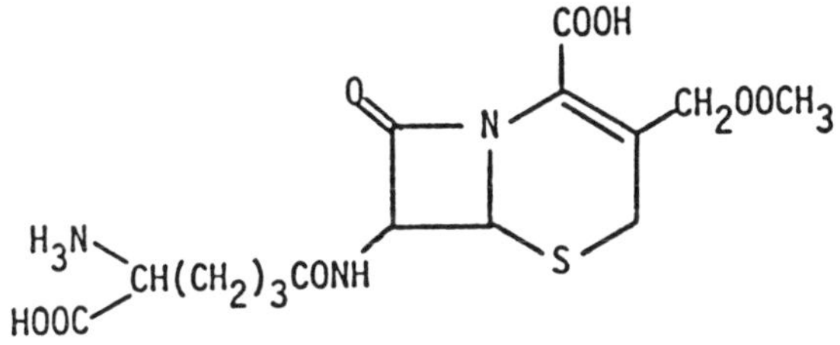

Figure 3.69. Chemical structure of Cephalosporin C.

Penicillin is not usually recovered using ion exchange resins because of the high degree of instability of penicillin once it has been produced in fermentation (128). Penicillin is unstable in water solution as evidenced by a 1% solution degrading by 10% in one hour at room temperature. Penicillin is also very sensitive to enzymatic decomposition by the enzyme *penicillinase*, produced by many organisms, which may contaminate the harvested fermentation product during processing. More stable derivatives of penicillin and cephalosporins have met with more success in ion exchange recovery processes.

Cephalosporin C is hydrophilic and chemically

amphoteric. It cannot be extracted directly from the fermentation broth with ion exchange resins. It is first necessary to acidify, filter and perform preliminary purification using as adsorbent resin as shown in Figure 3.70 (129-131). The eluate from the adsorbent resin contains over 80% of the antibiotic obtained by fermentation and less than 20% of the initial impurities. This eluate becomes the feedstream for a medium base anion resin in the acetate form. The flow rate may be as high as 5 bed volumes per hour. Following the loading phase of the cycle, Cephalosporin C may be eluted with a buffer solution of pyridine acetate at a pH of 5.5 to 5.6. After elution, the resin is regenerated with caustic soda and reconverted to the acetate form for the next loading cycle. Over 95% of the treated antibiotic is recovered in the eluant from the ion exchange resin.

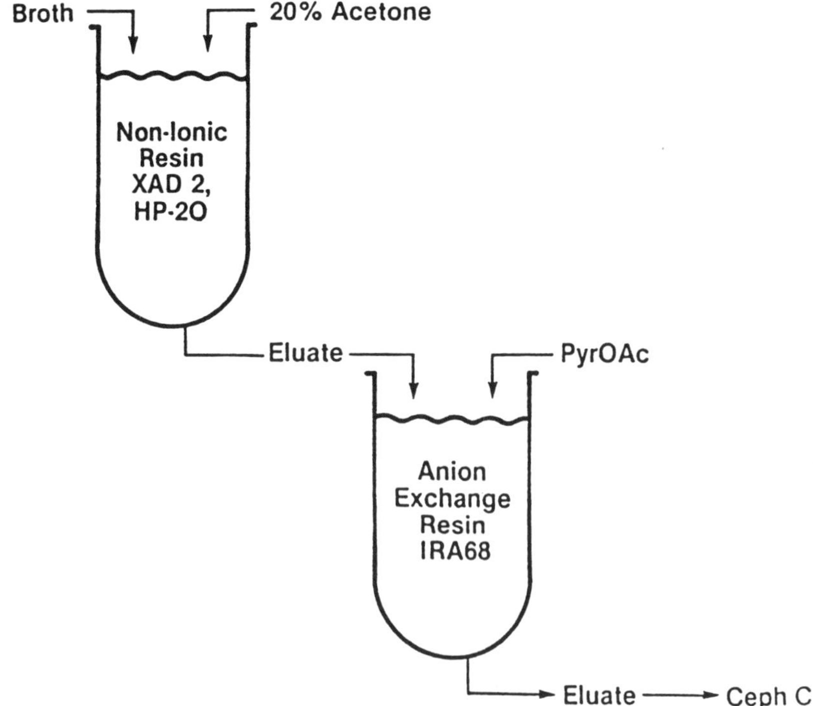

Figure 3.70. Non-ionic and anionic resin purification of Cephalosporin C (Reference 129).

Other combinations of resin columns have been proposed for the recovery of Cephalosporin C (132-134). A general scheme for these is shown in Figure 3.71. A cation exchange resin column in the hydrogen ion form is used to lower the pH of the feedstream without adding additional anions. A lead anion exchange column is used to adsorb the stronger anions, while a second anion exchange column adsorbs the antibiotic. The adsorbed antibiotic is eluted with potassium acetate buffer solution. Variations on this arrangement switch the positions of the cation and the first anion column or use a mixed bed of cation and anion resin followed by the bed of anion resin for cephalosporin adsorption.

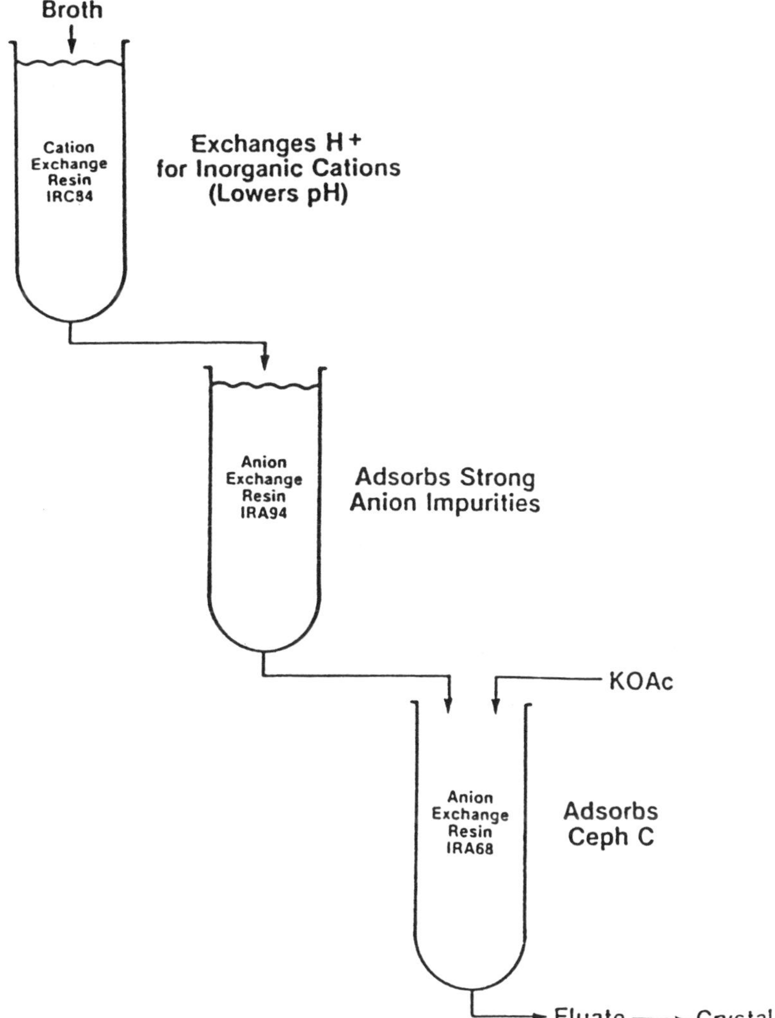

Figure 3.71. Cation and two anion resin purification of Cephalosporin C (References 132-134).

Since it is zwitterionic, Cephalosporin C may also be adsorbed on a strong acid cation resin if the pH of the feedstream is first adjusted to 2.5. However, the strong acidity of the resin results in significant amounts of degradation of the antibiotic to its lactone. A related antibiotic, Cephamycin C, which contains a carbamoyl ester in place of the acetyl ester at C-3, is more stable. This antibiotic has been successfully purified using a low crosslinked strong acid cation resin (135). Elution was performed with sodium acetate buffer solutions.

An alternate approach (136), after adsorbing the Cephalosporin C onto the cation resin, is to wash the column with 2 bed volumes of water, one bed volume of 10% NaCl, recirculating these effluents and then recovering the antibiotic by eluting with cold water (0°C to 5°C) when the effluent reaches a pH greater than 5.5. This results in a yield of over 90%.

More recently another scheme (shown in Figure 3.72) has been developed that uses two anion resin columns and two adsorbent columns to recover Cephalosporin C (137). Using this process results in a yield of 360 mg from the original 60 liters of fermentation broth.

The macrolides are characterized by a cyclic lactone structure to which one or more substituted sugars are attached. Examples of this type are erythromycin, shown in Figure 3.73, oleandomycin, rosamicin and tylosin.

Erythromycin has been recovered from a filtered fermentation broth by passing it through a column of cation resin (138). The effect of resin crosslinkage and acid strength on the amount of antibiotic adsorbed is shown in Table 3.23 for solutions that have an initial antibiotic concentration of 1000 micrograms per milliliter. Lower crosslinked strong acid resins had significantly higher antibiotic adsorption levels.

The erythromycin was eluted from the low crosslinked strong cation resin with the solvents shown in Table 3.24. The combination of 0.25N ammonia and certain aqueous organic solutions gave recovery of over 80% of the adsorbed antibiotic.

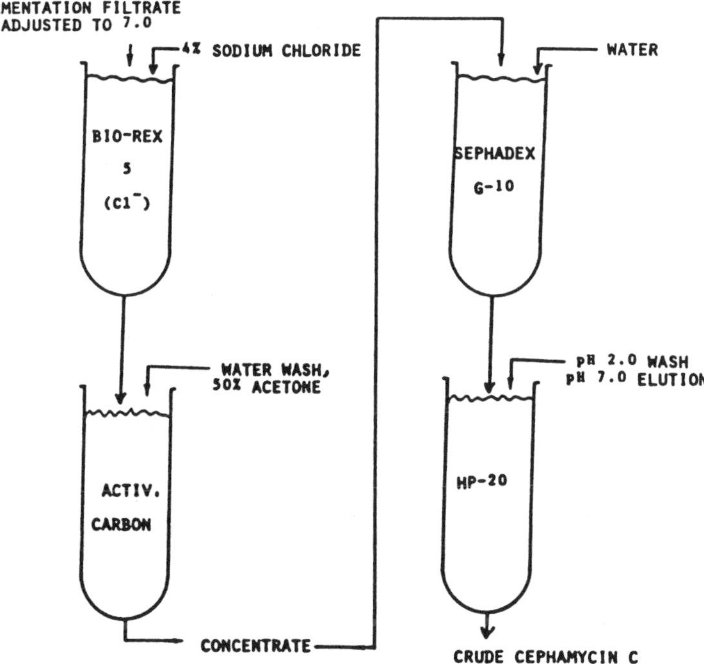

Figure 3.72. Alternate purification scheme for Cephamycin C (Reference 137).

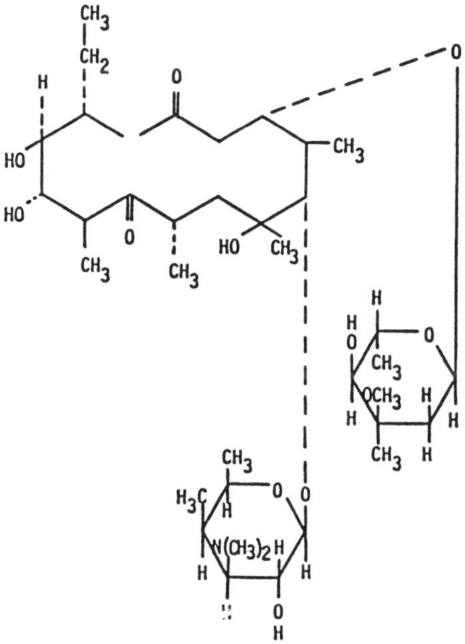

Figure 3.73. Chemical structure of Erythromycin.

Table 3.23: Effect of Resin Crosslinkage and Acid Strength on the Amount of Erythromycin Adsorbed (138)

Type of Ion Exchange Resin	Trade Name	Form	Amount Adsorbed (mg/ml)
Low crosslinked, strong acid cation	Diaion PK-204	NH_4^+	12.0
	Diaion PK-208	NH_4^+	11.6
	Dowex 50WX4	NH_4^+	11.7
	Amberlite XE-100	NH_4^+	7.5
	Diaion SK-104	NH_4^+	7.5
	Duolite C-25	NH_4^+	8.2
	Lewatit SP-100	NH_4^+	7.6
High crosslinked, strong acid cation	Dowex 50WX8	NH_4^+	0.86
Low crosslinked, weak acid cation	Duolite CS-101	NH_4^+	0
	Amberlite IRC-50	NH_4^+	0
	Amberlite IRC-50	pH 6.0	7.4
	Amberlite IRC-50	pH 6.5	5.1

Table 3.24: Elution Solvents for the Recovery of Erythromycin (138)

Elution Solvent	Erythromycin Recovered (%)
0.25 N ammonia	37.9
1.0 M NaCl	8.1
60% methanol	0
0.25 N ammonia/60% methanol	71.3
0.25 N ammonia/90% methanol	89.0
0.25 N ammonia/60% isopropanol	71.0
0.25 N ammonia/90% isopropanol	86.0
0.25 N ammonia/60% acetone	65.0
0.25 N ammonia/90% acetone	58.2

Rakutina and coworkers (139) reported that decreasing the particle size of microporous ion exchange resins would greatly increase its adsorption capacity for erythromycin and oleandomycin antibiotics when the particle size was less than 40 microns. This is shown in Figure 3.74. The adsorption was performed under batch conditions with the antibiotic initially at a concentration of 4000 units/mL. These differences in amounts of adsorption can be interpreted as changes in the selectivity coefficient as shown in Figure 3.75.

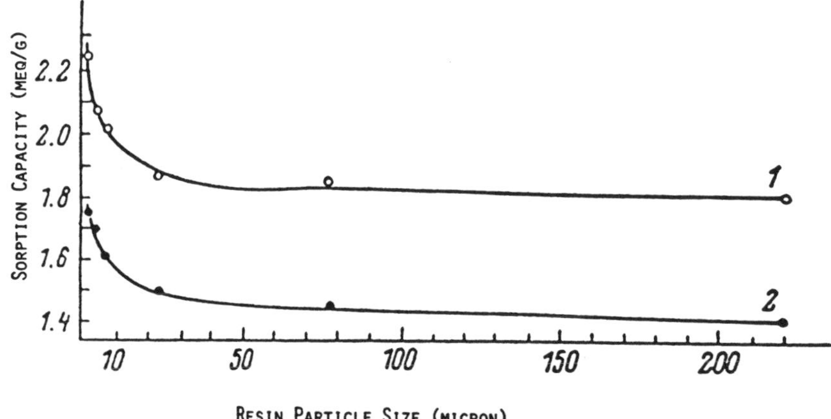

Figure 3.74. Dependence of the sorption capacity of the sulfonated naphthol cation exchange resin SNK-10E for erythromycin (curve 1) and oleandomycin (curve 2) on the resin particle size (Reference 139).

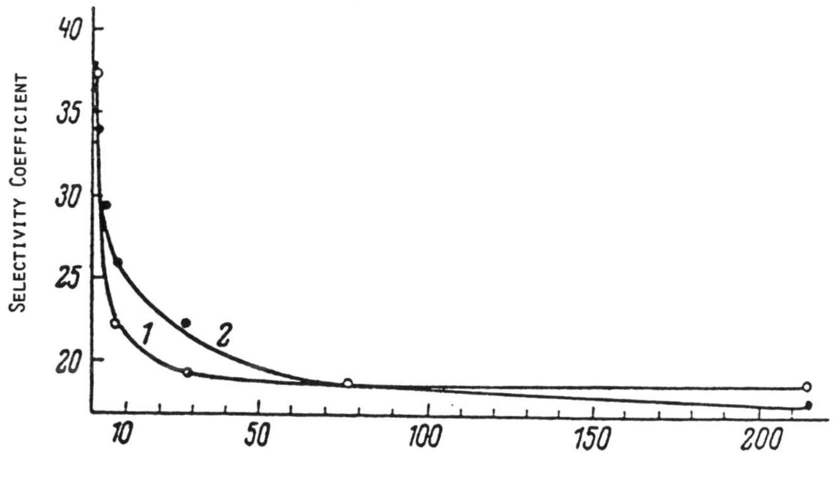

Figure 3.75. Dependence of the selectivity coefficient of erythromycin (curve 1) and oleandomycin (curve 2) sorption on the resin particle size of SNK-10E (Reference 139).

Samsanov and coworkers (140) have developed a system of finely divided ion exchange particles immobilized in a porous, highly permeable polymer matrix to adsorb the antibiotics erythromycin and oxytetracycline. When the ion exchange particles were only about 7 microns in diameter, the individual permeable polymer beads were approximately 500 to 600 microns in diameter and contained 20 to 75% of the resin particles (by volume).

This type of system allows a higher utilization of the ion exchange capacity and yields diffusion coefficients which are higher by an order of magnitude when compared to systems with ion exchange resins of 500 to 800 micron diameters. This is shown in Table 3.25. The effect of solution feed rate is shown in Figure 3.76 with different breakthrough curves for erythromycin adsorption.

The carbapenem antibiotics, thienamycin and olivanic acid, have a structure that is a β-lactam fused to a five member ring with a sulfur functionality attached. These antibiotics have been adsorbed on low crosslinked strong based anion resins in the chloride form (141). The resin would adsorb about 1,000,000 units of antibiotic activity per kilogram of resin in an agitated slurry. After transfer of the resin-antibiotic complex to a column, the antibiotic is eluted using a solution of 6% KCl with 50% acetone.

Thienamycin has also been recovered from its fermentation broth using a column of strong base resin in the HCO_3^- form (142). A fermentation broth containing about 12 mg/mL of antibiotic can be adsorbed with about 14 mL of resin per mg of antibiotic when the flow rate is 0.08 bed volumes per minute. This first step only yields a product with 5 to 10% purity. The additional steps necessary (anion resin column, adsorption column) are shown in Figure 3.77 which raise the purity to over 90%.

In addition, thienamycin has been recovered and purified by adjusting the fermentation broth to a pH between 3 and 5 and then passing it through a column of low crosslinked strong cation resin in the sodium form (143). A 2% aqueous pyridine solution can be used to elute 45% of the original antibiotic. Further purification is performed using chromatographic techniques.

Table 3.25: Influence of the Content of Cation Resin Microdispersion on the Capacity and Diffusion Coefficients for Erythromycin and Oxytetracycline (140)

Resin Type	Antibiotic	Total Exchange Capacity (meq/g)	Amount of Antibiotic Adsorbed (meq/g)	Capacity Utilized (%)	Diffusion Coefficient (cm^2/sec)
Whole Bead	Erythromycin	1.62	0.42	26	3.4×10^{-9}
Dispersions	Erythromycin	0.42	0.26	62	2.5×10^{-8}
		0.66	0.40	61	3.0×10^{-8}
		0.88	0.57	69	1.3×10^{-8}
Whole Bead	Oxytetracycline	5.11	2.35	46	1.3×10^{-8}
Dispersions	Oxytetracycline	0.76	0.42	55	1.5×10^{-7}
		1.60	0.83	52	1.2×10^{-7}
		2.37	1.36	58	0.8×10^{-7}

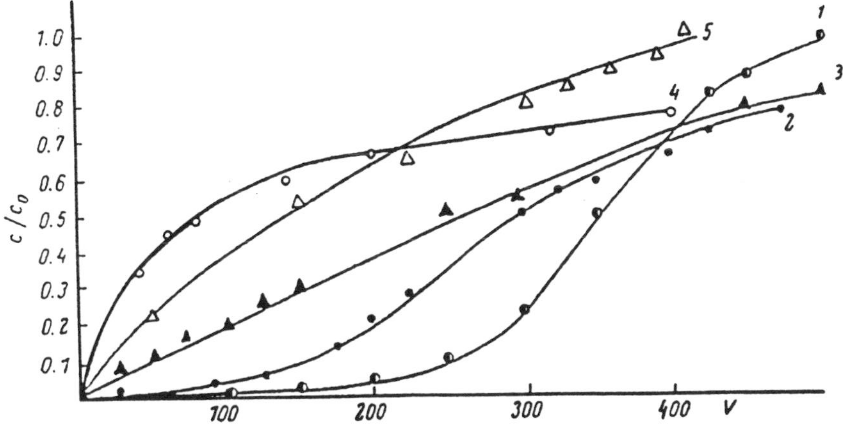

Figure 3.76. Breakthrough curves (effluent concentration fraction versus effluent volume) for the sorption of erythromycin by an immobilized SNK-30D dispersion at different solution flow rates. Curve (1): 50 ml/(cm^2 -hr); Curve (2): 300 mL/(cm^2/hr); Curve (3): 500 mL/(cm^2-hr); Curve (4): 1500 mL(cm^2-hr); Curve (5): 200 mL/(cm^2-hr) through a non-dispersed SNK-39 sample (Reference 140).

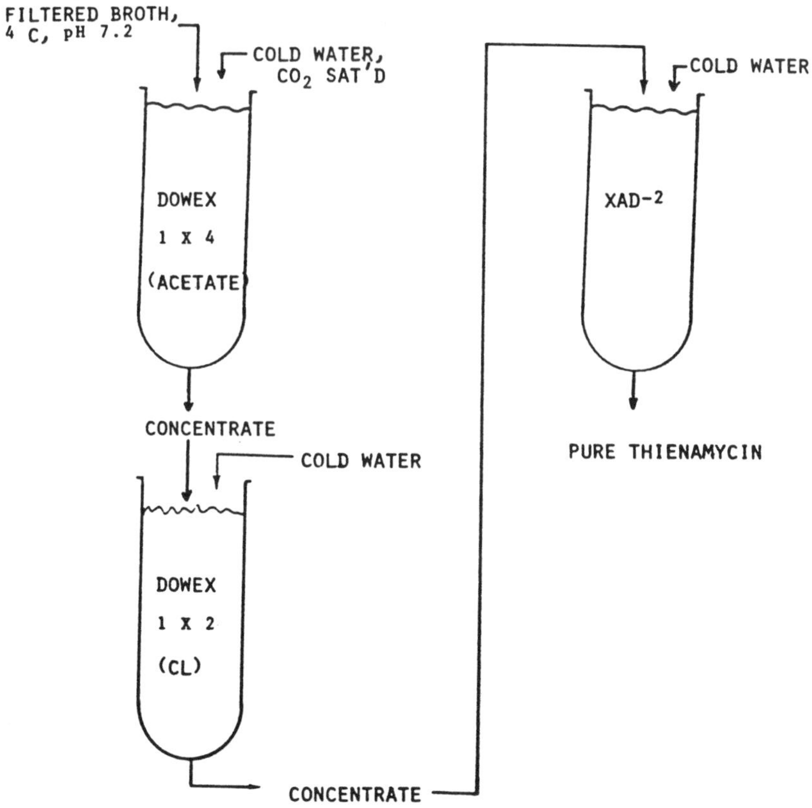

Figure 3.77. Purification sequence for thienamycin. The first anion column may be in either the carbonate or the acetate form (Reference 142).

The antibiotic A-4696 has been recovered from acidified, filtered fermentation broths by passing it through a column of strong acid cation resin in the sodium form (144,145). After loading, the resin is washed with three bed volumes of deionized water. The antibiotic is eluted with an aqueous solution of NaOH (pH 10.5), which is neutralized after elution. The eluate is further processed so that eventually an 80% yield is obtained.

3.8.3 Pharmaceuticals And Organic Acids

In addition to use in recovering antibiotics, ion exchange resins have been used to recover other pharmaceuticals and valuable organic substances from fermentation broths.

Khose and Dasare (146) have used the porous phenol-formaldehyde resins they developed to recover theophylline (Figure 3.78). The amount of adsorption was not as great as observed with the commercial Duolite resin, S-761, on a dried basis. However, since their resin has 20% less moisture content, the difference is not as large as it initially appears. Figure 3.79 shows the dependence of this adsorption on the flow rate of the feed stream.

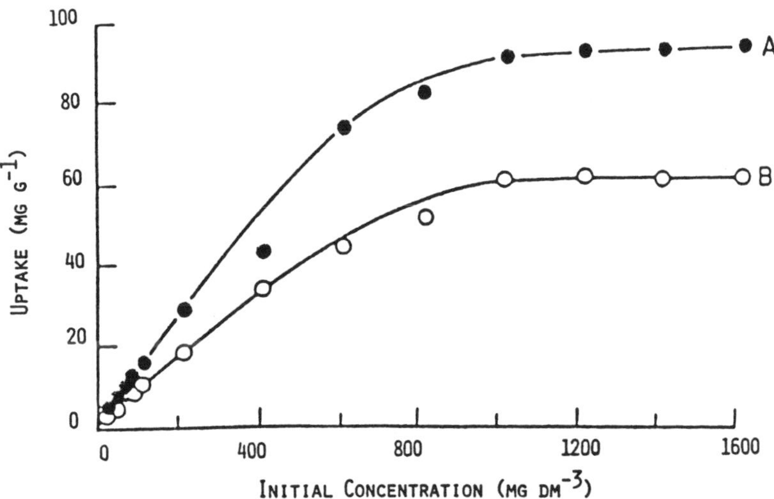

Figure 3.78. Dependence of the uptake of theophylline on the initial concentration of the equilibrated solution for Duolite S-761 (Curve A) and for a new synthetic resin (Curve B) (Reference 146).

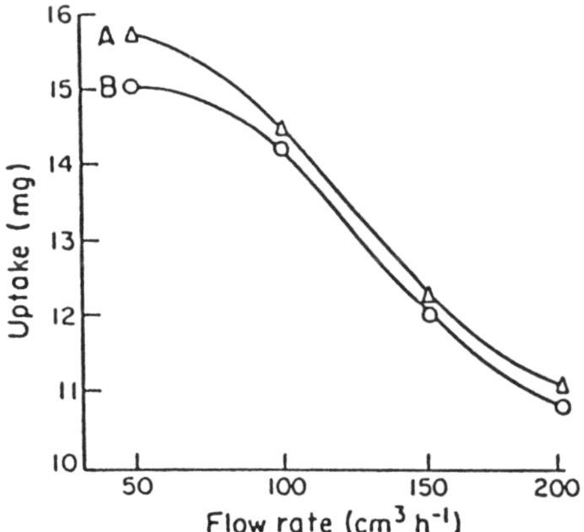

Figure 3.79. Dependence of the breakthrough capacity on the feedstream flow rate for Vitamin B-12 (Curve A) and theophylline (Curve B) (Reference 146).

The acetate forms of strong base anion exchangers have been used to purify acetylsalicylic acid from commonly occurring impurities (147).

Citric acid may be produced by the fermentation of molasses or dextrose. After fermentation, the citric acid is precipitated by the addition of lime at a temperature of 60°C. A precipitate of calcium citrate can then be separated from the mother solution. The calcium citrate must then be treated with sulfuric acid which dissolves the calcium citrate and produces a precipitate of calcium sulfate. The solution must be filtered and then treated with ion exchange resins to remove dissolved salts and organic impurities. The treated fluid can then be evaporated and crystallized to obtain the citric acid crystals.

Jacob (148) has described the recovery of xanthylic acid from fermentation broths. The acid is initially adsorbed onto a strong base anion resin and then eluted with an aqueous acid solution.

Tsao and coworkers (149) extracted itaconic acid from a molasses fermentation liquor by adsorbing the itaconic acid onto a strong base anion resin. The best results were obtained by first passing the fermentation liquor (diluted with three volume equivalents of water) through a column containing the ion exchange resin. The itaconic acid was eluted quantitatively from the resin using 3N NaOH.

Itaconic acid has also been purified after ultrafiltration by passing it through a cation resin in the potassium form (150). The effluent from this column is neutralized to pH 7.0 and passed through an ion exchange column membrane system. The dialyzate is then passed through a weak base resin column from which the itaconic acid is eluted with a concentrated solution of potassium bicarbonate.

3.8.4 Protein Purification

The adsorption of a protein by an ion exchange resin is dependent on the chemical composition of the resin, on the nature of the protein and on the other components in the solution. Girot and Boschetti (151) showed that a maximum protein adsorption can be obtained as one varies the pH of the solution and that the pH of the maximum differs from protein to protein. They showed that this maximum can be predicted using the equation:

$$S = K Q_A \left(1 - \frac{1}{10^{pH - pK_a} + 1}\right)\left(1 - \frac{pH}{pI}\right) \qquad (3.57)$$

where Q_A is the practical specific capacity, S is the amount of protein adsorbed, pI is the isoelectric point of the protein and K is a constant. This equation is based on the hypothesis that the protein sorption is proportional to the concentration of the ionized carboxyl groups and that the net charge of the protein is a linear function of pH in the range studied.

The curve in Figure 3.80 shows the relationship between the protein's isoelectric point and the pH corresponding to maximum protein adsorption calculated using Equation 3.57. The agreement with experimental results is very good. Such a curve can be used to select the pH of a buffer for maximum adsorption of a protein on a given ion exchange resin.

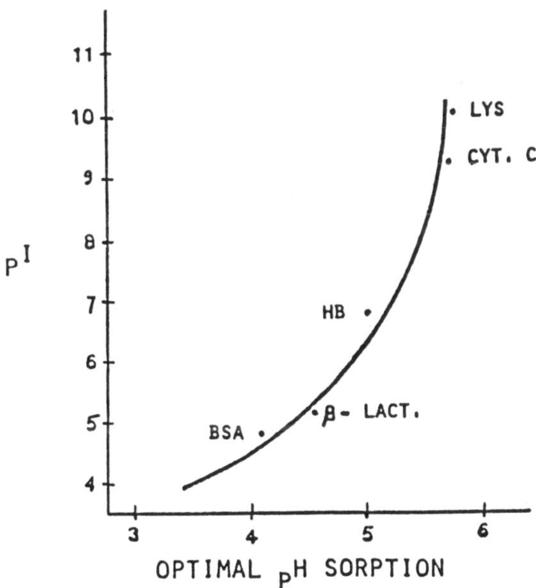

Figure 3.80. Relationship between protein isolectric point and optimal pH for protein sorption of the ion exchange resin. The curve is a theoretical model; the points are experimental results (Reference 151).

Keay (152) described the general conditions for the purification of a protein. His specific example was for the purification of neutral and alkaline proteases from *B. subtilis* by adsorption on a cation exchange resin.

The sulfonated phenol-formaldehyde resin was adjusted to neutral pH by washing it with the solution to be used as the enzyme solvent. This solution must have a low ionic strength (below 0.5 molarity) to enable neutral as well as alkaline proteases to be adsorbed. The pH should be between pH 6 and 7.5.

The maximum concentration of protein in the solution should be 2.5% with the best results when it is less than 1%. This corresponds to an enzyme level of about 50,000 enzyme units per milliliter.

After adsorption, the resin column is washed at neutral pH or neutral ionic strength so that the adsorbed protein is not removed while unadsorbed material is removed for subsequent treatment in the next cycle. Then the adsorbed enzymes are eluted with a 0.05 to 0.2M solution to remove the neutral proteases, followed by elution with a 0.2 to 1.0M solution to remove the alkaline proteases. These solutions are usually 0.1M phosphate buffer followed by 1.0M sodium chloride. The elution pH is usually slightly higher than the pH of the feed stream to avoid contamination of the product with any adsorbed color body impurities.

When DEAE cellulose ion exchange resins have been used, as was the case in the purification of dextranase (153), the process is much different. The filtrate, with its pH adjusted to 8.5, is mixed with 2 weight percent of DEAE cellulose which had been pre-equilibrated by being allowed to stand in 0.02M "Tris" buffer. This suspension is stirred for a minimum of one hour. Then the mixture is filtered and the filter cake is washed with the "Tris" buffer.

The washed filter cake is resuspended in a solution containing 0.3 mole of sodium chloride and 0.05 mole of "Tris" buffer to maintain the pH at 8.5. After stirring for one hour to assure complete elution of the dextranase enzyme, the DEAE cellulose is removed by filtration. The filter cake this time is washed with the same sodium chloride-"Tris" buffer solution. After precipitation with ammonium sulfate, a very pure dextranase (about 6,000 units per mg) is recovered from the filtrate.

In the isolation of a glucopyranose amino-sugar (154), strong acid cation resins and anion resins are added to the fermentation broth. After mixing for about one hour, the resin-amino sugar complexes are separated from the liquid and mycelium using a sieve screw centrifuge. After rinsing with deionized water, a dilute salt solution is used to elute the glucopyranose. The eluate is deionized by passing it through a strong cation and weak base resin. Separation into specific derivatives of the glucopyranose is carried out with column chromatography. These procedures result in the recovery of over 90% of the saccharase inhibitor activity in the original fermentation broth.

The amastatin tetrapeptides, which inhibit the activity of aminopeptidase A, have been isolated using a combination of adsorption, ion exchange and chromatography (155). The process is illustrated in Figure 3.81. After each step in the process, the eluate was concentrated under vacuum to a dry powder. Such drying would not necessarily be employed in a commercial process.

Kiselev (156) has described a preparative method for the isolation of pure fractions of di- and tri-phosphoionositides from ox brain. The petroleum ether lipid extract was passed through a Dowex 50 W (H^+) resin column to remove the divalent metal ions. Then sodium hydroxide in methanol solution was added to the effluent to convert the lipids to the sodium salt. The resulting solution was added to a DEAE-cellulose column. Gradient elution with 0-0.6M ammonium acetate in chloroform/methanol/water (20:9:1) allowed the separation of the lipids into fractions of di- and triphosphoinositides. The desired salt forms of the lipids were obtained by passing the ammonium salts through Dowex 50 W (H^+) and neutralizing with the appropriate base in methanol solution. One kg of wet ox brain tissue yields about 0.35 mmol of diphosphoinositide and 0.63 mmol of triphosphoinositide. Table 3.26 shows the phospholipid concentration as the ox brain extract progresses through the preparative purification.

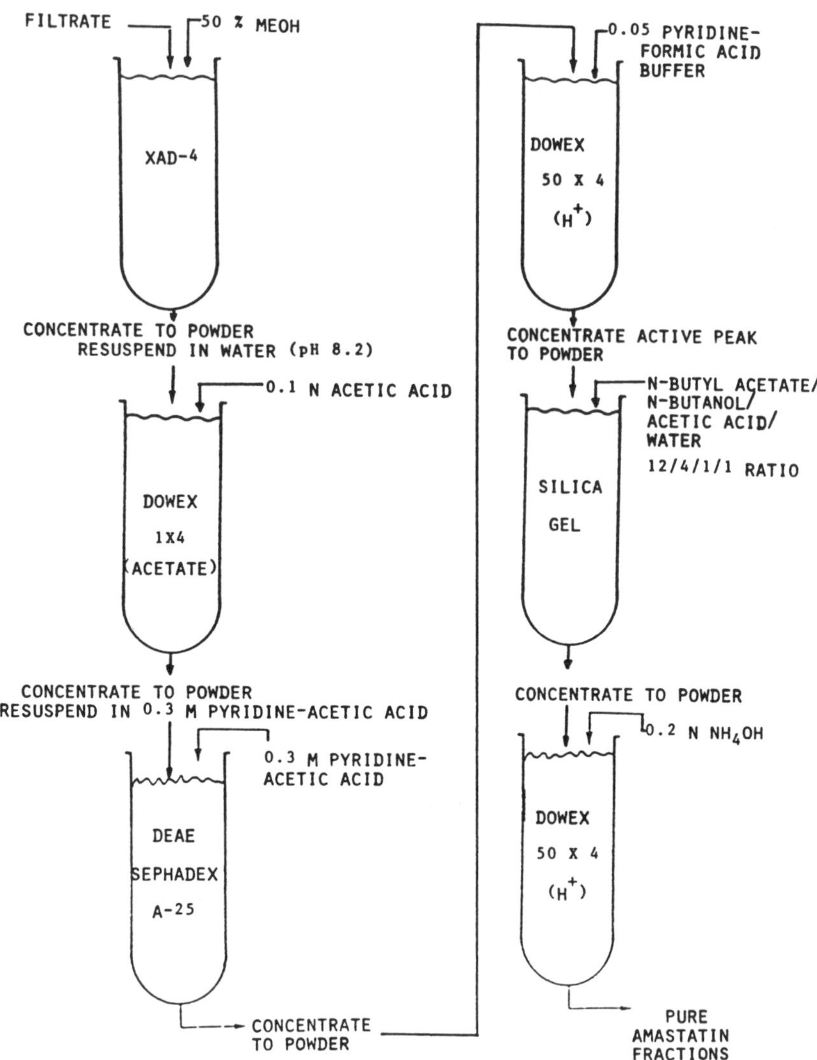

Figure 3.81. Purification scheme for preparing pure amastatin fractions (Reference 155).

Table 3.26: Phospholipid Yield (mmol/kg) from Wet Ox Brain Tissue (156)

Stage	Phosphoinositides		
	Tri-	Di-	Mono-
In Petroleum Ether Extract	0.84	0.43	1.68
After Dowex 50 Treatment	0.67	0.37	1.32
After DEAE Chromatography	0.63	0.35	1.29

The waste streams from food processors have been treated successfully with ion exchange resins to recover proteins (157). As Table 3.27 shows, the efficiency of protein removal is very good.

Table 3.27: Protein Recovery from Waste Streams (157)

Waste Stream	Protein Nitrogen (g/ℓ)		% Protein Removed
	Before	After	
Cheese Whey Effluent	0.60	0.02	93.3
Potato Effluent	0.78	0.035	95.5
Distillery Effluent	0.48	0	100.0
Animal By-Product	0.52	0.036	93.0

While whey has been treated by such simple operations as spray drying to concentrate the proteins, such processes unfortunately include the ash and lactose in the concentrate. Other processes, shown in Figure 3.82, have been devised to recover higher value products from cheese whey. As can be seen, the protein concentrates from these systems have ash and lactose contents which destroy their functional properties. These functional properties are foaming ability, heat gelation, water binding and fat emulsion.

A technique for the recovery of proteins from food processing operations has been developed by Bioisolates (158) which retains the functional properties of the proteins. As an example, the protein in the whey from milk is adsorbed onto a DEAE cellulosic ion exchange resin in an ion exchange bed with a sieve bottom, shown in Figure 3.83. The protein is eluted using 0.1N HCl to yield a dilute solution containing 1 to 5% protein (W/V). Further concentration using ultrafiltration provides solutions with 10 to 30% proteins (W/V) which can be dried to a product that contains less than 3% residual salts.

More recently another system has been developed (159) which allows the separation of cheese whey into its individual components. After removing the colloidal casein proteins by ultrafiltration, the clarified liquid is passed through a large pore (500 Å diameter), weak base resin to

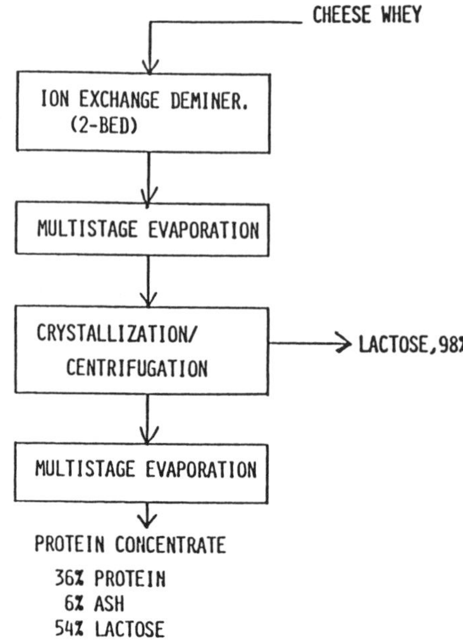

Figure 3.82. Purification scheme for proteins in cheese whey waste.

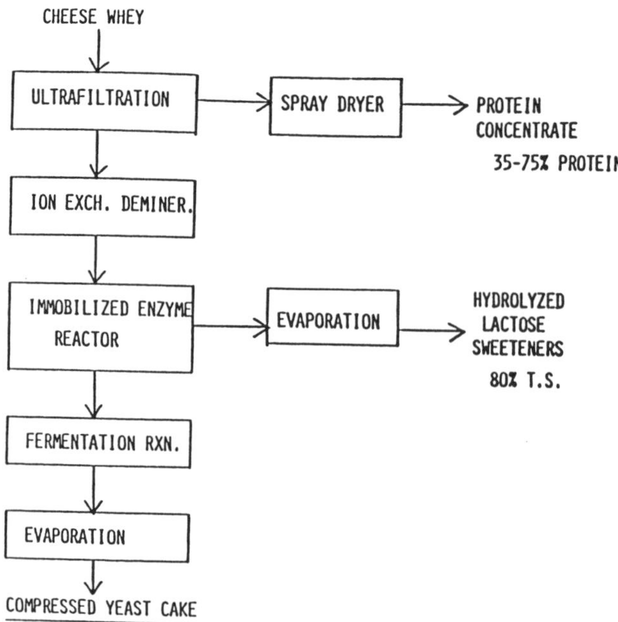

Figure 3.83. Alternate purification scheme for proteins in cheese whey waste (Reference 158).

adsorb about 90% of the soluble whey proteins. These are eluted from the resin using a dilute aqueous HCl solution. In these systems it is essential that the molarity of the eluting acid be kept low to prevent the irreversible binding of the proteins onto the resin.

3.8.5 Sugar Stream Processing

The largest use, on a volume basis, of ion exchange resins in the purification of biochemicals or organic feedstreams is the purification of sugar-bearing juices, liquors and syrups. The sugar purification operations may be either decolorization or demineralization processes.

3.8.5.1 Decolorization: Although the decolorization of liquid sugars almost always involves adsorption on some form of carbon, ion exchange resins are widely used in conjunction with these carbon beds. Anion exchange resins are particularly effective since most of the color impurities are organic substances with anionic characteristics. The resins are normally in the chloride form initially and the colored organic species are usually eluted from the resins using neutral brine solutions.

Figure 3.84 (160) shows the degree of colored impurity removal from cane and beet sugar solutions for anion resins with different degrees of crosslinking. As the degree of crosslinking is reduced, the amount of sugar which can be treated increases. It is also apparent from this figure that the efficiency of decolorization depends very much on the source of the sugar. This is shown for cane sugar from different locations in Figure 3.85 (161).

When strong base anion exchange resins (chloride form) are used for decolorization, the flow rate is typically 8 to 12 kL/hr for a space velocity between 2 and 4/hr. Approximately 80 to 85% of the color impurities are removed with units sized to treat 60 liters of fluid per liter of resin over a 12 or 24 hour loading cycle (162). The temperature is normally between 70°C and 80°C to decrease the viscosity of these 60°Brix solution. The pH must be maintained neutral in processing sucrose solutions since basic pH will cause the formation of undesirable colored impurities and acid pH will cause the inversion of sucrose to glucose and fructose.

If the strong base resin is regenerated to the chloride

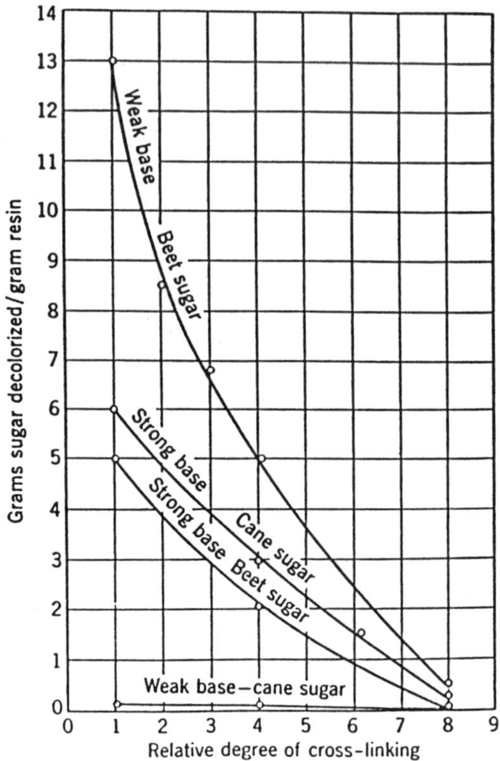

Figure 3.84. Relationship between the resin's degree of crosslinking and its decolorizing power for sugars from different sources (Reference 160).

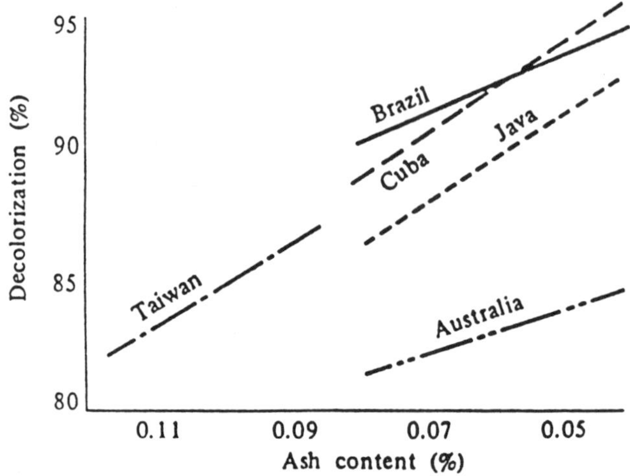

Figure 3.85. Ash content and degree of decolorization using an anion resin in the chloride form on sugars from different geographic regions (Reference 161).

form and then used to treat sucrose solutions, the pH of the solution will decrease, as shown in Table 3.28 (163), which will lead to inversion of some of the sucrose. This occurs because of the H_2CO_3 present in the sucrose solution after the carbonation step in the refining process, according to equation:

$$R\text{-}Cl + H^+ + HCO_3^- \longrightarrow R\text{-}HCO_3 + H^+ + Cl^- \quad (3.58)$$

The recommended method of preventing this pH drop is to pass a 0.1% soda ash solution (1.0 to 1.5 kg $CaCO_3/m^3$ resin) through the resin bed after the regeneration with NaCl solution.

Table 3.28: pH Drop of Sucrose Solutions Caused by Passing Through a Resin Bed (163)

Effluent (Bed Volumes)	pH Case 1	pH Case 2	pH Case 3
Original	6.7	8.05	5.3
10	4.9	5.4	4.0
20	4.9	5.25	3.9
30	4.9	5.25	3.85

It was shown (164) that the Type I anion resins have greater decolorizing ability then Type II resins. This is shown in Table 3.29. It is equally important that the regeneration efficiency shows the same preference for Type I resins over Type II resins even after twenty cycles, although there is a significant reduction in the about of adsorption after just ten cycles, as shown in Table 3.30.

Table 3.29: Decolorization with Type I and Type II Anion at Cycle 1 (164)

Resin	Type of Resin	Weight of Color In Influent	Weight of Color In Effluent	Weight of Color Adsorbed	Ratio of Decolor-ization
Amberlite IRA 401	Type I	65.93	6.47	59.46	90.19
Duolite A-101	Type I	65.12	8.28	56.84	87.29
Duolite	Type I	63.06	5.20	57.86	91.75
Duolite	Type II	62.89	8.97	53.92	85.74
Duolite A-102D	Type II	64.40	12.12	54.28	84.29

Table 3.30: Decolorization with Type I and Type II Anion Resins at Cycle 1, 10 and 20 (164)

Resin	Type of Resin	Cycle 1		Cycle 10		Cycle 20	
		Weight of Color Adsorbed	Ratio of Decolor-ization	Weight of Color Adsorbed	Ratio of Decolor-ization	Weight of Color Adsorbed	Ratio of Decolor-ization
Amberlite IRA 401	Type I	59.46	90.19	686.43	87.93	1515	85.6
Duolite A-101	Type I	56.84	87.29	669.9	86.19	1453	82.9
Duolite A-101D	Type I	57.86	91.75	688.8	88.46	1496	83.8
Duolite A-102	Type II	53.92	85.74	608.6	79.40	1308	75.1
Duolite A-102D	Type II	54.28	84.29	616.1	80.93	1333	75.8

Furukawa and Iizuka (165) have made an economic analysis of the cost of decolorizing sugar syrups with anion exchange resins. Other sugar decolorization studies using anion exchange resins have been reviewed by Parker (166) and Pollio and McGarvey (167).

3.8.5.2 Demineralization: In the beet and corn sugar industries, some demineralization of the sugar solution is normally required during the refining operation. In these cases, the sugar solutions are usually passed through a cation exchange resin bed in the hydrogen form followed by a weak anion exchange bed. The cation resin will exchange the cations in solution, primarily Ca^{++} and Na^{+}, with H^{+}, forming acids of the mineral salts. The weak base resin functions both to remove this acidity generated by the cation exchange resin and to adsorb colored organic species from the sugar solution.

For sugar beets, a thin sugar solution (<5%) is extracted from the beets in horizontal cylindrical diffusion units in which the beet slices circulate in an opposite direction to the hot water used to extract the sugar. This thin extract is treated with lime in a process known as carbonation. This results in a sugar solution that is about 90% pure sugar. However, the 10% impurities must also be removed since they prevent the crystallization of an equivalent amount of sugar.

The overall process is shown in Figure 3.86 for softening and demineralization of beet sugar. In the "softening" of sugar juice with ion exchange resins, the "hardening" salts are removed by conversion of Ca^{++} and Mg^{++} salts into Na^{+} salts, thus preventing the incrustation of

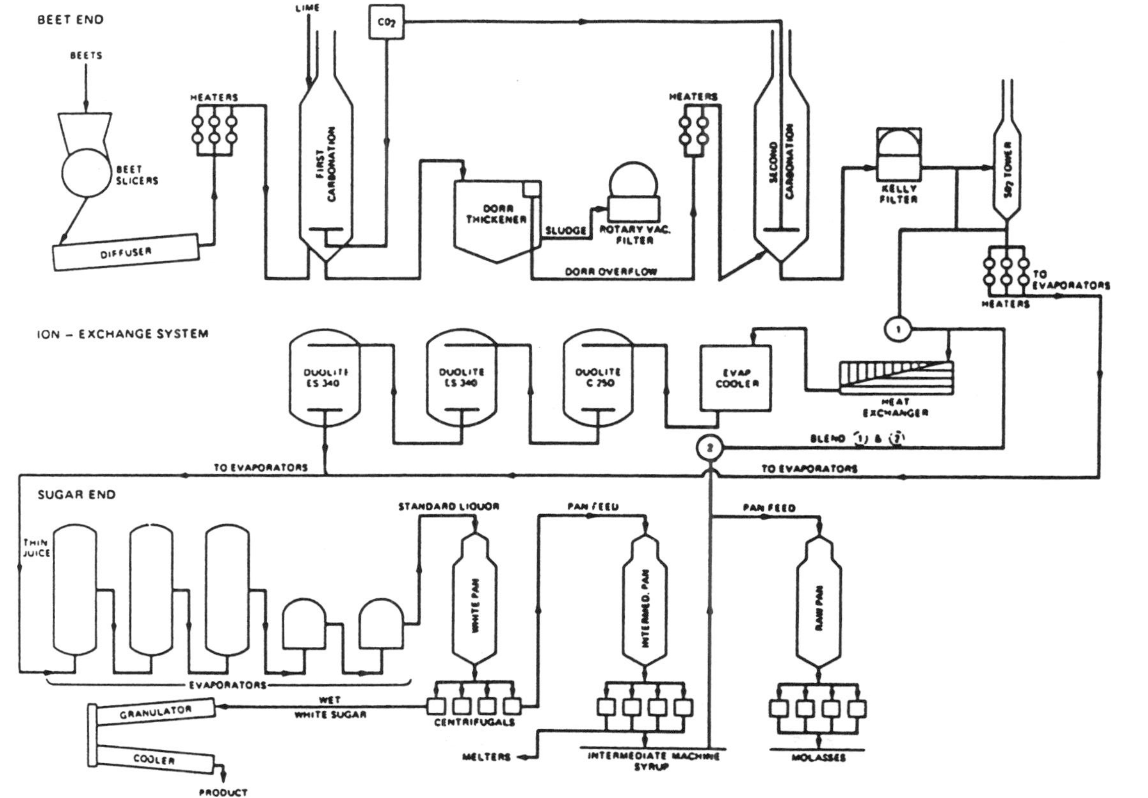

Figure 3.86. Simplified flow diagram of a beet sugar factory from a Duolite ion exchange resin brochure.

the evaporators and pipelines. In the demineralization of sugar juices, the occurrence of molasses is reduced by removal of salts from the clarified sugar juice, thereby increasing the sugar yield.

The starch from corn can be converted by acid or enzyme hydrolysis into glucose syrup (dextrose content between 30 and 68%) and into dextrose syrup (dextrose content over 96%). The dextrose can be isomerized with the enzyme glucose isomerase to produce a mixture of dextrose and fructose. The product, known as high fructose corn syrup (HFCS), contains 42% fructose and 55% dextrose.

It is necessary to have an ionically controlled environment for the optimum productivity of the glucose isomerase enzyme. Therefore, strong acid cation resin column and weak base anion resin columns are used to demineralize the dextrose solution prior to its passage through the enzyme reactor. This is shown in Figure 3.87. As the dextrose is introduced to the enzyme, the necessary magnesium sulfate cofactor, 3 millimolar, is added.

The feedstream is at a 30 to 50% dextrose concentration. This means that the viscous (3-6 cp) feedstream must travel at a relatively slow rate (0.5 to 1.0 $L/s\text{-}m^3$) to have maximum utilization of the ion exchange resin. This flow sensitivity is due, no doubt, to the slower diffusion rate even though macroporous resins are utilized. The operating temperature is normally about 50°C during the passage through the ion exchange columns and 60°C during the enzyme reaction. Using higher operating temperatures in the resin columns, particularly once the fructose is present, causes the formation of colored impurities.

The four bed resin system (2 cation and 2 anion) allows more complete utilization of the ion exchange capacity of each column. The lead cation and anion columns perform the bulk of the purification, while the second cation and anion act as "polishing" units to prevent the leakage of even minor amounts of ions of color impurities.

After a column is exhausted, it is necessary to remove the syrup solution from the column during a "sweetening-off" portion of the operating cycle. This step not only allows the recovery of the syrup, it also allows a gradual change in the osmotic strength of the solution. The syrup recovered during the "sweetening-off" may be used during the "sweetening-on" portion of the next cycle, which occurs just before the full-strength syrup feedstream is introduced.

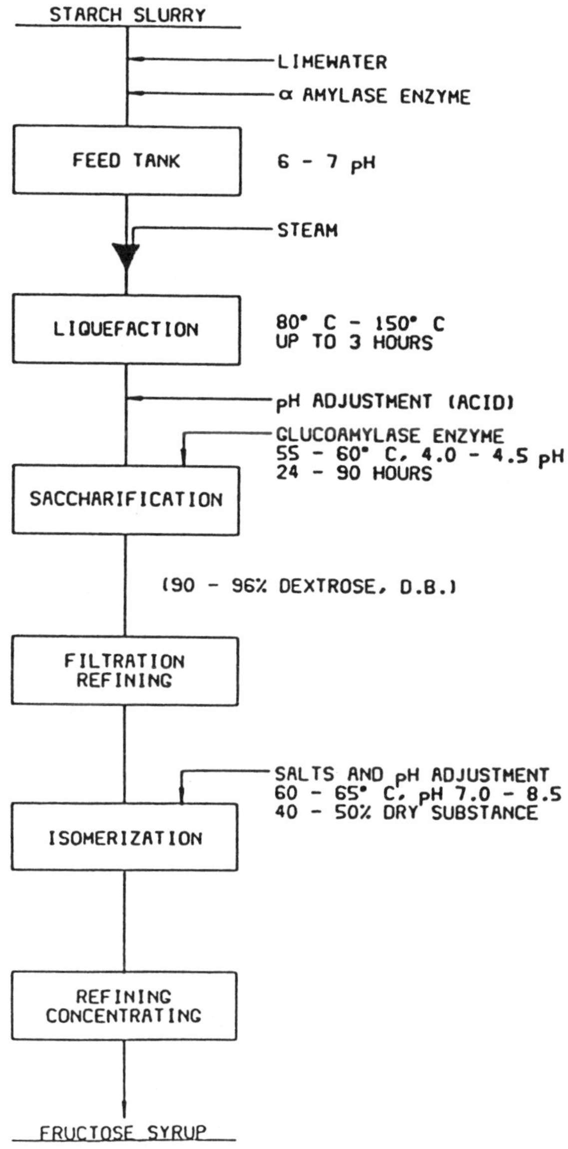

Figure 3.87. Generalized process scheme for corn starch liquefaction, conversion to dextrose (saccharification) and conversion to fructose syrup (isomerization).

The cation resin is normally regenerated with HCl solutions since it is much more efficient than H_2SO_4 as seen in Figure 3.88. The amount of regenerant required is typically 0.25 kg/L of resin. The anion resin may be regenerated with caustic, soda ash or ammonia. Table 3.31 shows how the operating capacity of a typical weak base resin depends on the regenerant used.

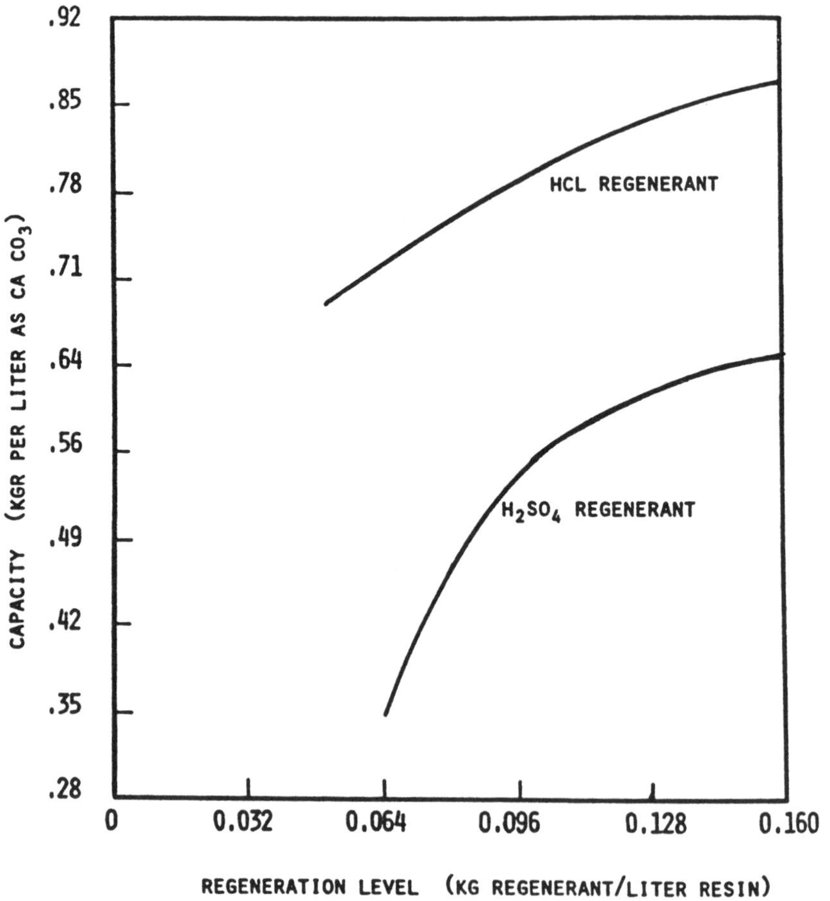

Figure 3.88. Operating capacity of a strong acid cation resin (Dowex 88) as a function of regeneration level for different regenerants.

Table 3.31: Operating Capacity of Dowex 66 Resin

Influent		Flow	Regenerant	Temper-	Capacity	
Exhaustant & Solids Level	Specific Acids	Rate (gpm/ft^2)	Dosage (lb/ft^3)	ature (°F)	(kgr/ft^3)	(meq/ml)
	(1,000 ppm)		NaOH			
Water	HCl/H_2SO_4	4	5	78	25.2	1.15
	HCl/H_2SO_4	6	5	78	24.1	1.10
	(250 ppm)		NaOH			
	HCl/H_2SO_4	4	5	100	27.4	1.26
	HCl/H_2SO_4	4	5	120	32.2	1.47
Dextrose Fructose (35% Dissolved Solids)	HCl/H_2SO_4	4	3	120	29.2	1.34
	HCl/H_2SO_4	4	5	140	35.5	1.63
	HCl/H_2SO_4	2	5	120	33.8	1.54
	HCl/H_2SO_4	6	5	120	24.0	1.10
	HCl	4	5	120	31.5	1.44
	H_2SO_4	4	5	120	32.1	1.47
	(250 ppm)		NH_4OH			
	HCl/H_2SO_4	4	4.35	120	26.6	1.22
	(250 ppm)		Na_2CO_3			
	HCl/H_2SO_4	4	6.6	120	27.7	1.27

3.8.6 Vitamins

When vitamins are obtained from natural sources or by fermentation, recovery processes using ion exchange resins have been found to be commercially desirable. The most notable example is vitamin B-12.

The isolation of vitamin B-12 using a carboxylic acid ion exchange resin was first described in 1953 (168). The adsorption proceeds best at pH 3 to 6. After the resin has been loaded with the vitamin, the resin-vitamin complex is washed with a 0.1N HCl solution. The vitamin may then be eluted using an aqueous organic solution. Since vitamin B-12 is a nonionic compound, the specific adsorption interactions with the carboxylic acid resin do not involve the exchange of ions. It is probably for this reason that less than 50 micrograms are adsorbed per gram of resin.

Recently porous phenol-formaldehyde cation resins have been developed (146) which have a significantly greater adsorption capacity for vitamin B-12. This is shown in Table 3.32 as a function of flowrate and in Figure 3.89 as a function of the initial concentration of vitamin B-12 in the feedstream. The elution was performed at 5 bed volumes per hour using a 1.0M HCl-60% aqueous acetone solution.

Table 3.32: Effect of Flow Rate on the Uptake of Vitamin B-12 (146)

Flow Rate (cm^3/h)	Volume Treated Until Breakthrough (Bed Volume)	Net Loading (mg/g resin)	Recovery (%)
50	349	1.72	86.14
100	313	1.54	85.43
150	280	1.38	87.24
200	240	1.18	88.30

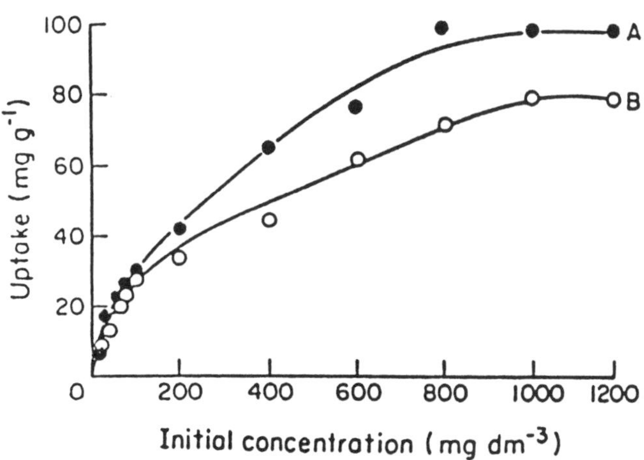

Figure 3.89. Dependence of the uptake of Vitamin B-12 on the initial concentration of the equilibrated solution on Duolite S-761 (Curve A) and on a new synthetic resin (Curve B) (Reference 146).

3.8.7 Wine Treatment

All wines contain naturally occurring potassium bitartrate, which is soluble in the original fermentation liquid but which precipitates as the wine develops (169). Classically, wines were matured for long periods and carefully decanted by the user to remove the precipitate. The greater consumption and the development of massive wine production facilities has led to the use of ion exchange techniques to overcome this problem. As Figure 3.90 shows, the wine was passed through a strong cation resin in the sodium form which converts the potassium bitartrate into the more soluble sodium hydrogen tartrate.

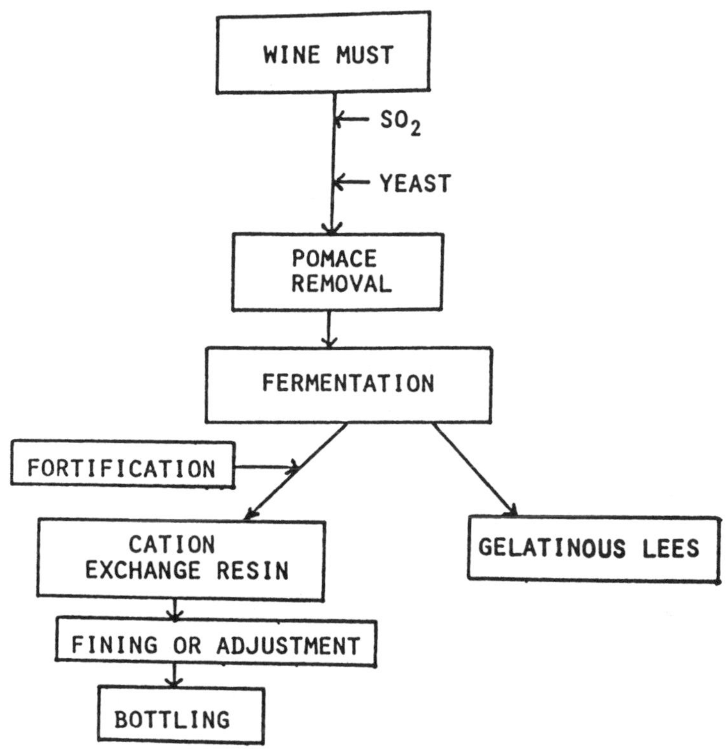

Figure 3.90. Generalized process scheme for the treatment of wine.

With the concern about sodium in prepared foods, alternate techniques have been developed for reducing the amount of potassium while, at the same time, insuring that the acidity of the wine is at the proper balance. Such process schemes are shown in Figure 3.91 (170-172).

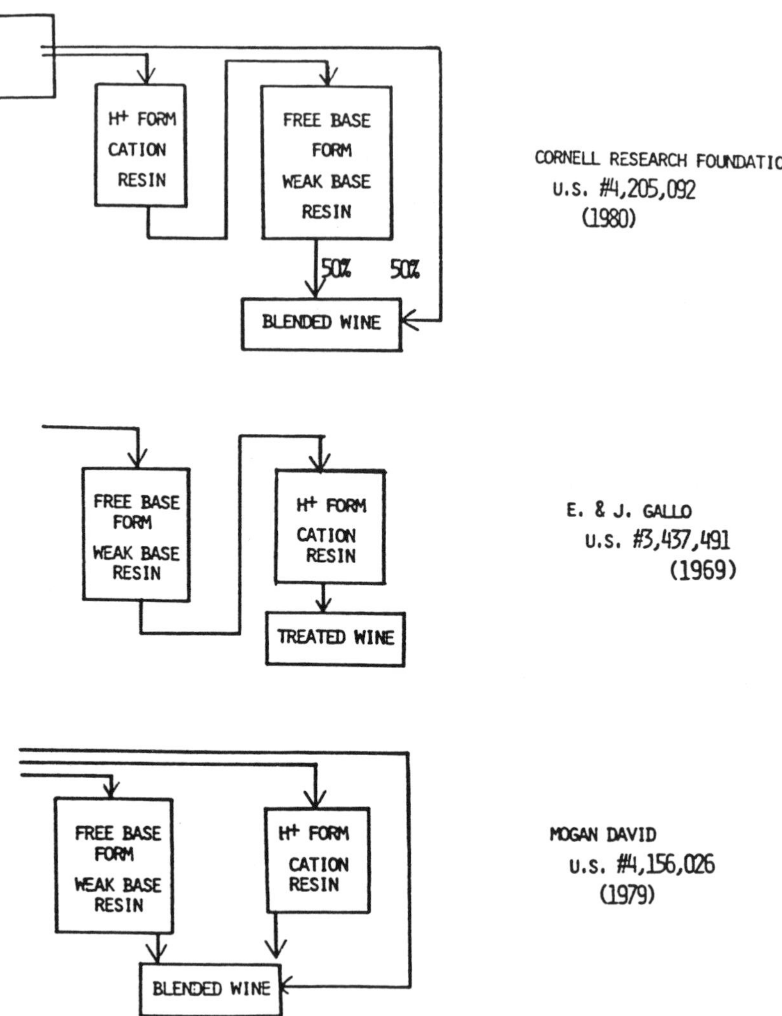

Figure 3.91. Process schemes for acid and tartrate adjustment of wine (References 170-172).

White wines produced during a poor growing season with inadequate sunshine tend to have an excess of acid which detracts from the taste of the wines (169). The acidity may be reduced to a lower level by passing the wine through a weak base resin in the free base form. Since the natural color and flavor ingredients are anionic in nature, the resin should have low porosity and be highly crosslinked to avoid removing these ingredients.

3.8.8 Enzyme Immobilization

It was in the 1950's that ion exchange resins began to be used seriously as carriers for immobilization of enzymes (173,174). These initial efforts resulted in immobilized enzymes that had very low activities of no commercial use. The results were somewhat better (175) for enzymes immobilized on cellulose derivatives. The macroporous resins of the 1970's allowed them to be used in commercial immobilized enzyme processes (176). The bonding specificity of the affinity chromatography type of resin with an enzyme has been employed to immobilize enzymes (177). A review of the use of these immobilized enzymes in the food industry has been prepared by Kilara and Shahani (178).

The advantages of immobilized enzymes over soluble enzymes are (179):

(1) Continuous processes become practical.

(2) The stability of the enzyme may be improved.

(3) A product of a higher purity may be possible.

(4) The sensitivity of the enzyme to changes in temperature or pH may be decreased.

(5) The enzyme activity may be increased.

(6) Effluent problems and material handing problems (separations) for feed and effluent are reduced.

Table 3.33 shows the features of a commercial process, mentioned in the same article, for the production of L-amino acids using an immobilized aminoacylase enzyme. More recently (180), the aminoacylase enzyme has been immobilized on a porous strong base resin of trimethylammonium-introduced porous silica with a pore size of about 1000 Å and a surface area of about 25 m^2/g. Additional examples of enzymes that have been immobilized on different carriers are given in Table 3.34 (181).

Tschang and Klefenz (182) have developed a series of macroporous crosslinked polystyrene resins which contain isocyanate, thioisocyanate or aldehyde functionalities for covalently binding proteins. These resins have capacities between 3.4 and 4.1 meq/g. The amount of enzyme bound and its activity are shown in Table 3.35.

Table 3.33: Commercial Process for the Production of L-Amino Acids (179)

Column capacity	:	1000 liter
Carrier	:	DEAE Sephadex
Amount of acylase bound	:	333 I.U./ml
Activity of bound enzyme	:	157 I.U./ml
Yield of enzyme activity	:	47%
Operating temperature	:	50°C
Operating pH	:	7.0
Activity loss	:	40% in 35 days
Column regeneration	:	once in 35 days
L-Methionine yield	:	715 kg/day @ SV = 2.0

Table 3.34: Commercially Available Immobilized Enzyme Technology (181)

Enzyme	Carrier	Level of Development
Glucose isomerase	DEAE cellulose, Alumina, Dowex MWA-1	Industrial
Glucoamylase	Dowex 1 x 10	Pilot
Lactase	DEAE cellulose, Alumina	Industrial
Invertase	Silica Gel, Alumina	Pilot Scale
Penicillin amidase	Polyacrylamide Gel	Industrial
Aminoacylase	Alumina, Diaion SA-11A	Industrial
Aspartase	Polyacrylamide Gel	Industrial
Fumarase	DEAE Cellulose	Pilot
Hydantoinase	DEAE Cellulose	Industrial

Table 3.35: Resin-Bound Enzymes and Activities (182)

Binding Functionality	Hydrophilic Group	Enzyme	Amount of Bound Enzyme (mg/g)	Activity
hexane-1,6-diamine/ glutarodialdehyde	$-SO_3Na(H)$	β-fructosidase	142	90%
hydrazine/hexane- 1,6-diisocyanate	$-SO_3Na(H)$	β-fructosidase	16.5	90%
hexane-1,6-diamine/ glutarodialdehyde	$-SO_2N(Prop)_2$	β-fructosidase	63	80%
hexane-1,6-diamine/ thiophosgene	$-SO_2NH(CH_2)_6NH_2$	β-fructosidase	--	80%
ethylene diamine/ toluylene diisocyanate	$-SO_2N(Prop)_2$	β-fructosidase	31.9	62%
4,9-dioxadodecane- 1,12-diamine/ glutarodialdehyde	$-SO_2NH(But)$	β-fructosidase	12.2	58%

Roy and Roy (183) have synthesized protein-compatible weak anion silica exchangers. These resins have been used to recover over 90% of the albumin and gamma globulin consistently with virtually no non-specific adsorption. However, even if kept in acidic media (pH 4 to 7), the useful life of the resin is only about three months.

Delin and coworkers (177) described an illustrative procedure for purifying and immobilizing penicillin acylase which could then be used to produce hypoallergenic penicillins from other penicillins. The penicillin acylase was purified by adsorption on a porous cation exchange resin from a solution adjusted to pH 4.0 - 5.0. The purified enzyme is eluted with 0.2M ammonium acetate buffer solution (pH 6.0 to 8.0). Sephadex G200 was treated with 5N NaOH and then reacted with cyanogen chloride at a cold temperature (0-3°C). After the reaction was complete, the resin was washed with ice water and a 0.1M borax solution. The wet polymer was added to the solution of purified penicillin acylase, more borax was added and the mixture stirred slowly for 24 hours to complete the enzyme immobilization.

Immobilizations such as this, which are due to the formation of a covalent bond, are more complex than the simple adsorption or ionic bonds which may be formed by simply passing the enzyme solution through the prepared ion exchange columns.

3.8.9 Catalytic Applications

As an early study showed (184), ion exchange resins can be used as catalysts for a variety of organic reactions including esterification, acetal synthesis, ester alcoholysis, acetal alcoholysis, alcohol dehydration, ester hydrolysis and sucrose inversion. Sucrose inversion was examined in detail by Bodamer and Kunin (185). They showed the effect of temperature, resin particle size, degree of crosslinking and resin type. The strong acid (sulfonic) functionality was much more effective than the weak (carboxylic) functionality. The rate of inversion increases with decreasing particle size, increasing porosity (lower crosslinking) and increasing temperature. These results are shown in Table 3.36.

Table 3.36: Effect of Particle Size, Porosity and Temperature on Sucrose Inversion (185)

Parameter	Value	Rate of Inversion ($k \times 10^4$)
Particle Size	0.24 mm	34.7
	0.45 mm	26.3
	0.63 mm	15.7
Degree of Crosslinking (Porosity)	1%	199.2
	4%	110.3
	10%	26.3
	15%	3.0
	20%	0.7
Temperature	25°C	0.7
	50°C	26.3
	75°C	117.0

A more recent study (186) has taken another look at the effect of these same parameters, the contact time of the sugar solution with the resin and the hydrogen ion concentration of the resin on sucrose inversion. The longer contact time and the higher hydrogen ion concentration caused greater rates of sucrose inversion. A temperature increase of only 5°C doubled the inversion rate. As the crosslinking was increased (porosity decreased), the rate of inversion decreased.

A similar catalytic inversion of lactose from whey into glucose and galactose has been demonstrated (187). The lactose solution was processed at a temperature of 95°C and a flow rate of 0.27 bed volumes/hour to give 80 to 90% conversion. It was necessary to remove the protein and ash from the lactose solution before processing to insure continued catalytic action by the cation resin. The concentration of lactose in the feedstream was determined to have a technical and economic optimum at 100 g/L. A column operation has also been developed by Holmberg (188).

Takahashi (189) combined 0.4% invertase enzyme with a strong acid cation resin to invert at 60°C a 50°Brix molasses solution. The contact time required was five hours. There was a 90% recovery of the sugars with 100% conversion to glucose and fructose. The color quality, however, was still that of raw sugar.

A strong resin has also been used to convert fructose into 5-hydroxymethyl-2-furancarboxaldehyde (190). This reaction required the presence of an organic solvent such as methyl isobutyl ketone and temperature between 80° and 90°C. The degrees of conversion for specific resins and times of reaction are shown in Table 3.37.

Table 3.37: Synthesis of 5-Hydroxymethyl-2-Furancarboxaldehyde from Fructose with Cation Resins (190)

Ion Exchange Resin	Reaction Time (hour)	Conversion Rates (%) [a]	Yield (%) [b]
Lewatit SC102	4	34	47
Amberlite IR118	4	35	58
Duolite C26	4	25	54
Amberlite A200C	4	13	42
Amberlyst A15	4	14	30
Lewatit SPC118	4	11	38
Lewatit SPC118	15	56	66
Lewatit SPC118	24	46	51

[a] The conversion rate is the ratio of the number of moles of obtained product to the number of moles of fructose introduced.

[b] The yield is the ratio of the number of moles of AMF produced to the number of moles of fructose consumed.

APPENDIX

Ion exchange resins are prepared by a number of manufacturers. During recent years there has been some consolidation among the producers of synthetic ion exchange resins. Rohm and Haas has purchased Duolite resins from Diamond Shamrock and Dow has purchased the Montecatini-Edison resin operation. These consolidations have caused some changes in designations for the resins now commercially available. The following tables include historical as well as current designations so that past literature references to specific commercial resins might be understood. It is recommended that the reader contact the resin manufacturer to learn about currently commercially available resins and their properties.

Table 3A.1: Ion Exchange Resins from U.S. Manufacturers

Cation Exchange Resins			
Moisture Content (%)	Percent DVB	Screen Mesh	Trade Name
I. Strong Acid Type			
a. Polystyrene, gel, $-SO_3H$			
75–85	2	20–50	Dowex 50WX2, BioRad AG-40WX2
		50–100	Dowex 50WX2, BioRad AG-50WX2
		80–200	Dowex 50WX2, BioRad AG-50WX2
65–75	4	20–50	Dowex 50X4, BioRad AG-50WX4
		40–80	Dowex 50X4, BioRad AG-50WX4
		60–140	Dowex 50X4, BioRad 50WX4
		100–200	Dowex 50X4, BioRad AG-50WX4
		200–400	Dowex 50X4, BioRad AG-50WX4
55–60	6	16–50	Duolite C-25D, Amberlite XE-100

(continued)

Table 3A.1: (continued)

Moisture Content (%)	Percent DVB	Screen Mesh	Trade Name
50-55	8	20-50	Dowex 50WX8, Dowex HCR AG-50WX8, BioRex RG-50WX8
50-55	8	40-80	BioRad AG 50WX8
		60-140	Dowex 50WX8, BioRad AG-50WX8
		100-230	Dowex 50WX8, BioRad AG-50WX8
		230-400	BioRad AG-50WX8
45-50	10	20-50	Dowex HGR, Dowex 50WX10, Amberlite IR-120, Amberlite IR-121, Amberlite IR-122
		100-500	BioRad AG-50WX10, Ionac C-240, Ionac C-249, Ionac C-253, Ionac C-257, Ionac Cl-295, Amberlite IR-120
		<325	Amberlite IR-120
40-45	12	20-50	BioRad AG-50WX12
		40-80	BioRad AG-50WX12
		60-140	Dowex 50WX12, BioRad AG-50WX12
		140-170	Dowex 50WX12, BioRad AG-50WX12
		230-400	BioRad AG-50WX12
35-40	16	20-50	Amberlite IR-124, Ionac C-250, Ionac C-258, Ionac C-255, Dowex 50WX16
b. Polystyrene, macroporous $-SO_3H$			
46-50		16-50	Amberlite 200, Dowex 88
			Amberlite 200C, Dowex MSC-1
			Amberlite 252
c. Sulfonated coal			
14-20		16-50	Ionac C-150
d. Cellulose or dextran, $-OC_2H_4SO_3H$			
89-90		120-400	SE-Sephadex C-25
96-98			SE-Sephadex C-50, Cellex SE

(continued)

Table 3A.1: (continued)

Moisture Content (%)	Percent DVB	Screen Mesh	Trade Name
e. Phenolic, $-CH_2SO_3H$			
—		20–50	BioRex 40, Duolite C-1
—		50–100	BioRex 40
—		100–200	BioRex 40
—		200–400	BioRex 40
II. Intermediate Acid Strength			
a. Polystyrene, phosphoric granular			
—		68–76	BioRex 63
b. Cellulose, phosphoric			
—		—	Cellex P

Polymer Type	Particle Type	Screen Mesh	Trade Name
III. Weak Acid Type			
a. Methacrylic acid-divinylbenzene, $-COOH$			
pK ~6.0	Spherical	20–50	Amberlite IRC-50
	Granular	100–500	Amberlite IRP-64
	Granular	<325	Amberlite IRP-64M
b. Acrylic acid-divinylbenzene, $-COOH$			
pK ~5	Spherical	20–50	Amberlite IRC-84 Dowex CCR-2, Duolite ES-80, BioRex 70
	Granular	50–100	BioRex 70
		100–200	BioRex 70
		200–400	BioRex 70
		<400	BioRex 70
c. Miscellaneous weak acid resins			
Condensation	Granules	16–50	Ionac C-265
Dextran, $-OCH_2COOH$	Spherical	120–400	Sephadex C-25
		120–400	Sephadex C-50
Cellulose, $-OCH_2COOH$	Fibrous	—	Cellex CM, Cellulose CM-22, Cellulose CM-23
		200–400	Cellulose CM-32, Cellulose CM-52

(continued)

Table 3A.1: (continued)

Cross-Link (%)	Moisture (%)	Particle Type	Screen Mesh	Trade Name
Anion Exchange Resins				
I. Strong Base				
a. Polystyrene—Type I, $CH_2\overset{+}{N}(CH_3)_3$				
2	70-80		20-60	BioRad AG-1X2, Dowex 1X2
			60-140	BioRad AG-1X2, Dowex 1X2
			80-200	BioRad AG-1X2, Dowex 1X2
			200-325	BioRad AG-1X1, Dowex 1X2
4	60-70		20-50	BioRad AG-1X4, Dowex 1X4, Amberlite IRA-900 (macroporous), Amberlite IRA-938-C (large pore), Duolite ES-111, Amberlite A-26 (macroporous), Amberlite IRA-401S
4	60-70	Spherical	40-80	BioRad AG-1X4, Dowex 1X4
			60-140	BioRad AG-1X4, Dowex 1X4
			100-230	BioRad AG-1X4
			230-325	BioRad AG-1X4
—	50-60		16-20	BioRad AG-21K, Dowex 21K, Amberlite IRA-401, Ionac A-540, Ionac A-544, Ionac A-548, Duolite A101D
			20-40	BioRad AG-21K, Dowex 21K, Dowex MSA-1 (macroporous)

(continued)

Table 3A.1: (continued)

Cross-Link (%)	Moisture (%)	Particle Type	Screen Mesh	Trade Name
8	40-50	Spherical	16-50	BioRad AG-1X8, Dowex 1X8, Dowex SBR, Dowex 11, Ionac A-546, Ionac A-000, Amberlite IRA-400, Amberlite IRA-400C
			40-80	BioRad AG-1X8, Dowex 1X8
			80-140	BioRad AG-1X8, Dowex 1X8
		—	140-325	BioRad AG-1X8, Dowex 1X8
		—	<230	BioRad AG-1X8, Dowex 1X8
—	40-50	Granular	100-500	Amberlite IRP-67
			<325	Amberlite IRP-67M
10	35-40	Spherical	40-80	BioRad AG-1X10
			60-140	BioRad AG-1X10
			140-325	BioRad AG-1X10
			<230	BioRad AG-1X10
b. Polystyrene—Type II, $CH_2-\overset{+}{N}(CH_3)_2(CH_2CH_2OH)$				
4	55-60	Spherical	20-50	BioRad AG-2X4, Dowex 2X4, Amberlite IRA-910 (macroporous)
	53-60		20-50	Dowex MSA-2 (macroporous)
	47-50		16-50	Ionac A-550, Ionac A-5
7	40-45		20-50	Amberlite IRA-410
8	34-40		20-50	BioRad AG-2X8, Dowex SAR
			40-80	BioRad AG-2X8
			60-140	BioRad AG-2X8
			140-235	BioRad AG-2X8
10	28-36	Spherical	40-80	BioRad AG-2X10
			60-140	BioRad AG-2X10
			140-235	BioRad AG-2X10

(continued)

Table 3A.1: (continued)

Cross-Link (%)	Moisture (%)	Particle Type	Screen Mesh	Trade Name
c. Polystyrene, $CH_2\overset{+}{N}$(cyclohexyl ring)				
—	46-54	Spherical	16-50	BioRex 9
		Granular	50-100	BioRex 9
			100-200	BioRex 9
			200-325	BioRex 9
—	—	Spherical	16-50	Ionac A-580
			10-20	Ionac A-590
d. Dextran				
—	—	Spherical	140-400	QAE Sephadex A-25
			140-400	QAE Sephadex A-50
II. Weak and Intermediate Base Strength, Quaternary and Ternary				
—	60-66	Spherical	16-50	Duolite A-57
—	50-60	Spherical	20-50	BioRex 5
		Granular	50-100	BioRex 5
			100-200	BioRex 5
			200-400	BioRex 5
—	—		16-50	Ionac A-300, Ionac A-302
III. Weak Base				
a. Polystyrene				
—	50-60	Spherical	20-50	Dowex 66, Dowex MWA-1 (macroporous)
—	46-55	Spherical	20-50	Amberlite IRA-93 (macroporous)
—	40-45	Spherical	20-50	Amberlite IRA-45, Amberlite IRA-21
—	25-35		16-40	BioRad AG-3X4
			60-140	BioRad AG-3X4
			140-325	BioRad AG-3X4
b. Acrylic				
—	57-63	Spherical	20-50	Amberlite IRA-68
c. Phenolic-polyamine				
—	—	Granular	100-500	Amberlite IRP-58
			<325	Amberlite IRP-58M
d. Condensation polymer				
—	—	Granular	16-50	Amberlite IRA-401, Dowex WGR, Dowex WGR-2, Ionac A-260

(continued)

Table 3A.1: (continued)

Cross-Link (%)	Moisture (%)	Particle Type	Screen Mesh	Trade Name
e. Dextran and Cellulose, DEAE–$OCH_2CH_2N(C_2H_5)_2$				
—	—	Spherical	160–400	DEAE Sephadex A-25
			140–400	DEAE Sephadex A-50
—	—	Fibrous	—	Cellex D, Cellulose DE-22, Cellulose DE-23
—	—	Spherical	100–200	Biogel DM-2
			100–200	Biogel DM-30, high cap.
			100–200	Biogel DM-30, low cap.
			100–200	Biogel DM-100, high cap.
			200–400	Cellulose DE-32, Cellulose DE-52

Resin Trade Name	Manufacturer
Dowex	Dow Chemical Co., Midland, MI
Amberlite, Duolite	Rohm and Haas Co., Philadelphia, PA
Ionac	Ionac Chemical Co., Birmingham, NJ
Sephadex	Pharmacia Fine Chemicals, Piscataway, NJ
Cellex, BioRad, BioRex, Biogel	BioRad Labs, Richmond, CA

Table 3A.2: Foreign Producers of Synthetic Ion Exchange Resins

Company	Country	Trade Name
Bayer	Federal Republic of Germany	Lewatit
Chemolimfex	Hungary	Varion
Mitsubishi	Japan	Diaion
Ostion	Czechoslovakia	Ostion
Permutit	United Kingdom	Zeocarb, Deacidite, Zerolit
Permutit, A.G.	Federal Republic of Germany	Orzelith, Permutit
Resindion	Italy	Relite
Wolfen	German Democratic Republic	Wofatit
	USSR	AW-, AV-, KB-, KU-

Table 3A.3: Corresponding Foreign and U.S. Ion Exchange Resins

U.S. Resin	Foreign Resins
Dowex 50 W X8	Diaion SK-1B Kationit KU 2 Lewatit S-100 Varion KS Wofatit KPS 200
Dowex CCR-1	Diaion WK-10 Wofatit CP 300
Dowex MSC-1	Diaion PK-228 Lewatit SP-100
Dowex MWC-1	Diaion WK-20 Lewatit CNP
Dowex SBR	Diaion 10A Lewatit M 500
Dowex SAR	Diaion 20A Lewatit M 600
Dowex MSA-1	Diaion PA 316 Lewatit MP 500 Wofatit SZ-30
Dowex MSA-2	Diaion PA 416 Lewatit MP 600
Dowex MWA-1	Diaion WA-30 Lewatit MP-62

Table 3A.4: Ion Exchange Process Designers and Manufacturers

Company	Location
Aqua Media	Sunnyvale, CA
Bateman Uranium Corporation	Lakewood, CO
Belco Pollution Control Corp.	Parsippany, NJ
Chemical Separations Corp.	Oak Ridge, TN
Chem Nuclear Systems	Columbia, SC
Cochrane Division of Crane Corp.	King of Prussia, PA
Downey Welding & Manufacturing Co.	Downey, CA
Envirex Water Conditioning Div.	Waukeska, WI
Epicor	Linden, NJ
Ermco, Inc.	St. Louis, MO
Graver, Division of Ecodyne	Union, NJ
Himsley	Toronto, Canada
Hittman Nuclear & Development	Columbia, MD

(continued)

Table 3A.4: **(continued)**

Company	Location
Hungerford & Terry, Inc.	Clayton, NJ
Hydro-Max Corporation	Milwaukee, WI
Illinois Water Treatment	Rockford, IL
Industrial Filter and Pump	Cicero, IL
Infilco Degremont, Inc.	Richmond, VA
Intensa	Mexico City, Mexico
Kinetico, Inc.	Newbury, OH
Kurita	Japan
L.A. Water Treatment	City of Industry, CA
Liquitech, Div. of Thermotics, Inc.	Houston, TX
Mannesmann	Federal Republic of Germany
Mitco Water Labs, Inc.	Winter Haven, FL
Morgan Company	Houston, TX
Permutit, Division of Zurn	Paramus, NJ
Rock Valley Water Conditioning	Rockford, IL
Smith, F.F. & Associates	Houston, TX
Technichem, Inc.	Belvidere, IL
United States Filter, Fluid System Corp.	Whittier, CA
Wynhausen Water Softener Company	Los Angeles, CA

3.9 REFERENCES

1. Thompson, H.S., J Roy Agr Soc, Eng, 11:68 (1850)
2. Bersin, T., Naturwissenschaften, 33:108 (1946)
3. Winters, J.C., Kunin, R., Ind Eng Chem, 41:460 (1949)
4. Polis, B.D., Meyerhoff, O., J Biol Chem, 169:389 (1947)
5. Carson, J.F., Maclay, W.D., J Am Chem Soc, 67:1808 (1945)
6. Nagai, S., Murakami, J Soc Chem Ind, Japan, 44:709 (1941)
7. Wieland, T., Berichte, 77:539 (1944)
8. Bergdoll, M.S., Doty, D.M., Ind Eng Chem, Arcl Ed, 18:600 (1946)
9. Lejwa, A., Biochem Z, 256:236 (1939)

10. Kingsburg, A.D., Mindler, A.B., Gilwood, M.B., Chem Eng Prog, 44:497 (1948)

11. Coke Cruz, E., Gonzales, F., Hulsen, W., Science, 101:340 (1945)

12. Jackson, W.G., Whitefield, G., DeVries, W., et al., J Am Chem Soc, 73:337 (1951)

13. Moore, S., Stein, W.H., J Biol Chem, 211:893 (1954)

14. Starobinietz, G.L., Gleim, J.F., Zh Fiz Khim, 39:2189 (1965)

15. Kunin, R., Ion Exchange Resins, Robert E. Krieger Publ. Co., Huntington, NY, p 26, (1972)

16. Wiklander, L., Ann Roy Agr Coll Sweden, 14:1 (1946)

17. Nachod, F.C., Wood, W., J Am Chem Soc, 66:1380 (1944)

18. Boyd, G.E., Schubert, J., Adamson, A.W., J Am Chem Soc, 69:2818 (1947)

19. Harned, H.S., Owen, B.B., The Physical Chemistry of Electrolytic Solutions, Reinhold Publishing Corp, (1943)

20. Gregor, H.P., J Am Chem Soc, 70:1293 (1948)

21. Argersinger, W., Davidson, A., Bonner, O., Trans Kansas Acad Sci, 53:404 (1950)

22. Reichenberg, D., Pepper, K., McCauley, D., J Chem Soc, p 493 (1951)

23. Savitskaya, E.M., Yakhontova, L.F., Nys, P.S., Pure & Appl Chem, 54(11):2169 (1982)

24. Wolf, F.J., Separation Methods in Organic Chemistry and Biochemistry, Academic Press, NY, p 144 (1969)

25. Peterson, S., Ann N Y Acad Sci, 57:144 (1953)

26. Ito, T., Kino Zairyo, 2(5):8 (1983)

27. Savitskaya, E.M., Yakontova, L.F., Nys, P.S., Ion Exchange, Nauka, Moscow, p 229 (1981)

28. Anderson, R.E., A I Ch E Symp Ser, 71(152):236 (1979)

29. Klein, G., A I Ch E Symp Ser, 81(242):28 (1985)

30. NATO Advanced Study Institute on Mass Transfer and Kinetics of Ion Exchange, Maratea, Italy (May 31-July 11, 1982)

31. Helfferich, F.G., Plesset, M.S., J Chem Phys, 28:418 (1958)

32. Cantwell, F.F., Ion Exchange Solvent Extrac, 9:339 (1985)

33. Helfferich, F.G., J Phys Chem, 66:39 (1985)

34. Helfferich, F.G., In: Mass Transfer and Kinetics of Ion Exchange, (Liberti, L., Helfferich, F.G., eds) Martinus Nijhoff Publ., The Hague, p 157 (1983)

35. Schogl, R., Helfferich, F.G., J Chem Phys, 26:5 (1957)

36. Glueckauf, E., Coates, J.I., J Chem Soc (London), 1315 (1947)

37. Helfferich, F.G., J Phys Chem, 69:1178 (1965)

38. Helfferich, F.G., Liberti, L., Petruzzelli, D., et al., Israel J of Chem, 26:3 (1985)

39. Tsai, F.N., J Phys Chem, 86:2339 (1982)

40. Vermeulen, T., Klein, G., Hiester, N.K., In: Perry's Chemical Engineers Handbook, (Perry, R.H., Chilton, C.H., eds) McGraw-Hill, NY, Section 16 (1973)

41. Helfferich, F.G., Ion Exchange, McGraw-Hill, NY, p 255 (1962)

42. Samsonov, G.V., Elkin, G.E., Ion Exchange and Solv Extract, 9:211 (1985)

43. Huang, T.C., Tsai, F.N., Proc Pac Chem Eng Cong 3rd (Seoul), 1:339 (1983)

44. Rosen, J.B., Ind Eng Chem, 46:1590 (1954)

45. Klein, G., In: Mass Transfer and Kinetics of Ion Exchange (Liberti, L., Helfferich, F.G., eds) Martinus Nijhoff Publ., The Hague, p 213 (1983)

46. Mathews, A.P., Weber, W.J., Jr., In: Adsorption and Ion Exchange with Synthetic Zeolites (Flank, W.H., ed) ACS Society, Washington, p 27 (1980)

47. Barba, D., Del Re, G., Foscolo, P.U., The Chem Eng J, 26:33 (1983)

48. Samuelson, O., Ion Exchangers in Analytical Chemistry, John Wiley, NY, p 52 (1953)

49. Friedlander, S.K., A I Ch E J, 3:381 (1957)

50. Marchello, J.M., Davis, M.W., Jr., I & EC Fund, 2(1):27 (1963)

51. Clearfield, A., Nanocollas, G.H., Blessing, R.H., Ion Exchange and Solvent Extraction (Marinsky and Marcus, eds), 5:1 (1973)

52. Millar, J.R., In: Mass Transfer and Kinetics of Ion Exchange (Liberti, L., Helfferich, F.G., eds.) Martinus Nijhoff Publ., The Hague, p 1 (1983)

53. Daniels, S.L., In: Adsorption of Microorganisms to Surfaces (Britton, G., Marshall, K.C., eds.), John Wiley and Sons, NY (1980)

54. Gold, H., Calmon, C., A I Ch E Symp Ser, 76(192):60 (1980)

55. Warshawsky, A., Die Angewandte Makromol Chem, 109/110:171 (1982)

56. Yotsumoto, K., Hinoura, M., Goto, M., Ger Offen 2,718,649 (1977)

57. Warshawsky, A., Fridkin, M., Stern, M., J Polym Sci Chem, 20(6):1469 (1982)

58. Freeman, D.H., Schram, S.B., Anal Chem, 53:1235 (1981)

59. Saldadze, K.M., Brutskus, T.K., Teploenergetika, 23(9):6 (1976)

60. Millar, J.R., Smith, D.G., Merr, W.E., et al., J Chem Soc, p 183 (1963)

61. Mindick, M., Svarz, J., U.S. Patent No 3,549,562 (Dec. 22, 1970)

62. Ikeda, A., Imamura, K., Miyaka, T., et al., U.S. Patent No. 4,093,570 (1978)

63. Miyake, T., Kunikiko, T., A., Ikeda, et al., U.S. Patent No. 4,154,917 (1979)

64. Abe, T., Ikeda, A., Sakurai, T., Jap. Patent No. 79-11,088 (Jan. 26, 1979)

65. Millar, J.R., In: Mass Transfer and Kinetics of Ion Exchange (Liberti, L., Helfferich, F.G., eds.) Martinus Nijhoff Publ., The Hague, p 23 (1983)

66. Messing, R.A., Immobilized Enzymes for Industrial Reactors, Academic Press, NY (1975)

67. Okuda, T., Awataguchi, S., U.S. Patent No. 3,718,742 (Feb. 27, 1973)

68. Crispin, T., Halasz, I., J Chromatogr, 239:351 (1982)

69. Martinola, F., Meyer, A., Ion Exchange and Membranes, 2:111 (1975)

70. Puente, A., Microbiol Espan, 14:209 (1961)

71. Ishida, M., Sugita, Y., Hori, T., et al., U.S. Patent No. 3,565,951 (Feb. 23, 1971)

72. Gordienko, S.V., J Appl Chem USSR, 39:10 (1966)

73. Shirato, S., Miyazaki, Y., Suzuki, I., Fermentation Industry (Japan), 45(1):60 (1967)

74. Wolf, F.J., Separation Methods in Organic Chemistry and Bicohemistry, Academic Press, NY, p 153 (1969)

75. Nagase, T., Hirohare, H., Nabeshima, S., Brit. Patent No. 1.597,436 (Sept. 6, 1981)

76. Commerical inert resins for this application include Dowex Buffer Beads (XFS-43179), Ambersep 359 and Duolite S-3TR.

77. Higgins, I.R., U.S. Patent No. 3,580,842 (May 25, 1971)

78. Himsley, A., Canadian Patent No. 980,467 (Dec. 23, 1975)

79. Brown, H., U.S. Patent No. 3,549,526 (Dec. 22, 1970)

80. Cloete, F.L.D., Streat, M., U.S. Patent No. 3,551,118 (Dec. 29, 1970)

81. Streat, M., J Separ Proc Technol, 1(3):10 (1980)

82. Mallon, C., Richter, M., ZfI-Mitteilungen, Leipzig, 86:39 (1984)

83. Cloete, F.L.D., Streat, M., British Patent No. 1,070,251 (1962)

84. George, D.R., Ross, J.R., Prater, J.D., Min Engin, 1:73 (1968)

85. Buijs, A., Wesselingh, J.A., Polytech Tijdschrift Procestechniek, 36(2):70 (1981)

86. Naden, D., Willey, G., Bicker, E., et al., Ion Exchange Technology (Nader, D., Streat, M., eds.) Horwood, Chichester, UK, p 690 (1984)

87. Naden, D., Bandy, M.R., presented at SCI meeting, Impact of SX and IX on Hydrometallurgy, University of Salford, UK (March, 1978)

88. Slater, M.J., Canadian J Chem Eng, 52:43 (Feb. 1974)

89. Chase, H.A., Ion Exchange Technology (Nader, D., Streat, M., eds.) Horwood, Chichester, UK (1984)

90. Wolf, F.J., Putter, I., Downing, G.V., Jr., et al., U.S. Patent No. 3,221,008 (Nov. 30, 1965)

91. Flemming, H.C., Vom Wasser, 56:215 (1981)

92. Spinner, I.H., Simmons, P.J., Brown, C.J., Proceed 40th International Water Conf, Pittsburgh (1979)

93. Powdex[R] is a registered tradename of the Graver Water Treatment Company.

94. Calmon, C., A I Ch E Symp Ser, 80(233):84 (1984)

95. Rodrigues, A.E., In: Mass Transfer and Kinetics of Ion Exchange (Liberti, L., Helfferich, F.G., eds.) Marinus Nijhoff, The Hague, p 259 (1983)

96. Passino, R., In: Mass Transfer and Kinetics of Ion Exchange (Liberti, L., Helfferich, F.G., eds.) Marinus Nijhoff, The Hague, p 313 (1983)

97. Boari, G., Liberti, L., Merli, C., et al., Env Prot Eng, 6:251 (1980)

98. Liberti, L., Passino, R., Ind Eng Chem, Pros Des Dev, 21(2):197 (1982)

99. Belter, P.A., Cunningham, F.L., Chen, J.W., Biotech and Bioeng, 15:533 (1973)

100. Slater, M.J., Effluent Water Treat J, 461 (Oct. 1981)

101. Snowdon, C., Turner, J., Proc Int Symp "Fluidization", Netherlands University Press, Amsterdam, p 599 (1967)

102. Ergun, S., Chem Eng Progr, 48:89 (1952)

103. Tan, H.K.S., Chem Eng, 91(26):57 (Dec. 24, 1984)

104. Klein, G., Computers and Chem Eng, 8(3/4):171 (1984)

105. Klein, G., Nassiri, M., Vislocky, J.M., A I Ch E Symp Ser, 80(233):14 (1984)

106. Erskine, D.G., Schuliger, W.G., Chem Eng Progr, 67:11,41 (1971)

107. Weber, W.J., Jr., Thaler, J.O., In: Scale-Up Water, Wastewater Treatment Processes (Schmidthe, N.W., et al., eds.) Butterworth, Boston, p 233 (1983)

108. Pelosi, P., McCarthy, J., Chem Eng, p 125 (Sept. 6, 1982)

109. Wirth, L.F., Jr., Feldt, C.A., Odland, K., Ind Eng Chem, 58:639 (1961)

110. Bolto, B.A., Eldridge, R.J., Eppringer, K.H., et al., Reactive Polymers, 2:5 (1984)

111. Kysela, J., Brabec, J., Jad Energ, 27(12):445 (1981)

112. Khodyrev, B.N., Prokhorov, A.F., Teploēnergetika, 25(3):60 (1978)

113. Stein, W.H., Moore, S., Sci American, 184:35 (1951)

114. Kunin, R., Ion Exchange Resins, Robert E. Krieger Publ. Co., Huntington, NY, p 290 (1972)

115. Yamada, K., Kinoshita, S., Tsunda, T., et al., The Microbial Production of Amino Acids, John Wiley and Sons, NY, p 228 (1972)

116. Nakayama, K., Hagino, H., U.S. Patent No. 3,594,279 (July 20, 1971)

117. Noguchi, Y., Nakajima, J., Uno, T., et al., U.S. Patent No. 3,684,655 (Aug. 15, 1972)

118. Okumura, S., Yoshinaga, F., Kubota, K., et al., U.S. Patent No. 3,716,453 (Feb. 13, 1973)

119. Ishida, M., Sugita, Y., Hori, T., et al., U.S. Patent No. 3,565,951 (Feb. 23, 1971)

120. Nakayama, K., Hagino, H., U.S. Patent No. 3,595,751 (July 27, 1971)

121. Abe, S., Takayama, K., U.S. Patent No. 3,700,556 (Oct. 24, 1972)

122. Nakayama, K., Kase, H., U.S. Patent No. 3,692,628 (Sept. 19, 1972)

123. Nagai, T., Yamaura, I., U.S. Patent No. 3,639,467 (Feb. 1, 1972)

124. Noguchi, L., Maruyama, H., Motoya, Y., et al., U.K. Patent No. 2,095,232 (Sept. 29, 1982)

125. Bartels, C.R., Kleiman, G., Korzun, J.N., et al., Chem Eng Prog, 54(8):49 (Aug. 1958)

126. Wolf, F.J., U.S. Patent No. 3,000,873 (Sept. 19, 1961)

127. Belter, P.A., A I Ch E Symp Ser, 80(233):110 (1984)

128. Denkewalter, R.G., Kazal, L.A., In: Ion Exchange Technology (Nachod, F.C., Schubert, J., eds.) Academic Press, NY, p 579 (1956)

129. Voser, W., U.S. Patent No. 3,725,400 (April 3, 1973)

130. Fujisawa, Japanese Patent No. 77 128 294 (1977)

131. Yamanouchi, Japanese Patent No. 7911298 (1979)

132. Abraham, E.P., Newton, G.G.F., Hale, C.W., U.S. Patent No. 3,184,454 (May 1965)

133. McCormick, U.S. Patent No. 3,467,654 (Sept. 1969)

134. Stables, H.C., Briggs, K., U.S. Patent No. 4,205,165 (May 27, 1980).

135. Goegelman, R.T., Miller, T.W., U.S. patent No. 3,709,880 (Jan. 9, 1973)

136. Schubert, P.F., U.S. Patent No. 4,196,285 (April 1, 1980)

137. Kawamura, Y., Shoji, J., Matsumoto, K., U.S. Patent No. 4,256,835 (March 17, 1981)

138. Fujita, S., Takatsu, A., Shibuya, K., U.S. Patent No. 3,629,233 (Dec. 21, 1971)

139. Rakutina, N.S., Shapenyuk, T.A., El'kin, G.E., et al., Zh Prikl Khim (Leningrad), 55(5):1171 (1982)

140. Rakutina, N.S., Borisova, V.A., Papukova, K.P., et al., Zh Prikl Khim (Leningrad), 55(3):540 (1982)

141. Huber, G., Schindler, P., European Patent No. 80,166 (June 1, 1983)

142. Treiber, L.R., Gullo, V.P., U.S. Patent No. 4,198,338 (April 15, 1980)

143. Geogelman, R.T., Kahan, F.M., U.S. Patent No. 4,000,161 (Dec. 28, 1976)

144. Debono, M., Merkel, K.E., Weeks, R.E., et al., U.S. Patent No. 4,375,513 (March 1, 1983)

145. Debono, M., Merkel, K.E., Weeks, R.E., et al., U.S. Patent No. 4,322,406 (March 30, 1982)

146. Khose, N.M., Dasare, B.D., J Chem Tech Biotechnol, 31:219 (1987)

147. Olthoff, U., Pharmazie, 32(8-9):536 (1977)

148. Jacob, T.A., U.S. Patent No. 3,366,627 (Jan. 30, 1968)

149. Tsao, J.C.Y., Huang, T., Wang, P., et al., J Chinese Chem Soc (Peking) Ser II, 2(3) (1964)

150. Kobayski, T., Nakamura, I., Nakagawa, M., U.S. Patent No. 3,873,425 (March 25, 1975)

151. Girot, P., Boschetti, E., J Chrom, 213:389 (1981)

152. Keay, L., U.S. Patent No. 3,593,738 (July 13, 1971)

153. Sipos, T., Viebrock, F.W., U.S. Patent No. 3,627,643 (Dec. 14, 1971)

154. Rauenbusch, E., Schmidt, D., U.S. Patent No. 4,174,439 (Nov 13, 1979)

155. Umezawa, H., Aoyagi, T., Takeuchi, T., et al., U.S. Patent No. 4,167,448 (Sept. 11, 1979)

156. Kiselev, G.V., Biochem Biophys Acta, 712:719 (1982)

157. Palmer, D.E., In: Food Proteins (Fox, P.F., Condon, J.J., eds.) Applied Science, London, p 341 (1982)

158. Phillips, D.J., Jones, D.T., Palmer, D.E., British Patent No. 1,518,111 (Feb. 25, 1981)

159. Harmon, Z., Dechow, F.J., U.S. Patent No. 4,543,261 (Sept. 24, 1985)

160. Kunin, R., Ion Exchange Resins, Robert E. Kreiger Publish. Co., Huntington, NY, p 183 (1972)

161. Obara, K., Proc Res Soc Japan Sugar Ref Tech, 4(4):8 (1955)

162. Sasaki, K., Naito, T., Tsukui, K., Seito Gijutsu Kenkyu Kaishi, 28:39 (1978)

163. Susuki, K., Seito Gijutsu Kenkyu Kaishi, 9:73 (1960)

164. Iwashina, S., Proc Res Soc Japan Sugar Ref Technol, 9:113 (1960)

165. Furukawa, N., Iizuka, T., Proc Res Soc Japan Sugar Ref Technol, 11:63 (1962)

166. Parker, K.J., Chem Indus, 21:782 (Oct. 1972)

167. Pollio, F.Y., McGarvey, F.X., Sugar Y Azucar, 63 (May 1978)

168. Shive, W., U.S. Patent No., 2,628,186 (Feb 10, 1953)

169. Arden, T.V., Rowe, MC., Process Biochem, 4 (April 1977)

170. Mattick, L.r., Gogel, E.V., U.S. Patent No. 4,205,092 (May 27, 1980)

171. Peterson, R.G., Fujii, G.R., U.S. Patent No. 3,437,491 (April 8, 1969)

172. Gogel, E.V., U.S. Patent No. 4,156,026 (May 22, 1979)

173. Brandenberger, H., Rev Fermentation et Indus Aliment, 11:237 (1956)

174. Barnett, L.B., Bull, H.B., Biochim Biophys Acta, 36:244 (1959)

175. Tosa, T., Mori, T., Fuse, N., et al, Enzymologia, 31:214 (1966)

176. Fujita, Y., Matsumoto, A., Nishikaji, T., et al., Japan Patent No. 50-9417 (July 26, 1975)

177. Delin, P.S., Ekstrom, B.A., Nathorst-Westfield, L.S., et al., U.S. Patent No., 3,736,230 (May 29, 1973)

178. Kilara, A., Shahani, K., CRC Critical Reviews in Food Sciences and Nutrition, 12(2):161 (1979)

179. Vieth, W.R., Venkatasubramanian, K., Chem Tech, p 677 (Nov 1973)

180. Chibata, I., Tosa, T., Mori, T., et al., Brit. Patent No. 2,082,188 (March 3, 1982)

181. Sweigart, R.D., In: Applied Biochem Bioeng (Wingard, L.B., Jr., Katchalski-Katzir, E., Goldstein, L.G., eds.) Academic Press, NY, p 209 (1979)

182. Tschang, C.J., Klefenz, H., U.S. Patent No. 4,266,030 (May 5, 1981)

183. Roy, A.K., Roy, S., LC, Liq Chromatogr HPLC Mag, 1(3):182 (1983)

184. Sussman, S., Ind Eng Chem, 38:1228 (1946)

185. Bodamer, G., Kunin, R., Ind Eng Chem, 43:1082 (1951)

186. Oikawa, S., Proc Res Soc Japan Sugar Ref Technol, 16:64 (April 1965)

187. Demaimay, M., LeHeneff, Y., Printemps, P., Proc Biochem, 3 (April 1978)

188. Holmberg, H., European Patent Appl No. 83 325 (July 6, 1983)

189. Takahaski, Y., Seito Gijutsu Kenkyu Kaishi, 31:76 (1982)

190. Rigal, L., Gaset, A., Gorrichon, J.P., Ind Eng Chem Proc Res Des, 20:719 (1981)

191. Hino, T., Ito, K., Hayashi, K., Nippon Nogei-Kagaku Kaishi, 35:773 (1961)

4

Column Chromatography Processes

4.1 INTRODUCTION

In most ion exchange operations, an ion in solution is replaced with an ion from the resin and the former solution ion remains with the resin. In contrast, ion exchange chromatography uses the ion exchange resin as an adsorption or separation media which provides an ionic environment, allowing two or more solutes in the feedstream to be separated. Beads of non-ionic resins, gels or molecular sieves may also be used as the solid stationary phase in column chromatography. The feed solution is added to the chromatographic column filled with the separation beads and eluted with solvent, often water for fermentation products. The resin beads selectively slow some solutes while others are eluted down the column (Figure 4.1). As the solutes move down the column, they separate and their individual purity increases. Eventually, the solutes appear at different times at the column outlet where each can be drawn off separately.

Tswett (1,2) was the first to identify correctly the nature of the separation of colored vegetable pigments in petroleum ether when the solution was passed through a column of fine particle calcium carbonate. The process was called a chromatographic separation because of the separation of the pigments into bands of different colors. Following Tswett have been numerous scientists who have developed ion exchange chromatography into a sophisticated analytical technique used in many scientific areas.

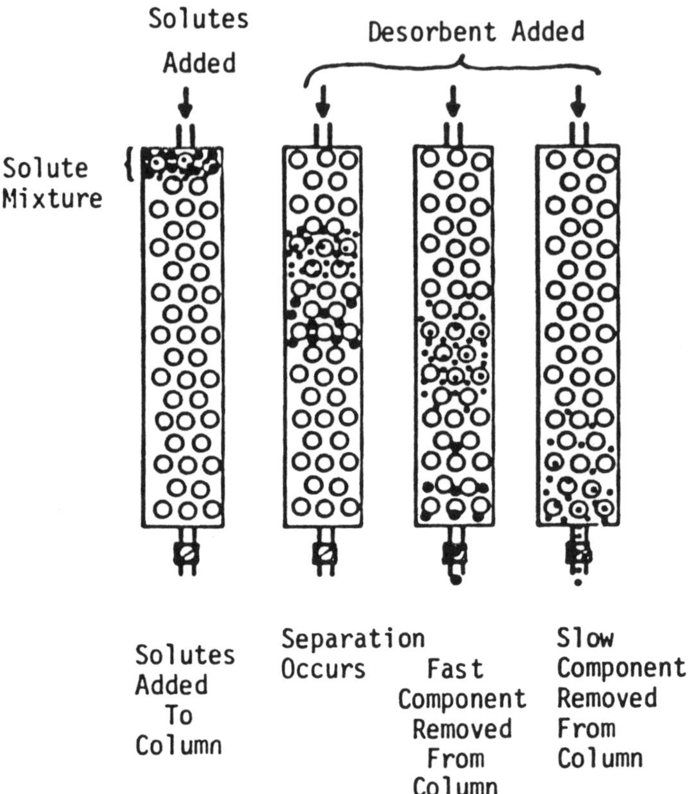

Figure 4.1. The steps of chromatographic separation are: addition of the mixed solutes to the column, elution to effect separations and removal of the separated solutes.

The first commercial use of ion exchange chromatography occurred during World War II at Iowa State University where Spedding and his coworkers (3,4) isolated transuranium elements as their contribution to the Manhattan Project. Several years later, the commercial preparation of individual rare earths was performed using ion exchange chromatography (5,6).

Despite these initial successes with chromatography in preparing pure materials in large quantities, other commercial applications were not developed until the mid to late 1960's. It was in the 1970's before industrial separation problems were identified which could not be solved by conventional, well-tried methods such as distillation or crystallization. At present, there are industrial ion exchange chromatographic units which separate amino acids, hydrocarbon isomers, glucose from fructose, sucrose from molasses, monosaccharides from di- and polysaccharides and ion exchange regenerants from salts.

A literature search by Sitrin and coworkers (7), covering the period from 1980 to 1983, revealed that out of the 7,000 citations for liquid chromatography, only 100 discussed preparative work. Few of these involved separations larger than 100 mg. That distribution remains typical of the published activity in the chromatography area.

There are two advances that have contributed strongly to the recent development of industrial ion exchange chromatography: (1) improvements in chromatographic equipment with recycle control systems and pseudo-moving bed control systems; and (2) improvements by resin manufacturers in developing resins with narrow, controlled size distributions and improved osmotic resilience. General resin considerations and commercial ion exchange resins have been described in Chapter 3. Descriptions of chromatographic systems will be covered in Section 7 of this chapter.

4.2 CLASSIFICATIONS OF CHROMATOGRAPHY

4.2.1 Chromatographic Methods

The literature is replete with various column chromatographic methods: adsorption chromatography, ion exchange chromatography, gel filtration chromatography, gel permeation chromatography and affinity chromatography. These methods are differentiated on the basis of the retentive ability of the stationary phase, the type of eluent employed and the material used as the separation phase.

In adsorption chromatography, it is mainly physical surface forces which are involved in retaining the solute. This method is the oldest of the chromatographic techniques. Although it does not have the specificity of the other methods, adsorption chromatography is still one of the simplest and most effective techniques of separating mixtures of non-polar substances and compounds of low volatility. Silica, alumina, activated carbon and macroporous non-ionic polymers are the adsorbents used most frequently as the separation media. Low molecular weight organic solvents are the most common eluents.

In ion exchange chromatography, true hetero-polar chemical bonds are formed reversibly between ionic components in the mobile and the stationary phase. The ion exchange beads may have either hydrophobic or hydrophilic matrices which have been functionalized with ionizable groups, as discussed in the last chapter. The eluent for the chromatographic separation of fermentation products at the preparative or higher level is usually water or an aqueous buffer.

The stationary phase may also serve as a molecular sieve to separate solutes on the basis of molecular size. This method is known as gel filtration chromatography or gel permeation chromatography.

Gel filtration chromatography originated in Sweden in 1959 (8) when columns packed with crosslinked polydextran gels, swollen in an aqueous solution, were used to separate water-soluble macromolecules on the basis of size differences. These gels are still used extensively for separating water soluble biological compounds for further characterization studies (9). Gel permeation chromatography was developed at Dow Chemical in 1964 (10) using crosslinked polystyrene gels swollen in organic solvents to separate synthetic polymers on the basis of size. Column chromatography with ion exchange resins also can utilize size differences for separation but more frequently relies upon adsorption for affinity differences to effect separations.

Partitioning of a compound between a hydrophobic stationary and a polar aqueous mobile phase is called reverse phase liquid chromatography. Reverse phase chromatography is usually performed on columns packed with silica gel to which a C-8 to C-18 hydrocarbon has been covalently attached. For such systems, the strength of an eluting solvent increases as its polarity decreases and compounds of similar structure can be expected to elute in order of decreasing polarity.

Polar, charged fermentation products, such as peptides and glycoproteins, are well suited for recovery and purification with reverse phase chromatography. This is due to the removal of very polar, often colored contaminants at the solvent front, ahead of the desired material. Using this technique early in a purification scheme can simplify the number of steps one must employ.

Affinity chromatography, which originated in 1968 at the National Institutes of Health (11), is based on a unique and fundamental property of biological macromolecules: their selective, high-affinity recognition of, and reversible interaction with, other molecules. This technique is sufficiently distinct from the other chromatographic methods that it will be discussed separately in the next chapter.

Most literature references to column chromatography are concerned strictly with analytical applications. While some of the information is directly relatable to the preparative and commercial recovery of fermentation products, one must be cautious due to the differences in ranges of operating parameters, throughput and the addition of components to aid in analytical resolution.

The low crosslinked polymer beads used in analytical chromatographic applications have mean diameters of 70 to 150 microns and can only be used at low flow rates and pressures less than 17 bars. These beads collapse at higher pressures which restricts the flow rate, making separations impossible. Pilot plant and industrial chromatography use larger mean diameter beads (200 to 450 microns) with a higher degree of crosslinking. The necessary degree of separation is achieved by removing a specific cut of the chromatogram.

4.2.2 Types of Chromatographic Separations

Chromatographic separations can be classed into four types according to the type of materials being separated: affinity differences, ion exclusion, size exclusion and ion retardation chromatography. These types of separations may be described in terms of the distribution of the materials to be separated between the phases involved.

Figure 4.2 shows a representation of the resin-solvent-solute components of a column chromatographic system. The column is filled with resin beads of the solid stationary phase packed together with the voids between the beads filled with solvent solution. The phases of interest are (1) the liquid phase between the resin beads; (2) the liquid phase held within the resin beads; and (3) the solid phase of the polymeric matrix of the resin beads. When the feed solution is placed in contact with the hydrated resin in the chromatographic column, the solutes distribute themselves between the liquid inside the resin beads and that between the resin beads. The distribution for component i is defined by the distribution coefficient, K_{d_i}:

$$K_{d_i} = C_{r_i}/C_{l_i} \tag{4.1}$$

where C_{r_i} is component i's concentration in the liquid within the resin bead and C_{l_i} is the component i's concentration in the interstitial liquid. The distribution coefficient for a given ion or molecule will depend upon that component's structure and concentration, the type and ionic form of the resin and the other components in the feed solution. The distribution coefficients for several organic compounds in aqueous solutions with ion exchange resin are given in Table 4.1 (12).

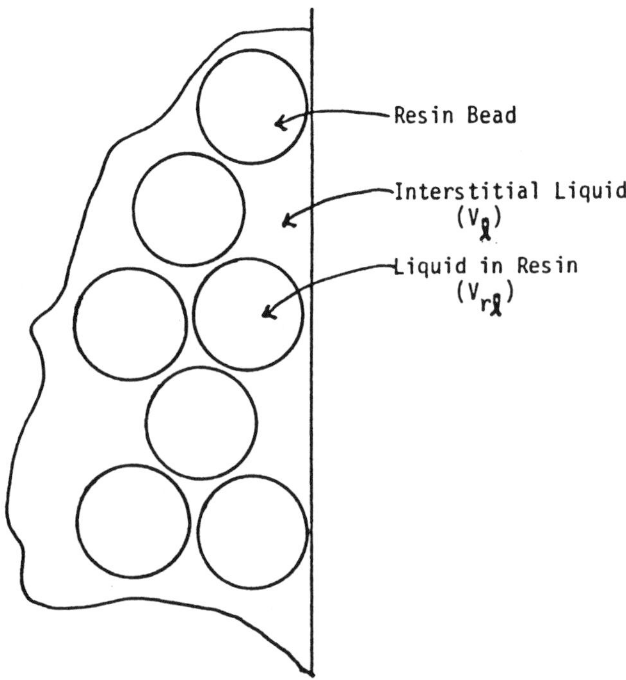

Figure 4.2. Representation of the three phases involved in chromatographic separation.

Table 4.1: Distribution Coefficients (12)

SOLUTE	RESIN	Kd
Ethylene Glycol	Dowex 50-X8, H^+	.67
Sucrose	Dowex 50-X8, H^+	.24
d-Glucose	Dowex 50-X8, H^+	.22
Glycerine	Dowex 50-X8, H^+	.49
Triethylene Glycol	Dowex 50-X8, H^+	.74
Phenol	Dowex 50-X8, H^+	3.08
Acetic Acid	Dowex 50-X8, H^+	.71
Acetone	Dowex 50-X8, H^+	1.20
Formaldehyde	Dowex 50-X8, H^+	.59
Methanol	Dowex 50-X8, H^+	.61
Formaldehyde	Dowex 1-X7.5, Cl^-	1.06
Acetone	Dowex 1-X7.5, Cl^-	1.08
Glycerine	Dowex 1-X7.5, Cl^-	1.12
Methanol	Dowex 1-X7.5, Cl^-	.61
Phenol	Dowex 1-X7.5, Cl^-	17.7
Formaldehyde	Dowex 1-X8, $SO_4^=$, 50-100	1.02
Acetone	Dowex 1-X8, $SO_4^=$, 50-100	.66
Xylose	Dowex 50-X8, Na^+	.45
Glycerine	Dowex 50-X8, Na^+	.56
Pentaerythritol	Dowex 50-X8, Na^+	.39
Ethylene Glycol	Dowex 50-X8, Na^+	.63
Diethylene Glycol	Dowex 50-X8, Na^+	.67
Triethylene Glycol	Dowex 50-X8, Na^+	.61
Ethylene Diamine	Dowex 50-X8, Na^+	.57
Diethylene Triamine	Dowex 50-X8, Na^+	.57
Triethylene Tetramine	Dowex 50-X8, Na^+	.64
Tetraethylene Pentamine	Dowex 50-X8, Na^+	.66

The ratio of individual distribution coefficients is often used as a measure of the possibility of separating two solutes and is called the separation factor, α.

$$\alpha = \frac{K_{d_1}}{K_{d_2}} \tag{4.2}$$

Example 4.1

From Table 4.1, the separation factors for acetone-formaldehyde separability would be 0.49, 0.98 and 1.54 for Dowex 50WX8(H^+), Dowex 1X8(Cl^-) and Dowex 1X8(SO_4^{-2}) resins, respectively. For comparison purposes, it may be necessary to use the inverse of α, so that the values would be 2.03 and 1.02 for Dowex 50WX8(H^+) and Dowex 1X8(Cl^-), respectively. When α is less than 1, the solute in the numerator will exit the column first. When α is greater than 1, the solute in the denominator will exit the column first. From this limited amount of data, the resin of choice would be Dowex 50WX8(H^+) and acetone would exit the column first. The separation factor is sometimes called the relative retention ratio.

The acetone-formaldehyde separation would be an example of affinity difference chromatography in which molecules of similar molecular weight or isomers of compounds are separated on the basis of differing attractions or distribution coefficients for the resin. The largest industrial chromatography application of this type is the separation of fructose from glucose to produce 55% or 90% fructose corn sweetener.

Ion exclusion chromatography involves the separation of an ionic component from a non-ionic component. The ionic component is excluded from the resin beads by ionic repulsion, while the non-ionic component will be distributed into the liquid phase inside the resin beads. Since the ionic solute travels only in the interstitial volume, it will reach the end of the column before the non-ionic solute which must travel a more tortuous path through the ion exchange beads. A major industrial chromatography application of this type is the recovery of sucrose from the ionic components of molasses.

In size exclusion chromatography, the resin beads act as molecular sieves, allowing the smaller molecules to enter the beads while the larger molecules are excluded. Figure 4.3 (13) shows the effect of molecular size on the elution time required for a given resin. An industrial chromatographic application using size exclusion is the separation of dextrose from di- and polysaccharides in the corn wet milling industry.

Example 4.2

The ion exclusion technique has been used for the separation of monosodium glutamate from other neutral amino acids (14). The amino acid solution, neutralized to pH 7.2 with sodium hydroxide, was passed through a column containing a sulfonic acid strong cation resin in the sodium form. The monosodium glutamate fraction was eluted first, as would be expected by ion exclusion, followed by the neutral amino acids, as shown in Figure 4.4.

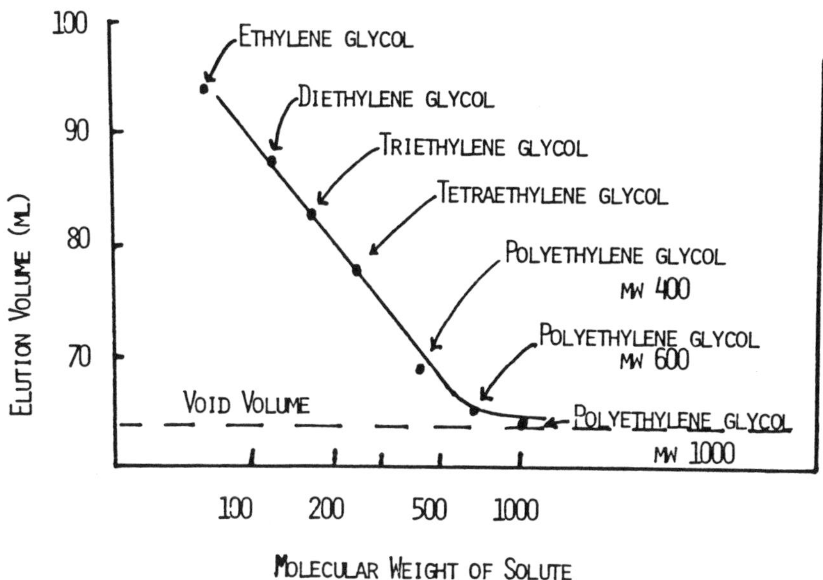

Figure 4.3. Effect of molecular weight on the elution volume required for glycol compounds (Reference 13).

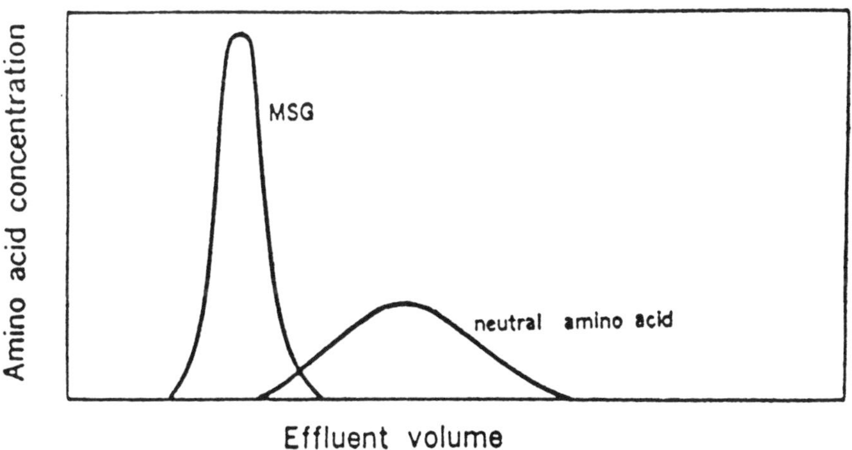

Figure 4.4. Separation of monosodium glutamate and neutral amino acids by ion exclusion at pH 7.2 through Dowex 50X8 (Na form) resin (Reference 14).

Ion retardation chromatography involves the separation of two ionic solutes with common counterion. Unless a specific complexing resin is used, the resin must be placed in the form of the common counterion. The other solute ions are separated on the basis of differing affinities for the resin. Ion retardation chromatography is starting to see use in the recovery of acids from waste salts following the regeneration of ion exchange columns.

4.2.3 Chromatographic Processing Techniques

Chromatography may also be categorized according to the techniques used to carry out the process. The chromatogram may be developed using frontal analysis, elution development or displacement development (15).

In frontal analysis, the sample is continuously fed into the column until "breakthrough" occurs. The least strongly adsorbed compound emerges first, as shown in Figure 4.5.

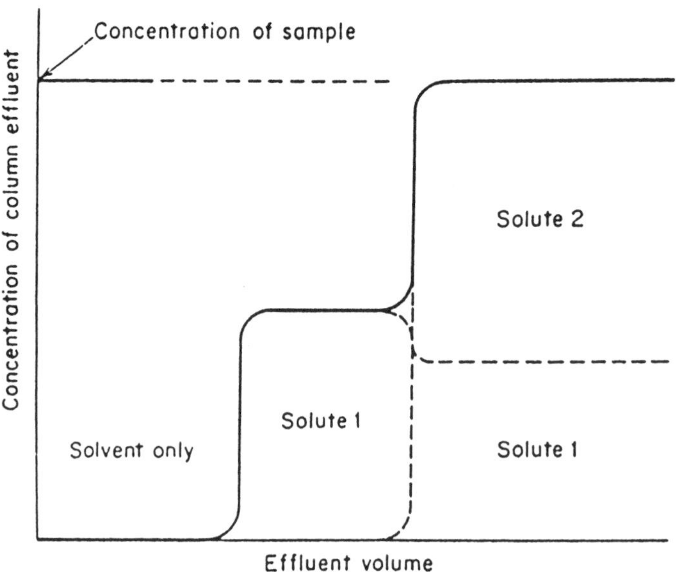

Figure 4.5. Concentration-volume relationship for frontal analysis of two-component system (Reference 15).

Each additional plateau indicates the emergence of another solute in the effluent. This technique only leads to the resolution of the least strongly adsorbed solute. It may be used to remove small amounts of a compound which is used to remove small amounts of a compound which is strongly adsorbed from a feedstream which contains other weakly adsorbed solutes. In such instances, the technique may be used as a preparative technique.

In elution development, a small amount of the sample mixture (at most a few percent of the adsorbent capacity) is fed into the column, then the eluent, which has no affinity for the adsorbent, is introduced. Separation is achieved in the form of "bands" of the individual solutes on a background of the mobile phase, as shown in Figure 4.6. This is the most widely used chromatographic technique due to its common occurrence in analytical methods.

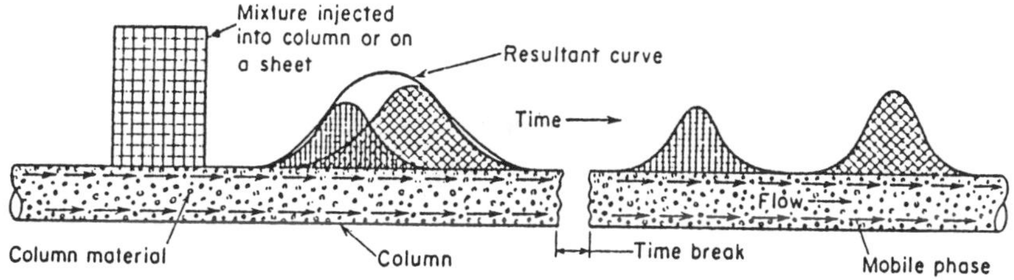

Figure 4.6. Schematic of the elution development of a chromatogram (Reference 15).

Displacement development is like elution development in that a small sample is fed to the column before elution begins. However, in displacement development, the eluent has a higher affinity for the adsorbent material than does the sample. All of the solutes in the sample are forced from the sorption sites and move ahead of the front produced by the eluent. In turn, the individual solutes displace one another in order of increasing sorption strength to give the type of chromatogram shown in Figure 4.7. When used as a preparative method, it is possible to obtain individual concentrated components from a diluted mixture.

4.3 CHROMATOGRAPHIC MATERIALS

The adsorption materials, described in Chapter 2, and the ion exchange resins, described in Chapter 3, are also used as the solid phase separation media in column chromatography. In addition, hydrophilic and hydrophobic gels are used in size exclusion chromatography. Common

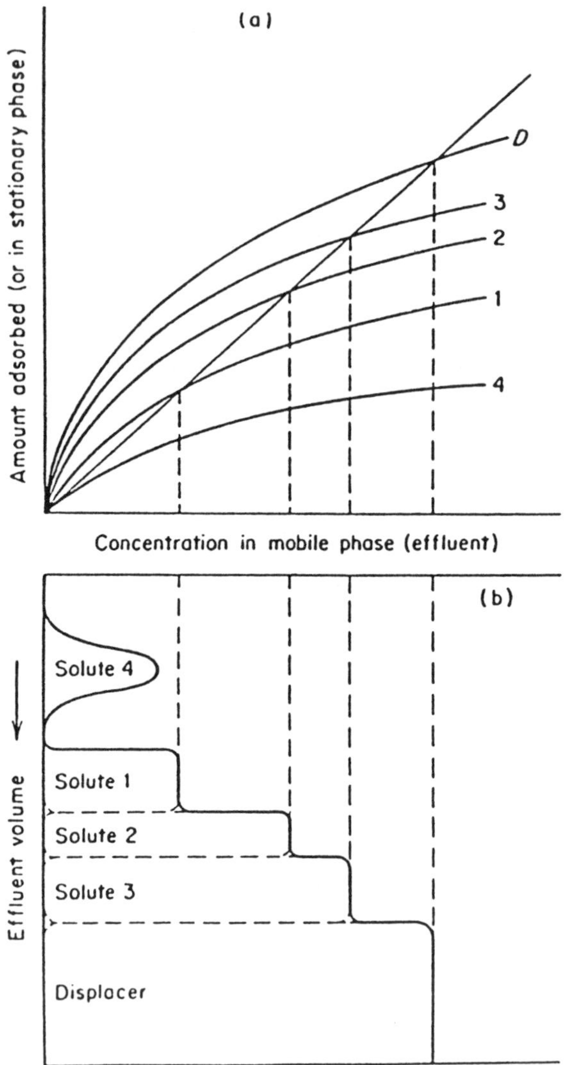

Figure 4.7. Displacement development. (a) sorption isotherms; (b) concentration-effluent volume profile (Reference 15).

hydrophilic gels are crosslinked dextrans, polyacrylamides and agarose, while the hydrophobic gels are usually based on polystyrene.

Dextran gels, under the tradename Sephadex, are composed of linear polyglucose chains of dextran crosslinked with epichlorohydrin. The polyacrylamide gels, under the tradename Bio-Gel, are prepared by the polymerization of acrylamide containing N,N-methylene bisacrylamide as the crosslinking agent. The agarose gels, under the tradenames Sepharose and Bio-Gel A, have a more fragile gel structure since it is due to hydrogen bonding, not chemical crosslinking. The agarose gels must be used at temperatures below 30°C and in a pH range of 5 to 8. Despite these limitations, agarose gels have the large pore structure needed to fractionate large macromolecules, viruses and subcellular particles.

The dextran and polyacrylamide gels are crosslinked sufficiently to render them insoluble in water. However, the gels are very hydrophilic and swell substantially in aqueous media. The gels are available in three size ranges of the hydrated beads: 40 to 80μ, 80 to 150μ and 150 to 300μ. The fractionation ranges for the various hydrophilic gels are shown in Table 4.2.

Table 4.2: Hydrophilic Type Gel Materials

Gel type		Fractional range in molecular-weight units	Hydrated bed volume, V_b, in ml per g dry gel	Water regain, ml per g dry gel
Bio-Gel	P-2	200 to 2,000	3.8	1.6
	P-4	500 to 4,000	6.1	2.6
	P-6	1,000 to 5,000	7.4	3.2
	P-10	5,000 to 17,000	12	5.1
	P-30	20,000 to 50,000	14	6.2
	P-60	30,000 to 70,000	18	6.8
	P-100	40,000 to 100,000	22	7.5
	P-150	50,000 to 150,000	27	9
	P-200	80,000 to 300,000	47	13.5
	P-300	100,000 to 400,000	70	22
Sephadex	G-10	up to 700	2.5	1.0
	G-15	up to 1,500	3.0	1.5
	G-25	100 to 5,000	5	2.5
	G-50	500 to 10,000	10	5.0
	G-75	3,000 to 70,000	13	7.5
	G-100	4,000 to 150,000	17	10
	G-150	5,000 to 400,000	24	15
	G-200	5,000 to 800,000	30	20
Agarose	0.5 m (10%)	10,000 to 250,000		
	1.0 m (8%)	25,000 to 700,000		
	1.5 m	10,000 to 1,000,000		
	2.0 m (6%)	50,000 to 2,000,000		
	5.0 m	10,000 to 5,000,000		
	15.0 m (4%)	200,000 to 15,000,000		
	50.0 m	100,000 to 50,000,000		
	150 m	500,000 to 150,000,000		

The hydrophobic gels, under the tradename Styragel, are prepared by the polymerization of styrene-containing divinylbenzene as the crosslinking agent. These beads are much more rigid than the hydrophilic gels. Unlike the hydrophilic gels which are normally used in an aqueous or very polar medium, the polystyrene gels require a solvent of low polarity, such as tetrahydrofuran, cyclohexanone or carbon tetrachloride. The fractionation ranges for Styragel are listed in Table 4.3

Table 4.3: Properties of Styragel Materials

Type		Fractionation range in molecular-weight units	Approximate exclusion limit in molecular-weight units (average porosity in Å)
60	Styragel	800	1,600
100		2,000	4,000
400		8,000	16,000
1×10^3		20,000	40,000
5×10^3		100,000	200,000
10×10^3		200,000	400,000
30×10^3		600,000	1,200,000
1×10^5		2,000,000	4,000,000
3×10^5		6,000,000	12,000,000
5×10^5		10,000,000	20,000,000
10×10^5		20,000,000	40,000,000

Table 4.4 shows the operational limits of these resins, along with additional resin suppliers.

Resins of very uniform size have been developed in several mean size ranges which could be very useful in the chromatographic separation of small biochemical products such as amino acids and short peptides. New FAST-FLOW Sepharose resins have been prepared to allow improved through-put, easier scale-up and simpler operation for industrial chromatography of proteins (16). Figure 4.8 illustrates the processing potential of this type of resin. The amount of protein processed per hour is 4.5 kg, with 2.3 kg albumin. The total processing time was only 60 minutes to treat 75 liters of plasma feed solution.

The use of porous glass which is silanized and functionalized (17) allows more protein-compatible resins to

Table 4.4: Packing Materials

Media	pH range	Max temp °C	Max flow (cm/h)	Max pressure drop (cm H_2O)	Makers*
Compressible Packings					
Dextran, cross-linked	>2	>120	<77	160-16	Ph
Dextran, bisacrylamide	3-11	>120	25-40	300-70	Ph
Polyacrylamide	2-10	>120	>2.8	100-20	BR
Agarose	4-9	40	--	100-30	Ph, BR, LK
Agarose, polyacrylamide	3-10	40	50-18	>15	PL, Si, Se, LK
Agarose, cross-linked	3-14	>120	30-15	120-50	Ph
Polyether, cross-linked	1-14	>120	--	--	EM, T
Cellulose					Me, Wh, Si
Rigid Packing Materials					
Surface-modified glass	<9	>120			Pi, EN
Surface-modified silica	<9	>120			Co, BR, Wa, Se, EN, EM
Cross-linked hydroxyethylmethacrylate	n/a	>120			LC, KL, Kn
See also Chapter 3 Appendix					

*BR = Bio-Rad; Co = Corning; EM = EM Science; EN = Electro-Nucleonics; KL = Koch-Light; Kn = Knauer AG; LC = LaChema; LK = LKB; Me = Merck; Ph = Pharmacia; Pi = Pierce; PL = P-L Biochemicals; Se = Serva; Si = Sigma; T = Toyo Soda; Wa = Waters; Wh = Whatman

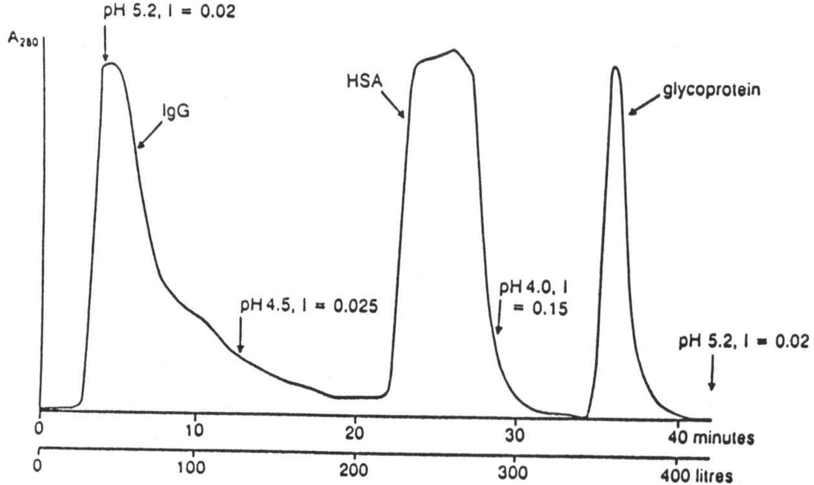

Figure 4.8. High performance industrial scale ion exchange chromatography on DEAE-Sepharose Fast Flow in a Sephamatic column (Reference 16).

be developed. Schomberg (18) has described the chemical modification of silica gels by polymer coating. These resins are especially useful in reverse phase chromatography applications.

4.4 THEORY

Mathematical theories for ion exchange chromatography were developed in the 1940's by Wilson (19), DeVault (20), and Glueckauf (21,22). These theoretical developments were based on adsorption considerations and are useful in calculating adsorption isotherms from column elution data. Of more interest for understanding preparative chromatography is the theory of column processes, originally proposed by Martin and Synge (23) and augmented by Mayer and Thompkins (24), which was developed analogous to fractional distillation so that plate theory could be applied.

One of the equations developed merely expressed mathematically that the least adsorbed solute would be eluted first and that if data on the resin and the column dimensions were known, the solvent volume required to elute the peak solute concentration could be calculated. Simpson and Wheaton (25) expressed this equation as:

$$V_{MAX} = K_d V_{rl} + V_l \qquad (4.3)$$

where V_{MAX} is the volume of liquid that has passed through the column when the concentration of the solute is maximum or the midpoint of the elution of the solute. K_d, defined in Equation 4.1, is the distribution coefficient of the solute in a "plate" of the column, V_{r1} is the volume of liquid solution inside the resin and V_1 is the volume of interstitial liquid solution.

The mathematical derivation of Equation 4.3 assumes that complete equilibrium has been achieved and that no forward mixing occurs. Glueckauf (26) pointed out that equilibrium is practically obtained only with very small diameter resin beads and low flow rates. Such restricting conditions may be acceptable for analytical applications but would severely limit industrial and preparative applications. For industrial and preparative chromatography, column processing conditions and solute purity requirements are often such that any deviations from these assumptions are slight enough that the equation can still be used as an approximation for the solute elution profile. An example of such an elution profile is given in Figure 4.9.

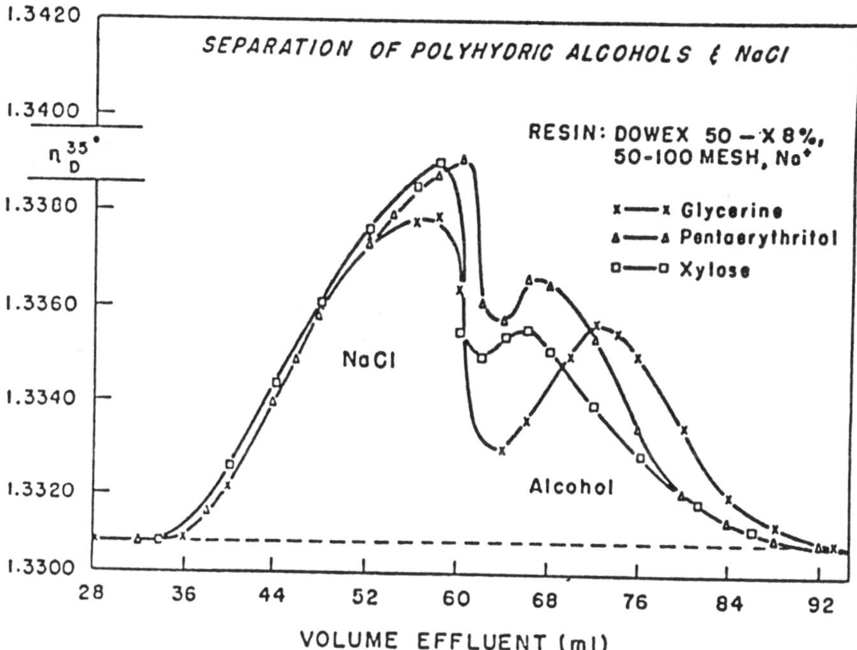

Figure 4.9. Elution chromatogram for the separation of polyhydric alcohols and NaCl using the sodium form of a cation resin (Reference 12).

4.4.1 Theoretical Plate Height

A second important equation for column processes is that used for the calculation of the number of theoretical plates, i.e., the length of column required for equilibration between the solute in the resin liquid and the solute in the interstitial liquid. If the elution curve approximates a Gaussian distribution curve, the equation may be written as:

$$P = \frac{2\,c\,(c + 1)}{W^2} \tag{4.4}$$

where P is the number of theoretical plates, c (= $K_d V_{r1}/V_1$) is the equilibrium constant, W is the half-width of the elution curve at an ordinate value of 1/e of the maximum solute concentration and e is the base of the natural logarithm. For a Gaussian distribution, W = 4σ, where σ is the standard deviation of the Gaussian distribution. The equilibrium constant is sometimes called the partition ratio.

An alternate form of this equation is:

$$P = \frac{2\,V_{MAX}\,(V_{MAX} - V_1)}{W^2} \tag{4.5}$$

Here W is measured in the same units as V_{MAX}. This form of the equation is probably the easiest to calculate from experimental data. Once the number of theoretical plates has been calculated, the height equivalent to one theoretical plate (H.E.T.P.) can be obtained by dividing the resin bed height by the value of P.

The column height required for a specific separation of two solutes can be approximated by (27):

$$\sqrt{H} = \frac{3.29}{c_2 - c_1}\left(\frac{c_2 + 0.5}{\sqrt{P_2}} + \frac{c_1 + 0.5}{\sqrt{P_1}}\right) \tag{4.6}$$

where H is the height of the column, P is the number of plates per unit of resin bed height, and c is the equilibrium constant defined above. Note that the number of plates in a column will be different for each solute. While this equation may be used to calculate the column height needed to separate 99.9% of solute 1 from 99.9% of solute 2, industrial and preparative chromatography applications typically make more efficient use of the separation resin by selectively removing a narrow portion of the eluted solutes as illustrated in Figure 4.10 (28).

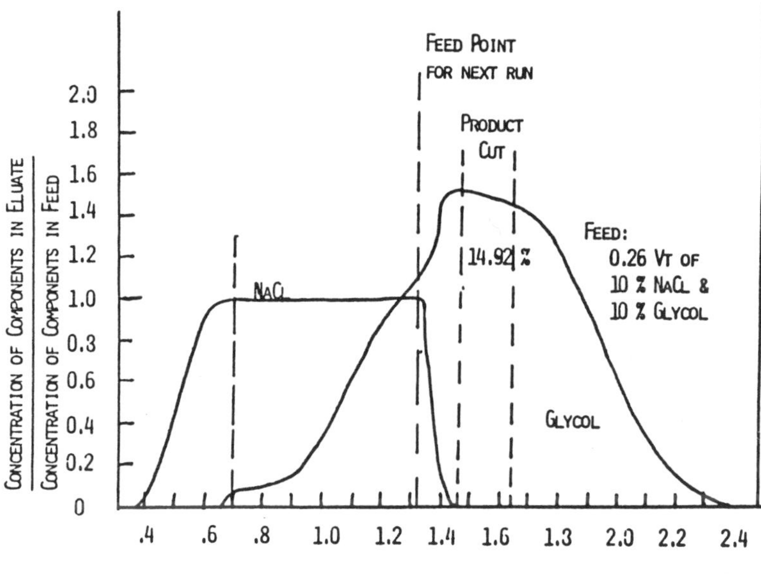

Figure 4.10. Distribution of eluate into fractions for product, recycle and waste for NaCl and glycol separation (Reference 28).

Table 4.5 shows how the theoretical plate number for a chromatography system may be calculated from various combinations of experimental data. The band variance, σ_t^2, is calculated from the experimental data and combined with the retention time, t_R, for a given solute. Figure 4.11 shows the different experimental values which may be used to calculate σ_t.

Table 4.5: Calculation of Plate Number from Chromatogram

Measurements	Conversion to Variance	Plate number
t_R and σ_t	------	$N = (t_R/\sigma_t)^2$
t_R and baseline width W_b	$\sigma_t = W_b/4$	$N = 16(t_R/W_b)^2$
t_R and width at half height $W_{0.5}$	$\sigma_t = W_{0.5}/\sqrt{8 \ln 2}$	$N = 5.54(t_R/W_{0.5})^2$
t_R and width at inflection points (0.607 h) W_i	$\sigma_t = W_i/2$	$N = 4(t_R/W_i)^2$
t_R and band area A and height h	$\sigma_t = A/h\sqrt{2\pi}$	$N = 2\pi(t_R h/A)^2$

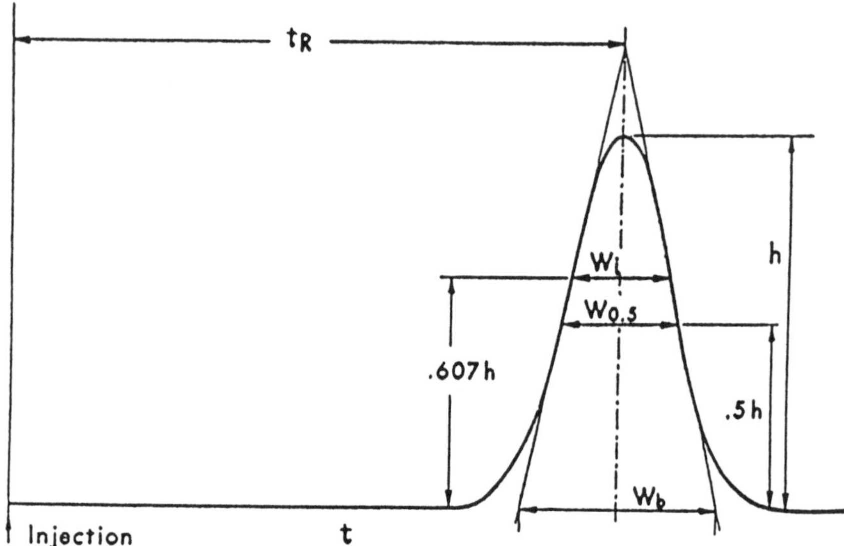

Figure 4.11. Identification of a chromatographic peak segments for the calculation of column performance.

4.4.2 Zone Spreading

The net forward progress of each solute is an average value with normal dispersion about the mean value. The increased band or zone width which results from a series of molecular diffusion and non-equilibrium factors is known as zone spreading.

The plate height as a function of the mobile phase velocity may be written as a linear combination of contributions from eddy diffusion, mass transfer and a coupling term:

$$H = \frac{B}{\nu} + E_S \nu + \frac{1}{1/A + E_M/\nu} \qquad (4.7)$$

where A is a constant, B is the coefficient for the diffusion term, ν is the linear flow rate, and E_M and E_S are the mass transfer coefficients for the mobile and stationary phases, respectively. The plate height is seen to depend on the partition ratio for each species. The smaller the partition ratio, the faster the respective zone moves and the larger is the respective plate height.

A plot of Equation 4.7 for any type of linear elution chromatography describes a hyperbola, as shown in Figure 4.12 (15). There is an optimum velocity of the mobile phase for carrying out a separation at which the plate height is a minimum, and thus, the chromatographic separation is most efficient:

$$\nu_{optimum} = \sqrt{D_M / \left[R_t(1 - R_t)\, d_p^2 / D_S \right]} \tag{4.8}$$

where D_M is the diffusion coefficient of the solute molecule in the mobile phase, D_S is the diffusion coefficient in the stationary phase, d_p is the diameter of the adsorbent particle and $R_t = L/\nu t$, where L is the distance the zone has migrated in time t.

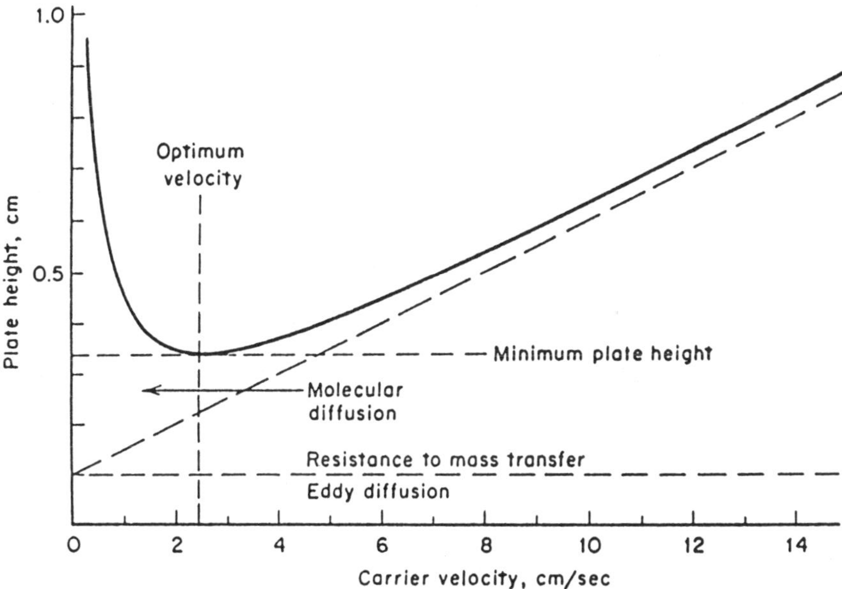

Figure 4.12. Relationship between plate height and velocity of the mobile phase (Reference 15).

4.4.3 Resolution

A variation on calculating the required column height is to calculate the resolution or degree of separation of two components. Resolution is the ratio of peak separation to average peak width:

$$R = \frac{V_{MAX,2} - V_{MAX,1}}{0.5(W_1 + W_2)} \quad (4.9)$$

The numerator of Equation 4.9 is the separation of the two solutes' peak concentration and the denominator is the average band width of the two peaks. This form of the equation is evaluating the resolution when the peaks are separated by four standard deviations, σ. If R = 1 and the two solutes have the same peak concentration, this means that the adjacent tail of each peak beyond 2σ from the V_{MAX} would overlap with the other solute peak. In this instance, there would be a 2% contamination of each solute in the other.

Resolution can also be represented (29) by:

$$R = \frac{\sqrt{P_2}}{4}\left(\frac{\alpha - 1}{\alpha}\right)\left(\frac{c_2}{1 + c_2}\right) \quad (4.10)$$

Resolution can be seen to depend on the number of plates for solute 2, the separation factor for the two solutes and the equilibrium constant for solute 2.

In general, the larger the number of plates, the better the resolution. There are practical limits to the column lengths that are economically feasible in industrial and preparative chromatography. It is possible to change P also by altering the flow rate, the mean resin bead size or the bead size distribution since P is determined by the rate processes occurring during separation. As the separation factor increases, resolution becomes greater since the peak-to-peak separation is becoming larger. Increases in the equilibrium constant will usually improve the resolution since the ratio $c_2/(1+c_2)$ will increase. It should be noted that this is actually only true when c_2 is small since the ratio approaches unity asymptotically as c_2 gets larger. The separation factor and the equilibrium factor can be adjusted by temperature changes or other changes which would alter the equilibrium properties of the column operations.

Equation 4.10 is only applicable when the two solutes are of equal concentration. When the solutes are of unequal concentration, a correction factor must be used:

$$(A_1^2 + A_2^2)/2A_1A_2$$

where A_1 and A_2 are the areas under the elution curve for solutes 1 and 2, respectively. Figure 4.13 shows the relationship between product purity (η), the separation ratio and the number of theoretical plates. This graph can be used to estimate the number of theoretical plates required to attain the desired purity of the products.

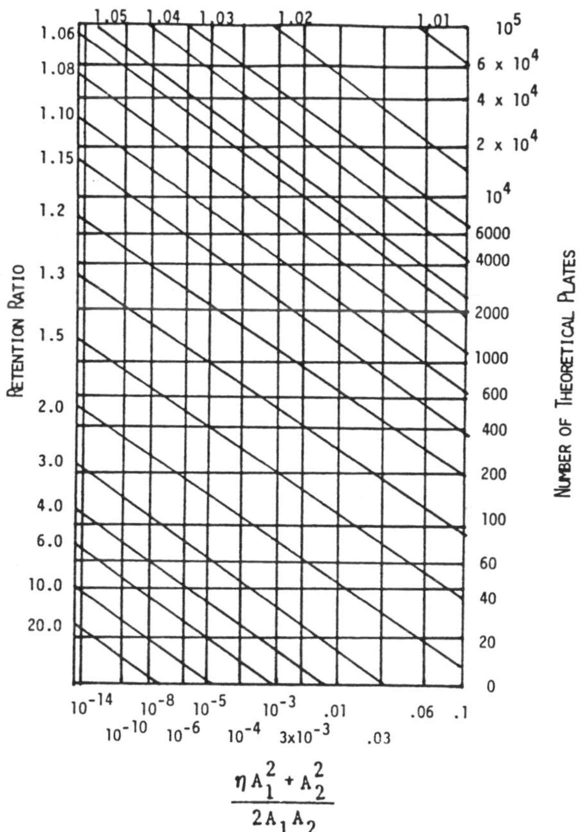

Figure 4.13. Relationship between relative retention ratio, number of theoretical plates and product purity (Reference 26).

Example 4.3

When the product purity must be 98.0%, then $\eta = \Delta m/m = 0.01$ when the amount of the two solutes is equal. If the retention ratio, α, is equal to 1.2, then the number of theoretical plates from Figure 4.13 is about 650. With a plate height of 0.1 cm, the minimum bed height would be 65 cm. In practice, a longer column is used to account for any deviation from equilibrium conditions.

Figure 4.14 shows the effect of relative concentration of two solutes on product purity and product recovery for a few resolution values. When R = 1.25, the solutes are effectively completely separated, with crosscontamination less than 1%. Rarely would this level of purity be required in current industrial chromatography applications. For R = 1.0, the cross contamination is less than 3%; however, the recovery of the more dilute component is significantly reduced when the solute concentration ratio is less than 1:4. When R = 0.8, the recovery and the pourity is still acceptable in industrial chromatography for solutes of approximately the same concentration.

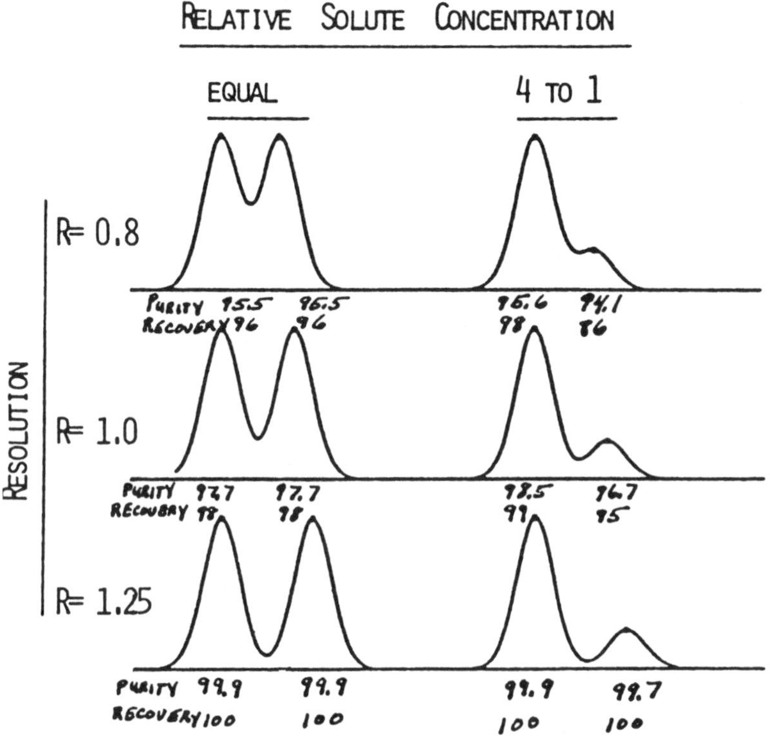

Figure 4.14. Effect of relative solute concentration on peak resolution.

As was mentioned earlier, both resin properties and process parameters will affect the numvber of theoretical plates and, therefore, the separation resolution. The mean size, the size uniformity and the crosslinking of the ion exchange resin are the critical resin parameters for rate processes.

The mean bead size is chosen so that the diffusion into and out of the resin beads does not become the rate controlling step and so that the pressure drop in the column does not become excessive. Figure 4.15 (25) shows the effect of particle size on the shape of the elution profile of ethylene glycol. Small beads adsorb and release the solute quickly for sharp, distinct elution profiles. Larger beads take longer to adsorb and release the solute so that the elution profiles are wider and less distinct. The resin with the smallest mean bead size has the greatest number of theoretical plates. If the separation ratio for the solutes is large enough, a resin with coarser beads may be used, while one with fine beads must be used for solutes with a separation ratio near 1.0. The practical lower limit on bead size is determined by the acceptable pressure drop limit of the column.

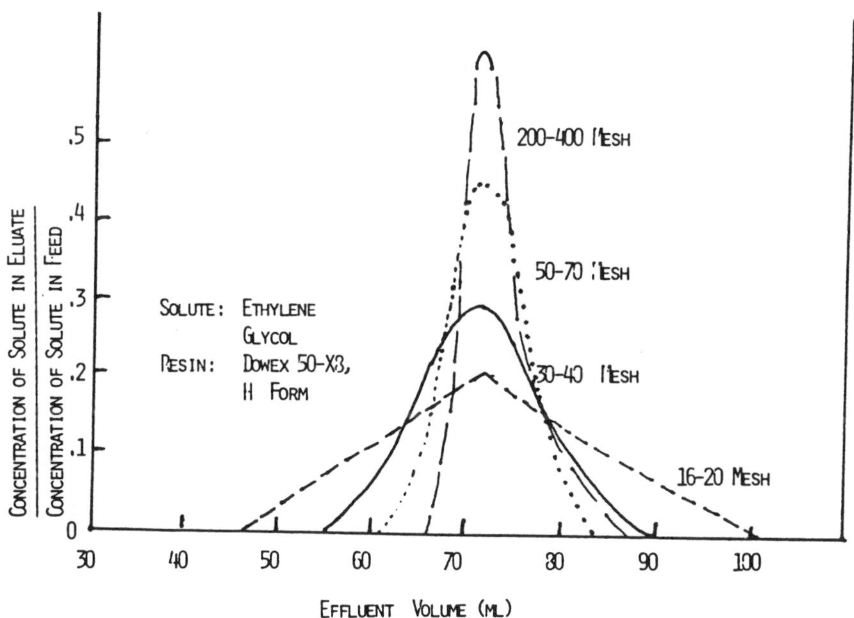

Figure 4.15. Effect of particle size on the elution profile for ethylene glycol (Reference 25).

When there is a large distribution of bead sizes for a resin of a given mean bead size, it becomes difficult to obtain proper elution profiles. The beads smaller than the mean will elute the solute more rapidly and those larger, less rapidly, to cause additional spreading or tailing of the elution profile. Ideally, the resin beads should all be of the same size to amplify the attraction differences of the solutes uniformly as they flow down the column. Separation resins are now available for industrial chromatography that have 90% of the beads within ±20% of the mean bead size. One resin manufacturer can even supply resins that have 90% of the beads within ±10% of the mean bead size (Figure 4.16). Table 4.6 shows the effect of size distribution on separation for glucose-fructose separation.

The crosslinkage of the resin will affect the position of the elution profile as is shown by Figure 4.17. This shift occurs because an increase in the crosslinking decreases the liquid volume inside the resin beads. As crosslinking is increased, the exclusion factor for ionic solutes is increased.

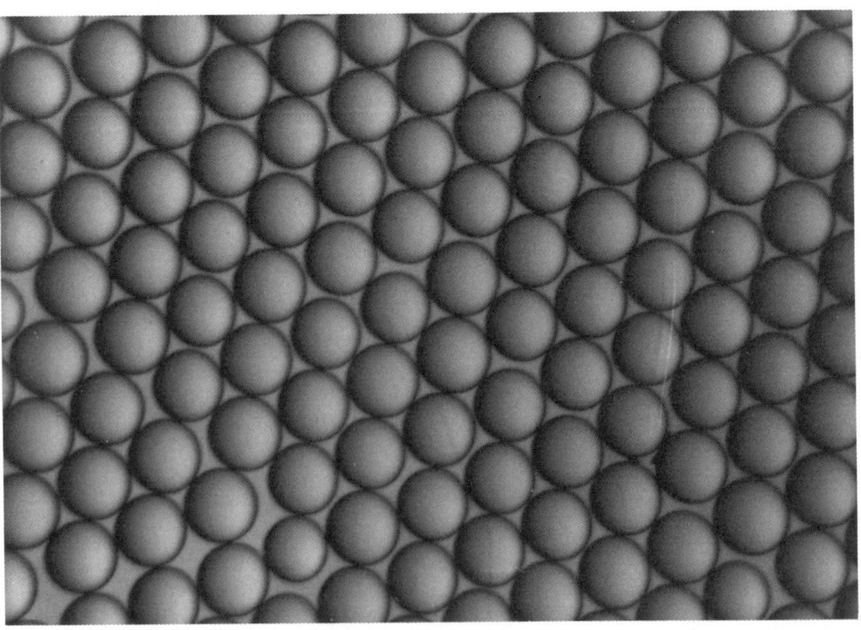

Figure 4.16. Dowex Monosphere 99 with a mean particle diameter of 390 microns. Dowex and Monosphere are registered trademarks of Dow Chemical.

Table 4.6: Effect of Bead Size Distribution on Glucose-Fructose Separation

Bead Size Distribution (Mesh)	Volume %	Glucose-Fructose Peak Separation (Bed Volumes)	Development Length (Elution Volume in Bed Volumes From Initial Glucose to Final Fructose)
1. 35-40 40-50	55 26	0.9	0.96
2. 40-50	80	0.10-0.12	0.81-0.87
3. 40-50	90	0.11-0.13	0.79-0.81
4. 40-50	95	0.12-0.15	0.76
5. 45-50	95	0.14-0.16	0.74

Likewise, increased crosslinking will decrease the rate of diffusion within the resin for non-ionic solutes. The mechanical strength and shrink/swell of organic polymeric resins is increased by increasing the resin crosslinking. The weight of the resin in the column combined with the pressure effect of flow may cuase deformation of the resin beads leading to even higher pressure drops if the resin crosslinkage is not sufficient. Industrial chromatographic separations of sugars have been achieved with resins of 3 to 8% crosslinking for resin bed depths of 3 to 6 meters (30).

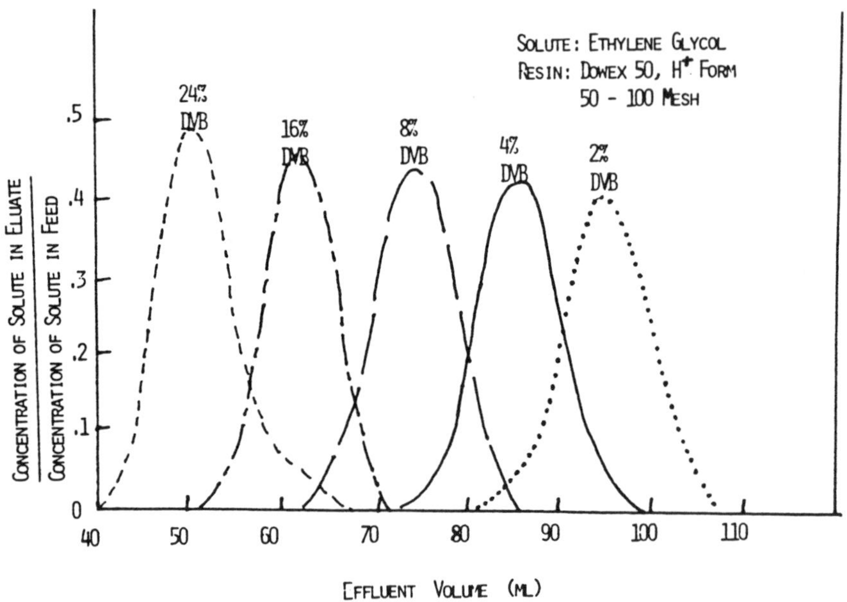

Figure 4.17. Effect of resin crosslinkage on the position of the elution profile for ethylene glycol (Reference 25).

The effect of porosity and resin crosslinking on glucose-salt separation was studied by Martinola and Siegers (31). Figure 4.18 shows that at the low crosslinkage normally used in spearations, the gel resins have a lower glucose loss and have less NaCl in the product. Macroporous resins are not manufactured at the low crosslink levels (<6%) needed for the glucose-salt separation.

Resin manufacturers have made significant advances in recent years to optimize and to customize resin properties for specific industrial applications. Often these resins are so

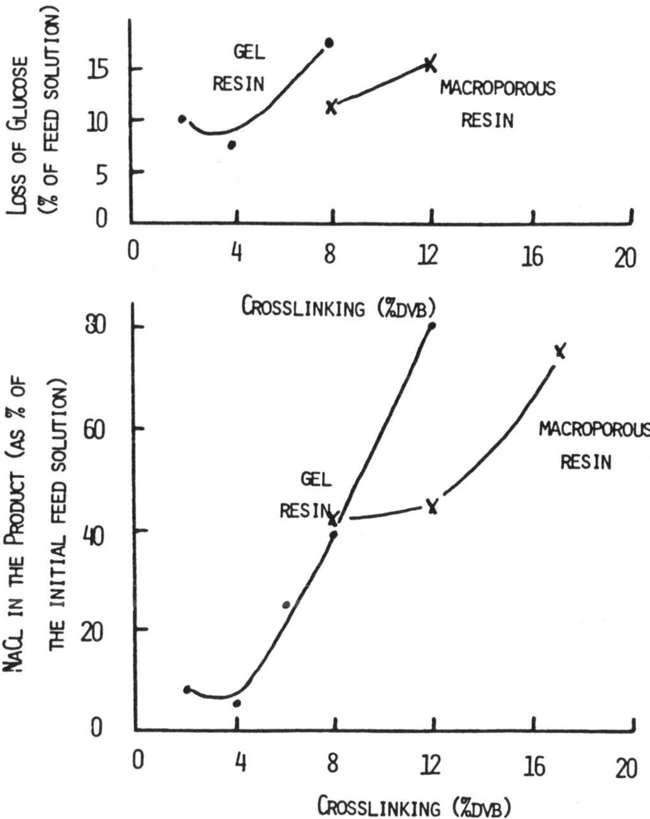

Figure 4.18. Chromatographic separation as a function of crosslinking and porosity (Reference 31).

tailored for the separation that data sheets and resin samples are only available to the customer with the application.

The flow velocity, the solute concentrations, the cycle frequency, the pressure drop and the column temperature are the process parameters to be controlled to achieve the desired separation.

The elution profile becomes sharper as the flow rate is decreased (Figure 4.19). The flow rate should be between 0.5 and 2.0 times the critical velocity (v_c) to insure that the profile "tailing", due to density and viscosity differences in the solution moving through the column, is avoided or at least minimized. The critical velocity is defined by (30):

$$v_c = \frac{g(\rho_2 - \rho_1)}{k(\eta_2 - \eta_1)} \quad (4.11)$$

where g is the gravity constant, ρ is the density, η is the viscosity, k (= $\Delta p/\eta vL$) is the permeability coefficient of the bed, Δp is the pressure drop in the resin bed, v is the linear flow rate of the solution and L is the height of the resin bed. The subscripts 1 and 2 in this equaiton refer to the low and high extremes of these values for the solution as it cycles from feed to solvent. The permeability coefficient for industrial chromatography of sugar and polyols has been found to be 1 to 4 x $10^{10}m^{-2}$.

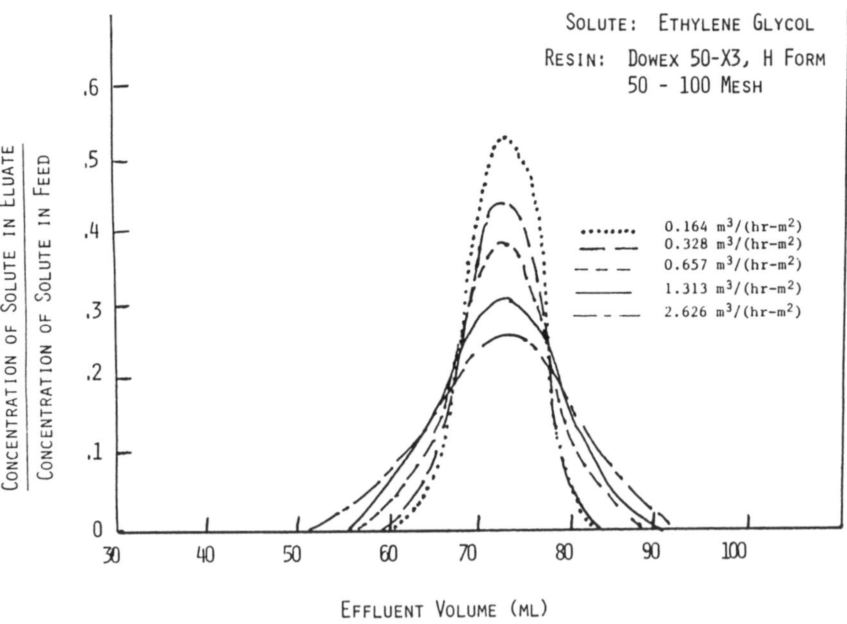

Figure 4.19. Effect of flow rate on the elution profile of ethylene glycol (Reference 25).

The effect of solute concentration on the elution profile is shown in Figure 4.20. The concentration changes will shift the point at which V_{MAX} occurs but does not change the appearance of the initial point when the solute emerges from the column. Therefore, the appearance of the solute in the lowest concentration may be the limiting factor for the total dissolved solute level. For industrial and preparative chromatography to be effective and economically justifiable, the solute concentration of the most dilute component should be greater than 3% dissolved solids. With a total dissolved concentration of 30 to 50%, this means the relative concentration for two solutes would range from 15 to 1 to 9 to 1. Solutions with as much as 60% total dissolved solids are being separated by industrial chromatography. Beyond that concentration level, the solution viscosity becomes excessive even at temperatures of 65°C.

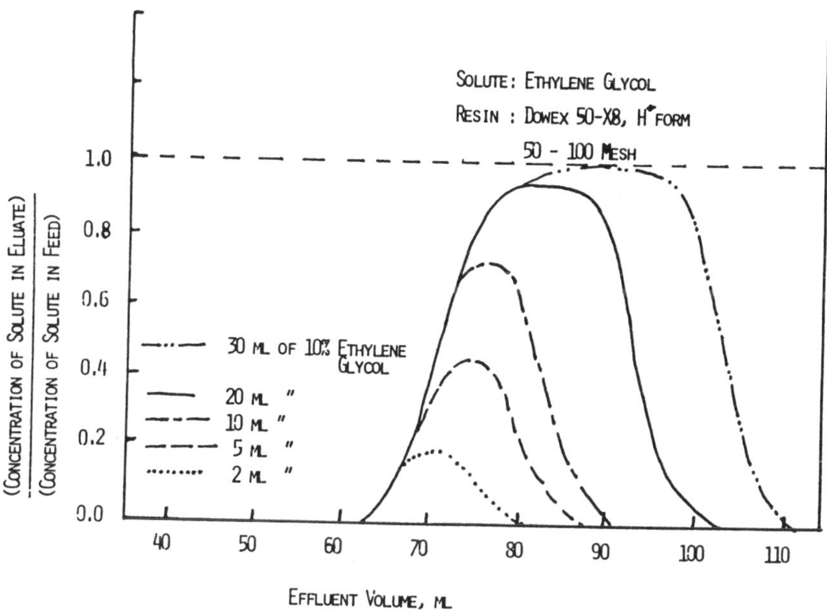

Figure 4.20. Effect of solute concentration on the elution profile of ethylene glycol (Reference 25).

As the resin beads are cycled between the feed solution and the eluting solvent, the beads will expand and contract in response to the relative osmotic pressure of the solution. If the feed/eluent cycle occurs too rapidly, the rapid expansion and contraction can cause the resin beads to "de-crosslink". As crosslinkage of the beads decreases, they lose strength and become more susceptible to compression under flow. A good indicator of the loss of crosslinkage is an increase in the bead water retention capacity. Figure 4.21 shows how cycle frequency affects bead crosslinkage with an increasing number of cycles. Typical cycle times for industrial chromatography are between 1 and 1.5 hours, which result in expected resin lifetimes of several years for resins of appropriate initial crosslinkage.

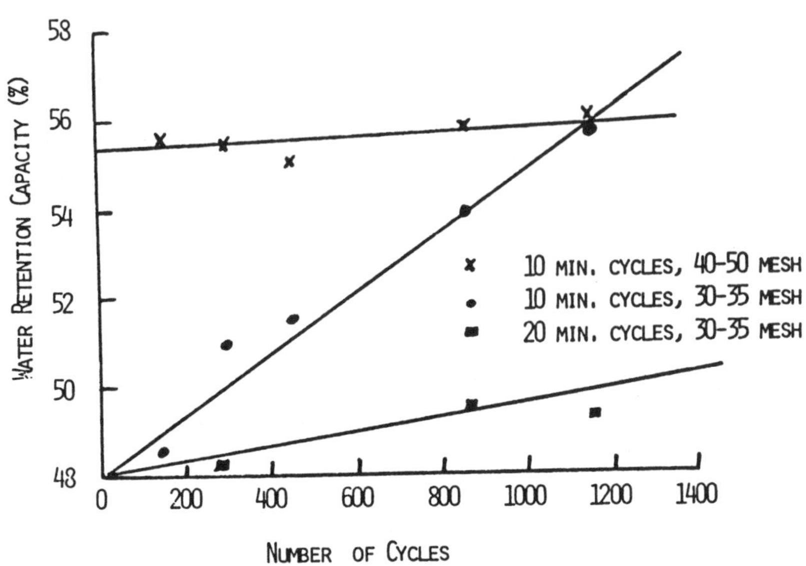

Figure 4.21. Effect of cycle frequency and bead size on resin crosslinkage changes.

Although separation efficiency normally decreases with increases in column temperature, there are two reasons for elevating the temperature. First, elevating the temperature may decrease the viscosity of the mixture making it possible to separate a mixture without excessive dilution. Second, elevated fluid temperature is often necessary to prevent microbial growth on the ion exchange beads or in the column. Generally, fouling due to microbial growth can be prevented with a column temperature greater than 60°C.

On a periodic basis it will be necessary to backwash the resin to remove small particles and to relieve compression from ion exchange beads. For columns with organic polymer beads, some method of allowing 50% freeboard volume must be provided for bed expansion during backwashing. Inorganic chromatographic materials have the advantage that they do not compress under flow and do not have to be backwashed. However, one must monitor the effluent to insure that silicates are not being sloughed from the inorganic beads.

Chromatographic processes are usually not based on the exchange of ions on the resin. At times the solutes are loaded onto the column and then eluted chromatographically by gradient changes in the elution medium. In this case, as well as in chromatography with a single eluent, control of the ionic concentration of the feed and eluent streams is critical. Traces of ionic material in the feed or eluent may eventually cause enough exchange of ions to necessitate the regeneration of the resin. Figure 4.22 (32) shows the effect even partial conversion of a chromatographic resin would have on the resolution of molasses constituents.

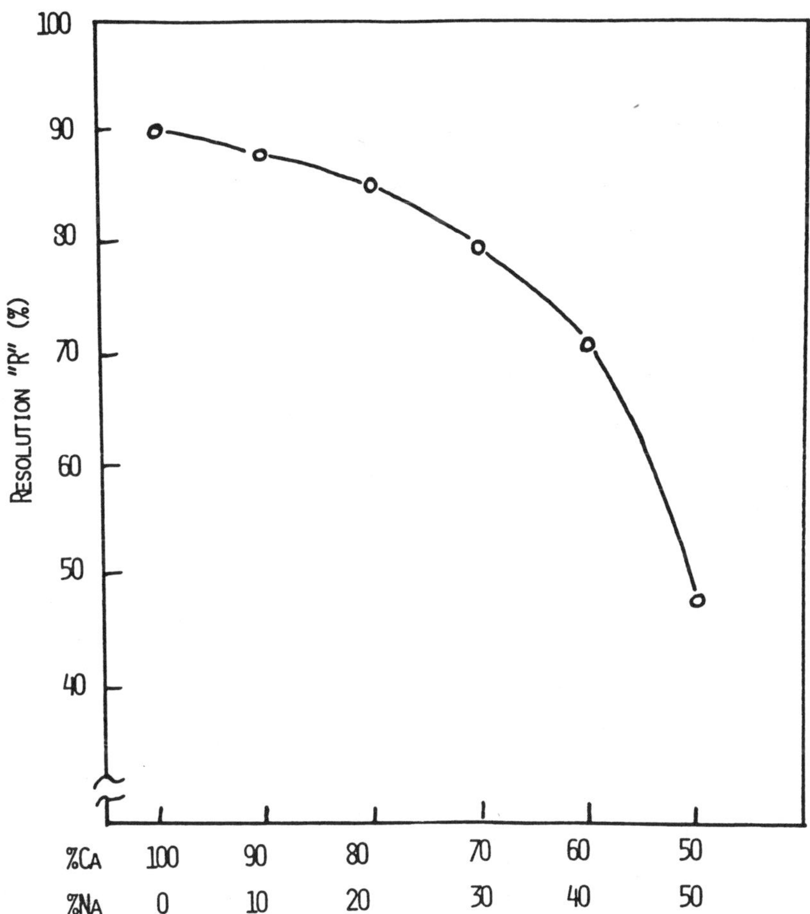

Figure 4.22. Effect of partial conversion of resin from Ca to Na form on the resolutions of sugar and salt in molasses (Reference 32).

4.5 LABORATORY TECHNIQUES

Many of the same laboratory procedures described in the Adsorpiton and the Ion Exchange chapters are useful in preliminary chromatographic studies. In those applications where fine adsorbent particles are being studied, it may be necessary to cover the sintered-glass bed support with a thin layer of sand to prevent its becoming clogged by the fine particles. A bed height-to-diameter ratio of 10 to 1 is recommended with a column height of approximately 20 to 30 cm for laboratory evaluations.

Prior to beginning any chromatography of biologically active molecules, it is necessary to establish the relative stability of the specific molecule in various eluants (33). Table 4.7 shows such a study of the stability of bovine liver acid phosphatase over a pH range from 4.0 to 8.5. The optimal phosphatase activity was maintained when the pH was approximately 5.6 The duration of such a stability test should correspond to the length of time the protein solute will be in the buffer solution during the proposed chromatography operation.

Table 4.7: Effect of Elution Buffer on Bovine Liver Homogenate Acid Phosphatase Activity (33)

Elution Buffer	Units Activity/ml (Zero Time)	Units Activity/ml (3 Hour Time)	% Decrease in 3 hours at Room Temp.
20 mM Na Acetate 0.5 N NaCl 1 mM EDTA, pH 4.0	57.35	18.87	67.1%
20 mM Na Acetate 0.5 N NaCl 1 mM EDTA, pH 5.6	60.02	57.68	3.9%
20 mM Na Acetate 0.5 N NaCl 1 mM EDTA, pH 8.5	54.35	24.39	55.1%

The adsorbent is formed into a stirred slurry using the initial elution buffer. After a mixing time of 20 minutes for most adsorbents, the slurry may be decanted into a tall graduate cylinder and allowed to settle for 45 minutes. The turbid supernatant is discarded. Additional buffer is added to the settled resin to transfer it as a slurry into the column.

When gel-type materials are used in chromatography, it is necesssary to let them swell in a large excess of initial eluent buffer for the period of time recommended by the gel manufacturer. The supernatant liquid should be replaced at least three times with fresh buffer solution. The fully swollen gel slurry is carefully poured into the column and allowed to settle for at least five minutes before the excess fluid is slowly drained away. An excessive flow rate during the draining will lead to restrictive packing of the swollen gel particles.

The feed solution should be applied to the column in the same buffer solution that will be used for the initial wash or elution. Flow rates are typically between 5 and 25 $mL/cm^2/hr$. Such low rates are required by the rate at which the adsorption equilibrium is attained with macromolecules. A pump should be used to maintain this low rate at a constant value. The column effluent should be collected in separate vials at small time increments.

For simple molecules with large differences in distribution coefficients, a single eluting solution may be used to develop the chromatogram. However, more complex materials, such as peptides and proteins, require a shift in the ionic strength of the eluent. This can be done stepwise or as a gradient. Semenza (34) has proposed the following rules for the proper choice of eluent:

(1) Use cationic buffers (Tris-HCl, piperazine-$HClO_4$, etc.) with anion resins and anionic buffers (phosphate, acetate, etc.) with cation exchange resins.

(2) With anion resins use decreasing pH gradients and with cation resins use rising pH gradients.

(3) Avoid using buffers whose pH lies near the pK of the adsorbent.

If the chromatographed solutes are to be isolated by solvent evaporation, the use of volatile buffers, such as carbonic acid, carbonates, acetates and formates of ammonium should be used.

If better resolution is required, it may be obtained by changing the type of gradient applied. A convex gradient may be useful in improving the resolution during the last portion of a chromatogram or to speed up separation when the first peaks are well separated and the last few are adequately spaced. A concave gradient can be used if it is necessary to imporve resolution in the first part of the chromatogram or to shorten the separation time whenpeaks in the latter portion are more than adequately spaced.

The relationship between elution behavior and logarithmic molecular weight for various substances has been plotted in Figure 4.23 for several Sephadex gels, in figure 4.24 for the Bio-Gel P series of materials and in Figure 4.25 for agarose materials. The linear rising portion of each curve defines the molecular weight range within which each gel gives optimum separation.

Example 4.4

An elution scheme was developed for the separation of a synthetic mixture of human serum albumin (HSA), bovine chymotrypsinogen (BCT) and chicken lysozyme (CL) on a column of Spheron phosphate (35). Elution was first carried out with a series of linear gradients of buffers by simultaneously increasing pH and ionic strength in the region from 0.05M sodium formate (pH 3.5) to 1M sodium acetate (pH 8). In the last gradient, 1M sodium acetate was used with the same buffer enriched to 1M sodium chloride.

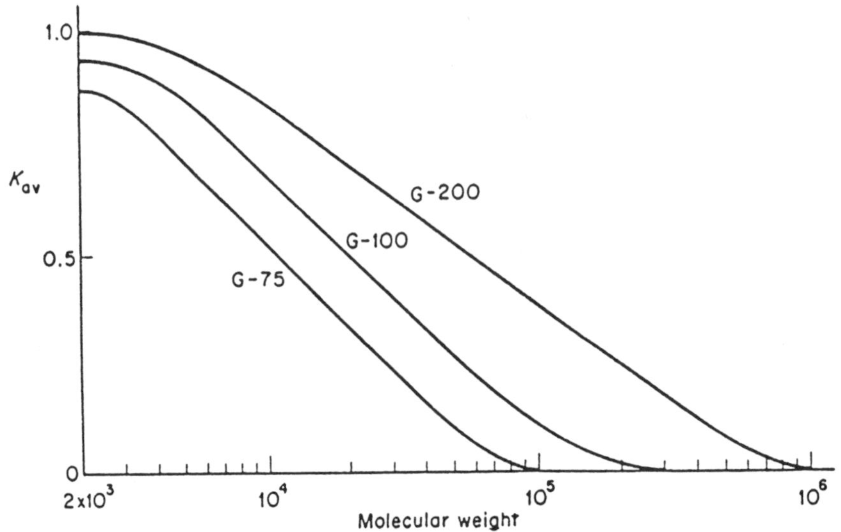

Figure 4.23. Relationship between K and molecular weight of globular proteins for Sephadex gels (Reference 15).

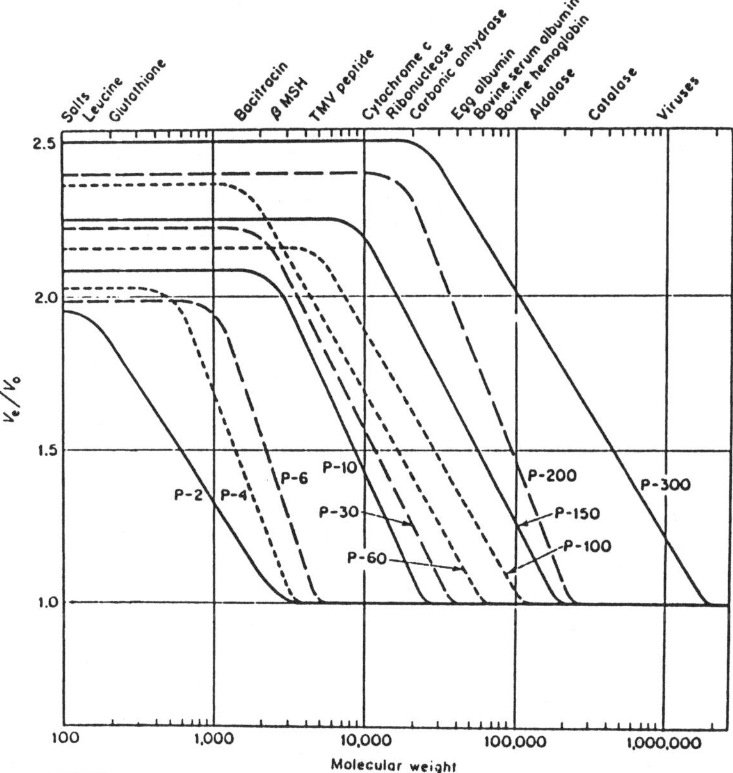

Figure 4.24. Relationship between elution volume and molecular weight on Bio-Gel P (Reference 15).

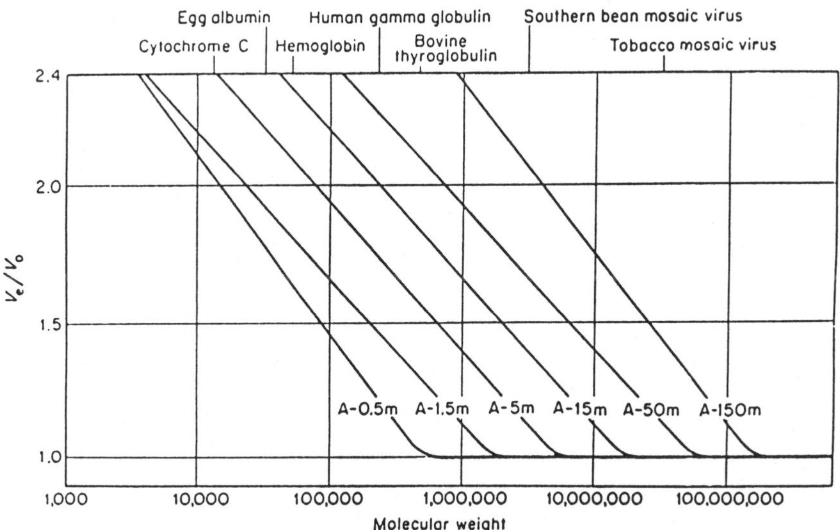

Figure 4.25. Relationship between molecular weight and elution volume on agarose materials (Reference 15).

Because HSA tended to be eluted with BCT under these conditions, a series of experiments were carried out with a gradient of ionic strength only, at each of the following pH values: 4, 5, 5.5, 6, 6.5 and 7.5. Only two peaks (HSA and BCT were combined) were observced in the experiments at low pH, whereas the proteins HSA, BCT and CL were separated at higher pH. A pH value of 6.5 was selected with the linear ionic gradient to give the separation shown in Figure 4.26.

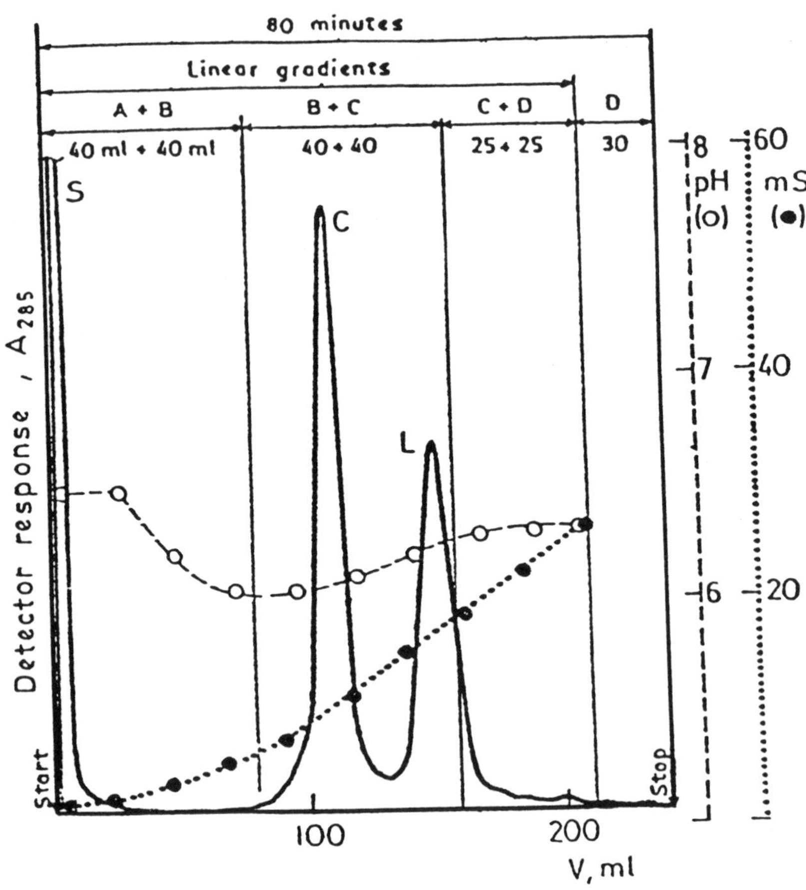

Figure 4.26. Separation of serum albumin (S), chymotrypsinogen (C) and lysozyme (L) on Sepheron phosphate 1000 of capacity 3.04 meg/g (Reference 35).

4.6 SCALE-UP PROCEDURES

The aim of scaling-up a chromatographic process is to

obtain the same yield and product quality in the same time period on laboratory, preparative and industrial scale. Laboratory analytical purifications tend to be optimized only for resolution of individual solutes. However, at the preparative and production scale, it is necessary also to maximize throughput. Table 4.8 (36) shows the effect of chromatographic operating parameters on resulution and throughput. While column length is a critical factor for resolution with gel filtation and isocratic elutions, it has little effect on resolution with gradient elution in adsorption chromatography. The wall effects on resolution are very noticeable with small radius columns, but decrease as the column length is increased.

Table 4.8: Effects of Process Parameters on Resolution and Throughput (36)

Paramter	Resolution varies with	Throughput varies with
Column Length (L)	L	1/L
Column Radius (r)	Some effect	r^2
Temperature (T)	Positive effect	T
Viscosity (η)	Negative effect	$1/\eta$
Sample Volume (V)	$1/\lvert V-V_{optimum}\rvert$	V
Flow Rate (J)	$1/\lvert J-J_{optimum}\rvert$	J

Voser and Walliser (37) viewed scale-up as a three step process involving selection of a process strategy, evaluation of the maximum and optimal bed height and finally column design. The selection of a process strategy involves choosing the direction of flow, frequency of backwashing, the operation with positive head pressure or only hydrostatic pressure, and the use of a single column or a series of columns in batch or semi-continuous operation. The maximum feasible bed height is determined by keeping the optimal laboratory-scale specific volume velocity (bed volume/hour) constant. The limiting factors will be either pressure drop or unfavorable adsorption/desorption kinetics since the linear velocity also increases with increasing bed height. Bed heights from 15 cm to as high as 12 m have been reported. The column diameter is then selected to give the required bed volume. The column design combines these column dimensions with the practical considerations of available space, needed flexibility, construction difficulty and flow distribution and dilution for the columns.

The scale-up considerations of column chromatography for protein isolation have been described by Charm and Matteo (38). When several hundred liters of a protein feedstream must be treated, the resin may be suspended in the solution, removed after equilibration by filtration and loaded into a column from which the desired proteins may be eluted. Adsorption onto a previously packed column was not recommended by them since they feared suspended particles would clog the interstices of the column, causing reduced flow rates and increased pressure drops across the column. The reduced flow rate may lead to loss of enzyme activity because of the increased time the protein is adsorbed on the resin.

It is important that all of the resin slurry be added to the column in one operation to obtain uniform packing and to avoid the formation of air pockets in the column. An acceptable alternative, described by Whatman (39), allows the addition of the adsorbed slurry in increments. When the resin has settled to a packed bed of approximately 5 cm, the outlet is opened. The next increment of the slurry is added after the liquid level in the column has dropped. It is important that the suspended adsorbent particles do not completely settle between each addition.

Stacey and coworkers (40) have used the relationships shown in Table 4.9 to scale up the purification from an 8 mg protein sample to a 400 mg sample. The adsorbent used in both columns was a Delta-Pak wide-pore C-18 material. When eluting the protein, the flow rate should change so that the linear velocity of the solvent through the column stays the same. The flow rate is proportional to the cross-sectional area of the column. The gradient duration must be adjusted so that the total number of column volumes delivered during the gradient remain the same. As with size exclusion chromatography, the mass load on the preparative column is proportional to the ratio of the column volumes. Figure 4.27 shows that the chromatogram from the 8 mg separation is very similar to that obtained for the 400 mg sample.

Many additional complicated equations have been developed for scaling up chromatography columns. For gel filtration, the same elution pattern is obtained with increases in the scale of operation when the dynamic similarity is

Table 4.9: Scale-Up Calculations (40)

	Small Scale	Preparative Scale
Column Dimensions	0.39 x 30 cm	3 x 25 cm
Flow Rate Scale Factor		
$\frac{(3.0)^2}{(0.39)^2} = 59$	1.5 ml/min	90 ml/min
Sample Load Scale Factor		
$\frac{(3.0)^2 \times 25}{(0.39)^2 \times 30} = 49$	8 mg	400 mg
Gradient Duration Calculation	40 min	33 min
	$\frac{1.5 \times 40}{(0.195)^2 \pi \times 30}$ = 16.7 Column Volumn	$\frac{90 \times (\text{Gradient Duration})}{(1.5)^2 \pi \times 25} = 16.7$ Grad. Duration = 33 min.

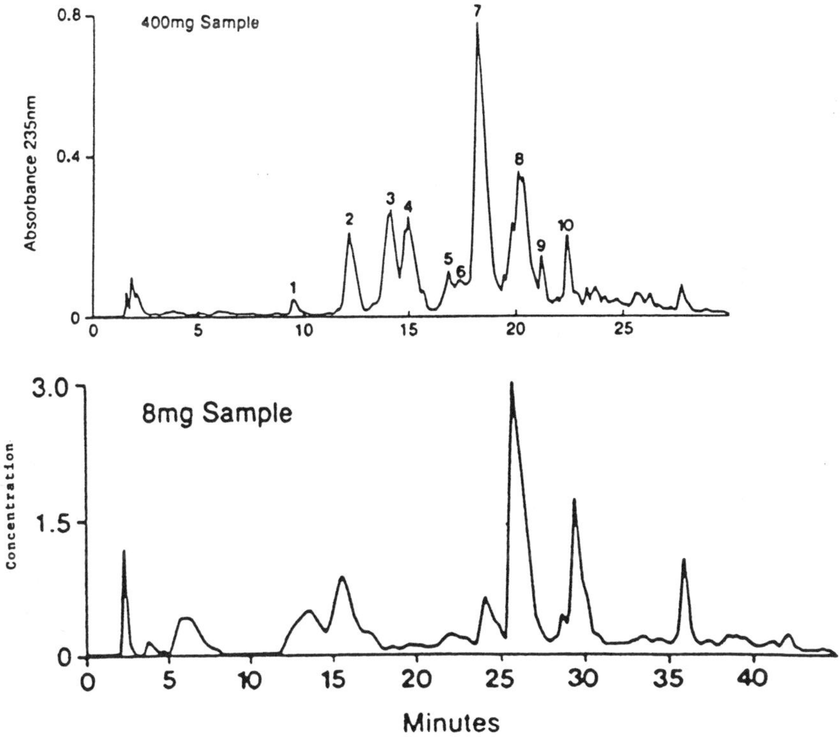

Figure 4.27. Laboratory and preparative scale separation of Cytochrome C digest (Reference 40).

maintained (41). This occurs when the small and scaled-up columns have the same L/D and $DV\rho/\mu$ ratios, where L is the column length; V is the flow rate per unit cross section; D is the column diameter; ρ is the liquid denisty; and μ is the liquid viscosity. It is also important that the volume of protein solution be kept at the same ratio to the volume of adsorbent.

Example 4.5 (38)

A Sephadex column, 1.5 cm in diameter and 40 cm in length, has been shown to purify 1 mL of protein solution at a flow rate of 100 mL/hr. The column dimensions and flow rate necessary to fractionate 200 ml of solution can be calculated as follows:

In the small column,

$$\text{Volume of adsorbent} = \frac{\pi\,(1.5)^2}{4}\,40 = 70.5\ \text{cm}^2 \quad (4.12)$$

$$\text{Cross section area} = 1.76\ \text{cm}^2 \quad (4.13)$$

$$\frac{\text{Volume of protein solution}}{\text{Volume of adsorbent}} = \frac{1}{70.5} \quad (4.14)$$

$$\frac{L}{D} = \frac{40}{1.5} = 26.6 \quad (4.15)$$

Since ρ/μ will be the same in both columns, it is only necessary that the values for DV be identical for both columns.

Therefore,

$$D\,V = 1.5 \times \frac{100}{1.76} = 85 \quad (4.16)$$

In the large column,

$$\frac{\text{Volume of protein solution}}{\text{Volume of adsorbent}} = \frac{1}{70.5} = \frac{200}{\text{Volume of adsorbent}} \quad (4.17)$$

$$\text{Volume of adsorbent} = 14{,}100\ \text{cm}^2 \quad (4.18)$$

$$\frac{L}{D} = 26.6 \quad \text{or} \quad L = 26.6\ D \tag{4.19}$$

Therefore:

$$\frac{\pi D^2}{4} \times 26.6\ D = 14{,}100 \tag{4.20}$$

$$20.9\ D^3 = 14{,}100 \tag{4.21}$$

$$D = 9\ \text{cm}, \quad L = 239\ \text{cm} \tag{4.22}$$

The flow rate is:

$$D\ V = 85 = \frac{(D)\ (\text{flow rate})}{\pi/4 \quad D^2} = \frac{9 \times \text{flow rate}}{\frac{\pi (9)^2}{4}} \tag{4.23}$$

$$\text{Flow rate} = 600\ \text{cm}^3 \text{ per hour} \tag{4.24}$$

Even though the cross-sectional area is 28 times greater in the scaled-up system, the flow rate only is increased by a factor of 6 to obtain the same elution pattern. Operating at a higher flow rate presents a risk of inferior protein separation.

Ladisch (42) has worked with a variety of column sizes ranging from 2 to 16 mm in diameter and 10 to 600 cm in length. His experience is that published semi-empirical scale-up correlations are useful in obtaining a first estimate on large scale column performance.

When scaling-up a chromatographic process, it may be necessary to change the order of certain steps from that used in the laboratory. Gel filtration, though a frequent first step at the laboratory scale, is not suitable for handling large scale feedstream volumes (43). When gel filtration is used to separate molecules of similar molecular weights, sample sizes may range from 1% to 5% of the total gel volume. Thus a 100 liter feedstream would require a gel filtration column of 2,000 to 10,000 liters. When one is separating a large molecule from a small molecule, as in desalting operations, the applied sample volume may be up to 30% of the total gel volume.

On the other hand, ion exchange chromatography is a very good first step because of its capacity of approximately 30 mg of protein per mL of resin. This capacity is relatively independent of feed volume. For the same 100 liter feedstream, only a 20 liter ion exchange column would be required.

4.7 INDUSTRIAL SYSTEMS

Packed bed, process scale high performance liquid chromatography (HPLC) equipment was first introduced by Millipore and Elf Aquitaine in 1982 (44). Since then, several companies, listed in Table 4.10, have entered the large scale HPLC market. The systems use packed beds at moderate pressures (30-140 bar). While there are substantial time savings in using these systems compared to other purification techniques, the short life of the packing material and its high cost continue to restrict this technique to applications that warrant the $100/kg separation cost.

Table 4.10: Manufacturers of Process Scale HPLC Equipment

Amicon	Danvers, MA
Dorr-Oliver	Stamford, CT
Elf Aquitaine (Varex in U.S.)	Rockville, MD
Millipore Corporation	Bedford, MA
Pharmacia	Piscataway, NJ
Separations Technology	Wakefield, RI
Y M C	Morris Plains, NJ

Example 4.6

The Waters Kiloprep Chromatography pilot plant is one example of the successful extension of an analytical chromatography process to the process scale. The ability to control the various operation parameters to scale up directly from the laboratory to the pilot plant and beyond to commercial production has been developed (45). Figure 4.28 illustrates how the performance of this larger system can be predicted from the data generated in an equivalent laboratory apparatus.

Voser and Walliser (37) have described the approaches different companies have devised so that fine and soft adsorbents may be utilized in large scale chromatography

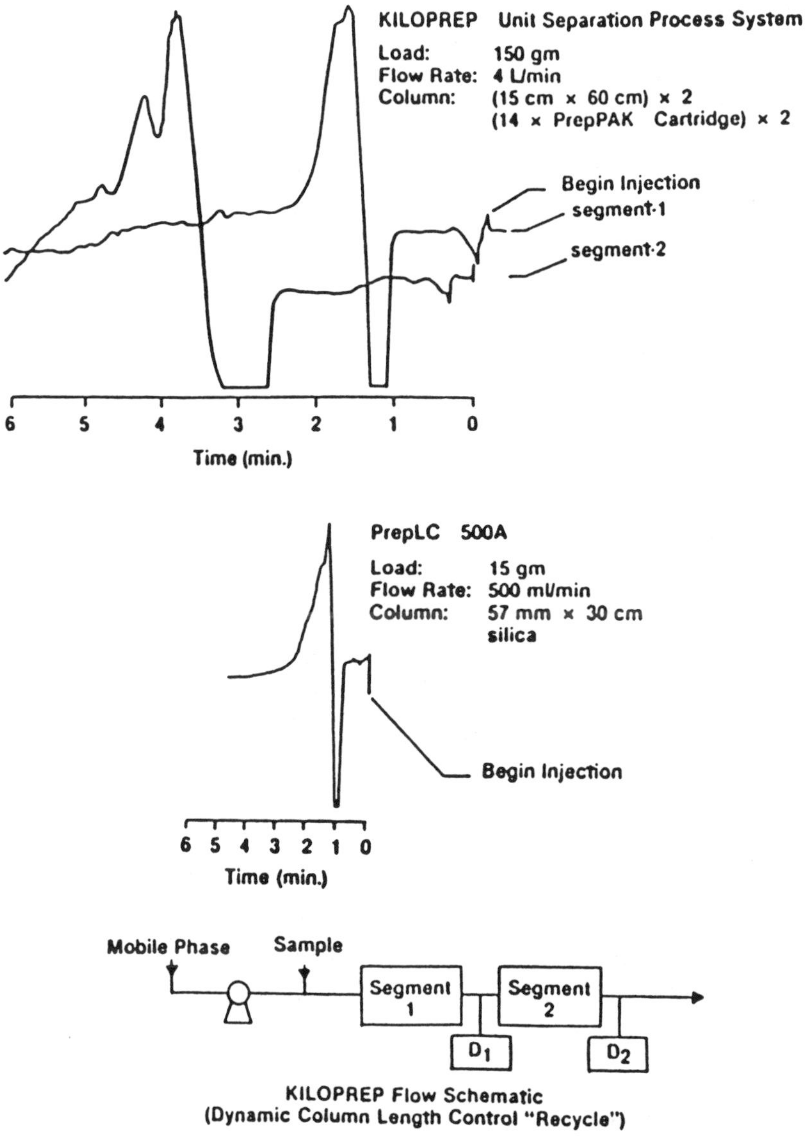

Figure 4.28. Comparison of column performance at equivalent loading (Reference 45).

operations. The scaling-up has usually been achieved by increasing the column diameter and using stack columns. The problem then becomes one of achieving uniform distribution of the feed solution over the entire resin bed surface, particularly when gradient elution is involved.

In the Pharmacia approach, the fluid input is split and distributed through six ports on the column end plates (46). At the entrance of each stream, an anti-jetting device spreads the liquid over a fine mesh net. A coarse net between the net and the end plate acts as both a support and a spacer. For the soft Sephadex gels G-50 to G-200, the maximum feasible bed height is 15 cm. Scale-up operations have used as many as six such squat columns in series (47). The drawbacks of this approach are the cumbersome adsorbent filling process, the possibility of air entrapment because of the split flow and the inability to regenerate the adsorbent in the column.

In Amicon/Wright columns, the flow is distributed through a carefully designed system of radial ribs cut out of the end plate. There is a single central port with a suitable anti-jetting device to reduce and divert the high velocity of the entering stream. The adsorbent bed is covered with a sintered plate. Sintered plates are claimed to be more efficient than mesh nets to achieve an even distribution. The larger the pore size of the sintered plate, however, the less efficient the system. Drawbacks of sintered plates are their tendency to adsorb substances on their very large surfaces and the possibility of fouling. These columns may also be stacked.

Whatman has developed a column with a new flow distribution system specifically designed for Whatman's cellulosic ion exchange resins. The bed height is 18 cm. The diameter of the first commercial unit was 40 cm and contained 25 liters of adsorbent. The resin bed is covered with a perforated plate with a high free surface. The slightly conical head plate covers an empty space and has a steep cone in the middle. The total empty head space is about 5% of the bed volume. The feed solution enters the steep cone tangentially in its upper part. The resulting rotary movement efficiently mixes the suprnatant liquid and allows gradient elution. It is claimed that filling and equilibration take only one hour.

AMF has developed an unconventional new approach with its ZETA-PREP cartridge. The cartridges consist of concentric polymer screens which bear the ionic groups and are supported by cellulosic sheets. The flow is radial from the outer rim toward a perforated central pipe. The available nominal cartridge lengths are from 3 cm to 72 cm with a constant diameter of about 7 cm throughout. Scaling up with this approach is quite straightforward. Single cartridges, each mounted in a housing, can be combined to a multi-cartridge system. For such a system, flow rates up to 12 L/min and bovine serum albumin capacities of 1400 g are claimed. The present ion exchange functionalities available are DEAE, QAE and SP.

The stepwise transition from high pressure liquid chromatography to medium pressure chromatography, such as described for the preparation of pectic enzyme (48), illustrate the progression toward large scale industrial application of the techniques developed in analytical laboratories. The pressure in these medium pressure chromatography applications is only 6 bar instead of the 100 to 150 bar associated with HPLC. The lower pressure results in longer processing times (about one hour) compared to the 5 to 20 minutes required for an analytical determination with HPLC.

Studies, such as the one by Frolik and coworkers (48), which examine the effect and optimization of variables in HPLC of proteins, can be expected to contribute to the implementation of this type of protein resolution technique into future commercial biotechnology processes.

The first chromatographic systems capable of handling more than 100 kg/day were merely scaled-up versions of laboratory chromatography (50,51). Even with some of these systems it was necessary to recycle a portion of the overlap region to have an economical process. A typical example of such a system would be the Techni-Sweet System of Technichem (52) used for the separation of fructose from glucose. The unique distributors and recycle system are designed to maximize the ratio of sugar volume feed solution per unit volume of resin per cycle while at the same time minimizing the ratio of volume of water requried per unit volume of resin per cycle.

The flow through the Technichem system is 0.56 $m^3/(hr\text{-}m^3)$ with a column height of 3.05 m. The feed solution contains 45% dissolved solids, and a feed volume equal to 22% of the volume of the resin is added to the column each cycle. The rinse water added per cycle is equal to 36% of the volume of the resin. This is much less rinse water than the 60% volume that was required by the earlier systems.

This technique is known as the stationary port technique since the feed solution and the desorbent solution are always added at the same port and the product streams and the recycle stream are always removed from another port. Technichem and Finn Sugar manufacture chromatography systems which utilize the stationary port technique.

One of the earlier attempts (53) at industrial chromatography used an adaptation of the Higgins contractor for the ion exclusion purification of sugar juices. The physical movement of the low crosslinked resin caused attrition as it was moved around the contactor. It was also difficult to maintain the precise control needed on flow rates because of the pressure drop changes and volume changes of the resin as it cycled from the mostly water zone to the mostly sugar solution zone.

An alternate approach (54) utilizes moving port or pseudo-moving bed techniques. With these techniques, the positions on the column where the feed solution is added and where the product streams are removed are periodically moved to simulate the countercurrent movement of the adsorbent material. At any given time the resin column can be segmented into four zones (Figure 4.29). Zone 1 is called the adsorption zone and is located between the point where the feed solution is added and the point where the fast or less strongly adsorbed component is removed. In this zone the slow or more strongly adsorbed component is completely adsorbed onto the ion exchange resin. The fast component may also be adsorbed, but to a much smaller extent. The second zone, Zone 2, is the purification zone and is located between the point where the fast component is removed and the point where the desorbent solution is added. Zone 3 is called the desorption zone and is between the point where the desorbent is added and the point where the slow component is removed. In this zone the slow component is removed from the resin and exits the column. The final zone, Zone 4, is called the buffer zone and is located between the point where the slow component is removed and the point where the feed solution is added. There is a circulating pump which unites the different zones into a continuous cycle.

Different sections of the column serve as a specific zone during the cycle operation. Unlike the stationary port technique, the liquid flow is not uniform throughout the column. Because of the variations in the additions and withdrawals of the different fluid streams, the liquid flow rate in each of the zones will be different.

With such a system one must slowly develop the chromatographic distribution pattern through the different zones. It may take from 8 to 36 hours for the pattern to be established. Another practical consideration is that the recirculation system must represent a small (<10%) portion of a single zone to prevent unacceptable backmixing which would alter the established chromatographic pattern.

The flow rate and the pressure drop per unit length of the chromatographic column are much lower for the stationary port compared to the moving port system. Also the stationary port method is much less caapital intensive. The moving port technique, however, is calculated to require only one third of the column volume and ion exchange volume and two thirds of the desorbent volume compared to the stationary port technique.

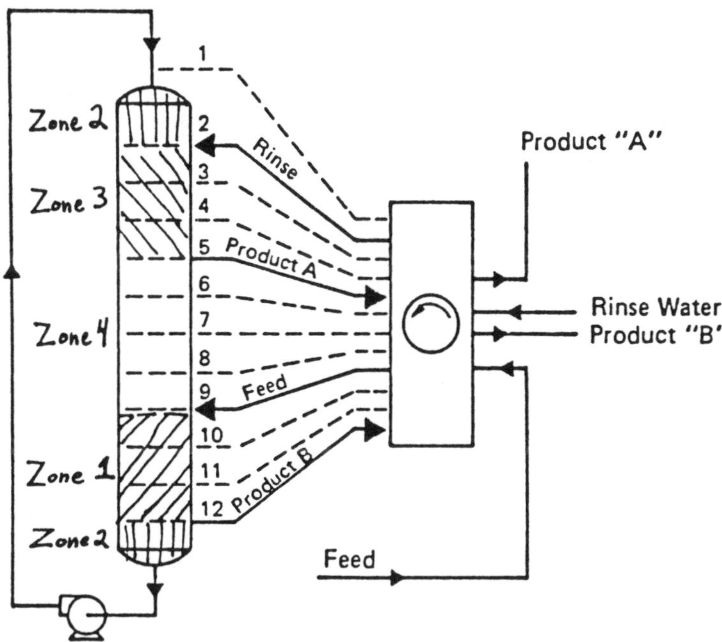

Figure 4.29. Moving port chromatographic column with four zones for continuous chromatographic separation (Reference 54).

After the expiration of the UOP patent involving the rotary valve, there have been several modifications to the moving port technique by Amalgamated Sugar (55), Illinois Water Treatment (56) and Mitsubishi (57). Each manufacturer has its own approach for the establishment and control of the chromatographic pattern. These are the subject of confidentiality agreements for specific applications.

Broughton and Gembicki have used a model of equilibrium theoretical trays with entrainment to describe mathematically the chromatographic separation of pseudo-moving bed operations (58). Two parameters are adjusted with this model to reproduce the experimental concentration profiles:

$$\text{Number of theoretical trays:} \quad n = KkH/Lz \qquad (4.25)$$

$$\text{Axial mixing ratio:} \quad e = E/L = Kk/DL^2 \qquad (4.26)$$

where

$$z = m \ln\left(\frac{m}{m-1}\right) \qquad (4.27)$$

K is the linear equilibrium constant; k is the mass transfer coefficient; H is the bed height; L is the liquid flow rate; m is the adsorption factor which is equal to the pore circulation rate K/L; E is the entrainment rate and D is the axial diffusion coefficient. Values of n and e are calculated for each zone since each zone is treated separately according to the component whose concentration is changing most rapidly. It should be noted that this model views both the adsorbent and the eluent as moving. Therefore it is necessary to adjust the flow rate of the eluent to account for the stationary position of the adsorbent when carrying out experiments based on the model results.

A comparative mathematical modeling of the pseudo-moving bed system with the conventional chromatography system has shown that the conventional system requries 3 to 4 times more adsorbent and twice as much desorbent solution. The reason for this is that in the pseudo-moving bed system, every part of the bed can be shown to be performing useful work at all times, with respect to the primary function of each zone. In the conventional system, on the other hand, some portion of the bed at all times is not contributing to the separation process.

Barker and Thawait (59) have used the theoretical plate model to show that the concentration of a component in a given plate element is given by:

$$C_n = C_{n-1}\left(1 - \exp\left(\frac{-Q\,\Delta t}{V_1 + K_d V_2}\right)\right) + C_n^o \exp\left(\frac{-Q\,\Delta t}{V_1 + K_d V_2}\right) \quad (4.28)$$

where C_n is the component concentration in the mobile phase leaving plate n; Q is the mobile phase flow rate; Δt is the time increment; V_1 is the mobile (fluid) phase volume of a plate; V_2 is the stationary (adsorbent) phase volume of a plate and K_d is the distribution coefficient. Each component can be considered individually assuming there is no interaction between components. The sequential nature of the switching valves is modeled by stepping the concentration profile backwards by one zone at the end of the switch period. The typical switch time is 5 to 15 minutes for commercial units. The one difficulty with using the model is the selection of a distribution coefficient since the measured value changes with concentration. Figure 4.30 shows the agreement between experimental results and those calculated using Equation 4.27.

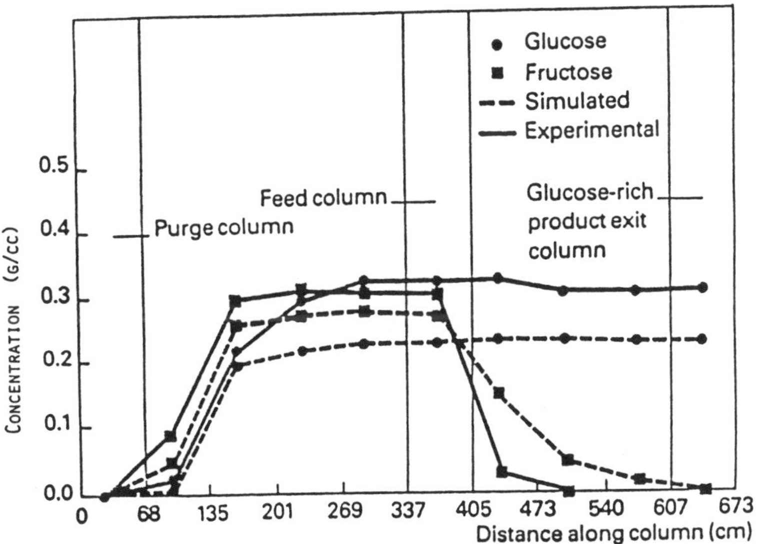

Figure 4.30. Experimental and simulated concentration profile for glucose-fructose separation (Reference 59).

Wankat (60) proposed a hybrid system which has some of the characteristics of both elution chromatography and the pseudo-moving bed system. During the feed pulse, the feed position was moved continuously up into the column at a velocity that lies between the two solute velocities. The eluting solvent was continuously fed into the bottom of the column. Elution development with solvent was used when the feed pulse was over. This method reduces irreversible mixing of solutes near the feed point. Wankat and Ortiz (61) have used this system for gel permeation chromatography and claim improved resolution, narrower bands and higher feed throughputs compared to conventional systems. McGary and Wankat (62) have had similar results applying it to preparative HPLC. His technique uses less adsorbent and produces more concentrated products compared to normal preparative chromatography, but more adsorbent and less concentrated products than pseudo-moving bed systems. Wankat (63) has proposed that his system will be of most value for intermediate size applications or when only one product is desired.

The key items to identify when considering an industrial chromatographic project are the capital for the equipment, yield and purity of the product, the amount of dilution of the product and waste stream, the degree of flexibility the computer controls allow, the expected life of the ion exchange material and whether the equipment allows for periodic expansion of the resin.

New techniques are continuing to be developed which can be expected to be used in future specialized industrial applications. Multi-segmented columns have been demonstrated for the preparative purification of urokinase (64). Begovich and coworkers (65,66) have developed a technique for continuous spiral cylinder purifications which may allow separation on the basis of electropotential in addition to the selective affinity of the adsorbent resin for the components in solution. A schematic of this device is shown in Figure 4.31.

Another new technology that offers promise for commercial biotechnology purifications is the use of parametric pumping with cyclic variations of pH and electric field. This has been described by Hollein and coworkers (67). They worked with human hemoglobin and human serum albumin protein mixtures on a CM-Sepharose cation

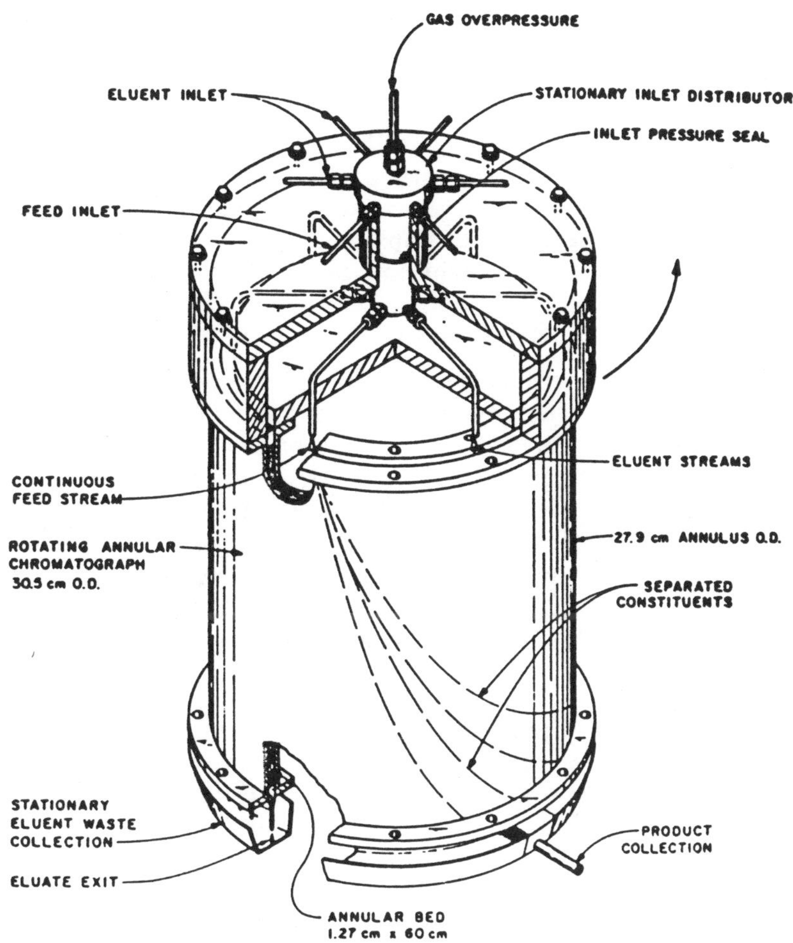

Figure 4.31. Schematic of the pressurized continuous annular chromatograph (Reference 65).

exchanger. The extensive equations they reported for parametric separations allow analysis of other systems of two or more proteins which may be candidates for this type of separation.

4.8 CHROMATOGRAPHY APPLICATIONS

Column chromatography techniques have been incorporated into the commercial purification scheme for fermentation products, biomaterials and organic chemicals. While the majority of these applications are on a small scale (less than 500 kg/month but greater than 10 g/month), several large industrial scale applications have arisen in the last decade. The extraction of sugar from molasses, the separation of glucose from fructose, the separation of polyhydric alcohols, the separation of xylene isomers and the separation of amino acids are carried out in industrial scale operations preparing thousands of metric tons of purified material each year. Examples of pilot scale and large industrial chromatography will now be given.

4.8.1 Amino Acid Separation

Preparatory scale separation of complex amino acid mixtures was reported by Moore and Stein (68). Today amino acid mixtures obtained from fermentation or from protein hydrolysis are separated by specific elution regimens after the amino acids are adsorbed on the ion exchange resin, as reported in Chapter 3. Although many of the commercial operations use standard ion exchange resins, such as Dowex HGR or Duolite C-25, product dilution and yield could be substantially improved by using the improved separation resins.

The use of ligand exchange with acrylic resins has been shown to provide advantages over conventional ion exchange resins in separating amino acids (69). Copper(II) carboxylate functionality provided the best kinetic performance and separation.

The separation of d- and l-isomers of amino acids may also be accomplished by using a derivatized copper form of the iminodiacetate ion exchange resin (70). Contrary to earlier assumptions, it is not necessary to have an optically active functionality on the ion exchange resin to separate optical isomers. With the increased importance of l-phenylalanine and l-tryptophan, it is expected that industrial chromatographic separations will soon include the separation of these amino acids from their d-isomers. Table 4.11 shows the results at the preparatory scale.

Hamilton (71) has measured the effect of resin crosslinking on the selectivity for amino acids. This is shown in Figure 4.32. Since commercial resins may vary ±0.5% in crosslinking from one batch to another, this could be an important factor in assuring reproducible separations when different batches of the same resin are used. This would especially be the case for the separation of alanine from glucosamine in a system designed for use with an 8% crosslinked resin.

Table 4.11: Separation and Recovery of d- and l-Isomers (70)

Amino Acid	Recovered in Pure Streams %d	%l	Residual Solids in Overlap Region % d,l
d,l-Tyrosine	82.4	79.2	19.2
d,l-Phenylalanine	100.0	93.0	3.6
d,l-Hystidine	97.0	100.8	1.1
d,l-Tryptophan	98.8	93.6	3.8
d,l-Cystine	97.0	100.0	1.5

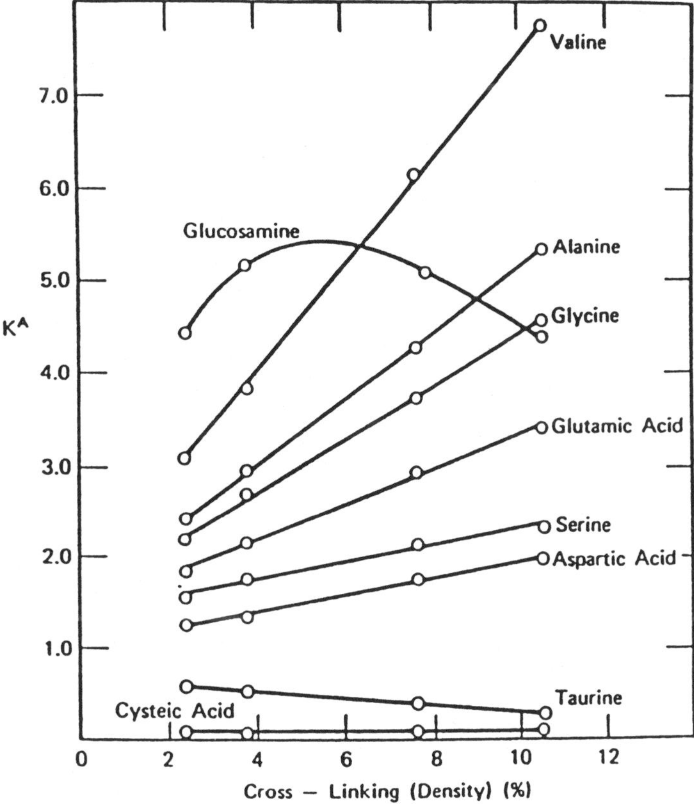

Figure 4.32. Selectivity coefficient ($K^{\text{Å}}$) for amino acid by a sodium form cation resin as a function of the resin crosslinking (Reference 71).

4.8.2 Antibiotics

One of the earliest chromatographic purifications of antibiotics was the separation of neomycin B from neomycin C using alumina (72). Weston and Putter (73) have used strongly hydrophilic anion resins (Sephadex A25), weak base resins (Amberlite IRA 68) and non-ionic crosslinked macroporous polystyrene polymer beads (XAD 2) to purify antibiotics in chromatography columns containing as much as 45 liters of adsorbent.

Aureomycin, an anti-tumor antibiotic, was isolated from a fermentation broth (74) by successive application of ion exchange chromatography with Amberlite IRA 93 and DEAE Sephadex, gel filtration on Sephadex G-50 and hydrophobic chromatography on Octyl-Sepharose CL-4B. The initial 64 liters of harvested broth was purified and concentrated to a 430 mg lyophilized powder.

Silica gel was used in the adsorption chromatography purification of carminomycin (75). This is an anthracycline antibiotic with known anti-tumor properties. The elution was carried out using chloroform and methanol at a 17 to 3 volume ratio. The fractions collected contained less than 1% impurities.

Cephalosporin antibiotics produced by fermentation of *Streptomyces* sp. S3907C were purified by chromatography with QAE Sephadex in the chloride form (76). A gradient elution of lithium chloride (0 to 10%) was applied in a linear manner. A crude fraction of 19 g from the 400 liter broth resulted in a purified active powder of 226 mg.

Sakuma and Motomura (77) have separated saikosaponins a, c and d in preparative quantities using ODS silica gel chromatography. These saponins, shown in Figure 4.33, have been used in anti-inflammatory medicinal applications. A three step gradient elution with acetonitrile and water gave a recovery of more than 90% with less than 1% impurity in each of the components. From the initial 10 g of crude saponin, 403 mg of saikosaponin c, 1210 mg of saikosaponin a and 1604 mg of saikosaponin d were obtained in a single preparative chromatogram.

4.8.3 Glucose-Fructose Separation

Glucose-fructose syrups may be obtained by the

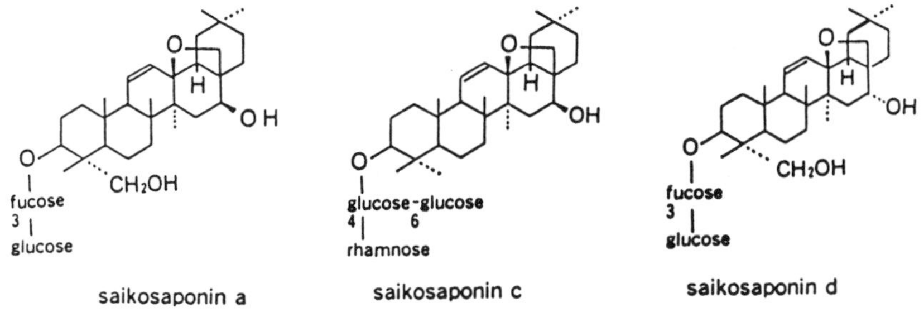

Figure 4.33. Chemical structure of saikosaponins (Reference 77).

inversion of sucrose or by the enzymatic conversion of starch. While rice and wheat starch have been used to make glucose-fructose syrups, the most common starch source is corn. The resulting gyrup is called high fructose corn syrup (HFCS). The enzymes convert the starch to a syrup that contains 55% glucose, 42% fructose and 3% oligosaccharides. While that level of fructose has sufficient sweetness for some applications, it is necessary to increase the fructose concentration to at least 55% of the dissolved solids in syrup to have a sweetness comparable to sucrose. The major industrial chromatographic application in the United States is this separation of HFCS into an enriched fructose stream and a glucose stream.

The separation of fructose from glucose was first reported by LeFevre (78,79) and Serbia (80) in 1962 using the Ba, Sr, Ag and Ca form of 4% crosslinked cation resin. The glucose and the higher saccharides do not form as strong a complex with these ions as the fructose does, so the glucose and higher saccharides appear first in the effluent, followed by the glucose-fructose mixture and then the relatively pure fructose. The Ca form of the resin is preferred in industrial chromatography.

Ghim and Chang (81) have reported the effect of different ionic forms of anions and cations in separating fructose from glucose using the moments of the elution curve. Table 4.12 shows the distribution ratios determined by them. It should be noted from the distribution coefficients that the order of elution would be fructose and then glucose for the anion resins and glucose followed by fructose for the cation resins. While it is possible to separate fructose from glucose using anion resins, the industrial chromatography applications only use cation resins. This is probably because the cation resins generally are more durable and less expensive than anion resins.

Table 4.12: Effect of Counterion on the Distribution Ratio (81)

(Dowex 1-X8 was used for anion resin and Dowex 50WX8 for the cation exchanger. Resin size was 200-400 mesh)

Ionic Form	Glucose	Fructose
SO_3^{-2}	0.45	0.27
CO_3^{-2}	1.43	1.00
$H_2PO_4^-$	0.16	0.08
Ca^{++}	0.30	0.80
Sr^{++}	0.30	0.80
Zn^{++}	0.20	0.30

Recently, zeolite and resin adsorbents have been compared in the chromatographic separation of glucose-fructose mixtures (82). The resin adsorbents were seen to have a higher equilibrium separation factor than the CaY zeolite. However, the resins offer greater resistance to mass transfer, resulting in little difference in overall performance between zeolites and resins in pseudo-moving bed and pulse chromatographic systems.

In spite of the great practical interest in the separation of glucose and fructose, there are only a few reports (83) covering their separation in industrial processes. Most available references deal with one or another specific chromatographic process (52,57,84) for carrying out the separation. These processes were covered in Section 4.7.

4.8.4 Glycerol Purification

One of the first industrial chromatographic separations proposed was for the purification of glycerol (85). In this case, elevated temperatures (80°C) were found to aid the separation of glycerol from salt since aqueous glycerol solutions are quite viscous at the concentrations necessary for practical operation. Figure 4.34 shows the separation profile for a solution of 10% NaCl and 36.3% glycerine.

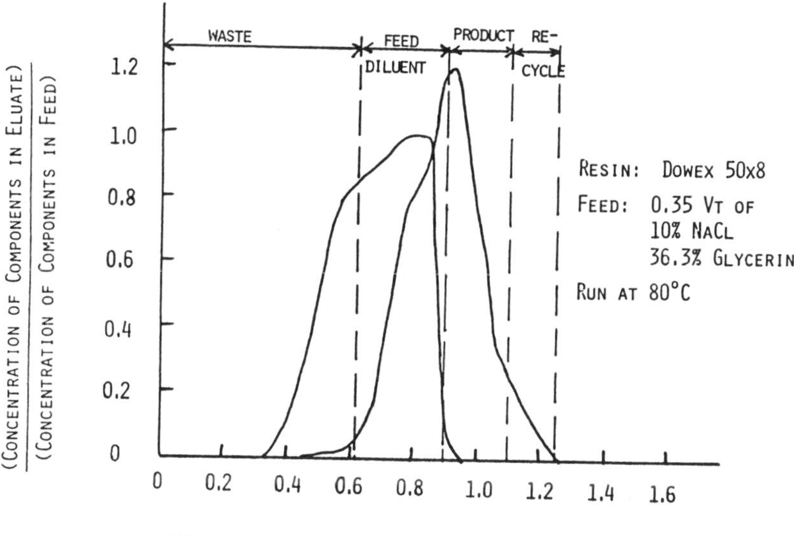

Figure 4.34. Separation profile for a solution of 10% NaCl and 36.3% glycerol (Reference 85).

Sargent and Rieman (86) proposed a chromatographic method for separating non-ionic organic solutes from one another by using aqueous electrolyte solutions as the eluent. As Figure 4.35 shows, this would lead to several changes per cycle in the composition of the eluent. This approach has only been feasible for laboratory or small scale preparatory separations. Recently, studies with ammonium carbonate solutions as the eluent for separating fermentation products containing amino acids showed that various strengths of the eluent were effective in amino acid separations and that $(NH_4)_2CO_3$ could be separated economically from the individual amino acid streams as the gaseous NH_3 and CO_2 and recycled for the next elution cycle.

4.8.5 Oligosaccharide Removal

The presence of 3 to 5% oligosaccharides in the high fructose syrup necessitates their removal during industrial chromatography of glucose and fructose since the

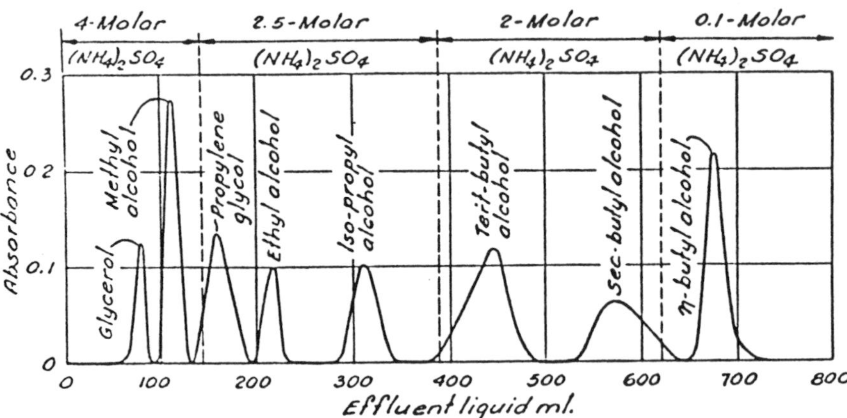

Figure 4.35. Elution profile and desorbent changes for the separation of alcohols (Reference 86).

oligosaccharides would decrease the sweetness of fructose or would hinder the enzymatic isomerization of the glucose. A process for removing oligosaccharides has been described by Hirota and Shioda (87). They used a strong cation resin in the calcium form to obtain a fructose stream that had 94 to 97% of the dissolved solids as fructose and less than 2% oligosaccharides. The glucose stream had a purity of 79 to 89% with an oligosaccharide concentration of 9 to 20%. The oligosaccharide stream had a dissolved solids oligosaccharide content of 64 to 76%. While this may not seem acceptable from an analytical chromatography standpoint, the low level of oligosaccharides in the feed solution (7 to 14%) and the high purity of the fructose product stream make the process acceptable for industrial chromatographic removal of a waste component. Such a process would also utilize the same resin and equipment that is used to separate the fructose from the glucose.

4.8.6 Peptides and Proteins

The now classic procedure of Moore and Stein (68) has been applied to the analytical separation of thousands of biochemical preparations and to the isolation of preparative quantities of biological materials (Table 4.13). Additional reviews of these activities have been described by Weaver (118), by Morris and Morris (119) and by Gordon (120).

Table 4.13: Proteins Purified by High Performance Ion Exchange Chromatography (88)

PROTEIN	COLUMN	RESIN TYPE	REFERENCES
Lactate dehydrogenase isoenzymes	DEAE-glycophase	weak base	89,90,91,92,93
	SynChropak AX300	weak base	91
Creatine kinase isoenzymes	DEAE-glycophase	weak base	92,93,94,95
	Synchropak AX300	weak base	91,96
Alkaline phosphatases	DEAE-glycophase	weak base	97
Hexokinase isoenzymes	Synchropak AX300	weak base	98
Arylsulfatase "	DEAE-glycophase	weak base	99
Hemoglobins	Synchropak AX300	weak base	100,101,102,103
	DEAE-glycophase	weak base	89,90
	IEX 545 DEAE	weak base	104
	IEX 535 CM	weak cation	104
	BIORex 70	weak cation	105
Cytochrome C	CM-polyamide	weak cation	106
Lysozyme	CM-polyamide	weak cation	106
Myoglobin	CM-polyamide	weak cation	106
	IEX 535 CM	weak cation	107
Soybean trypsin inhibitor	CM-glycophase	weak cation	90
Interferon	Partisil SCX	strong cation	108
Lipoxygenase	SynChropak AX300	weak base	109
Trypsin	DEAE-glycophase	weak base	90
Chymotrypsinogen	SP-glycophase	strong cation	90
	IEX 535 CM	weak cation	107
Immunoglobulin G	SynChropak AX300	weak base	110
Ovalbumin	SynChropak AX300	weak base	111
	IEX 545 DEAE	weak base	112
Albumin	SynChropak AX300	weak base	111
	DEAE-glycophase	weak base	90
	IEX 545 DEAE	weak base	107
Apolipoproteins	SynChropak AX300	weak base	113,114
Adenylsuccinate synthetase	SynChropak AX300	weak base	115
Insulin	Partisil SCX	strong cation	116
	IEX 535 CM	weak base	107
β-Lactoglobulin	Partisil SCX	strong cation	116
Carbonic anhydrase	Partisil SCX	strong cation	116
Monoamine oxidase	SynChropak AX300	weak base	117

DEAE and CM-glycophase are products of Pierce Chemical Company, Rockford, Illinois.

SynChropak AX300 is produced by SynChrom, Linden, Indiana.

IEX 535 CM and 545 DEAE are the products of Toya Soda Company, Yamaguchi, Japan.

Bio-Rex 70 is supplied by Bio-Rad, San Francisco, California.

Partisil SCX is manufactured by Whatman, Clifton, New Jersey.

Kelley and coworkers (121) have examined the use of gel chromatography for the industrial purification of proteins. Those applicatons reported only included those with column volumes greater than 4 liters (Table 4.14). The first seven examples (122-127) used soft, compressible gels, like Sephadex G-200 and Ultrogel AcA34. Productivities varied from 0.0016 to 0.045 liters feed solution per liter of gel for a 20 hour day (L/L/d). The average value was about 0.25 (L/L/d. The next three examples (128-130) used less compressible gels, such as Sepharose 4B, Sephadex G-75 and G-50. The productivities were on the order of 0.1 L/L/d, about 4 times higher than for the compressible gels. Finally, the productivity of a continuous annular chromatograph (131) was found to be approximately 0.37 L/L/d, 15 times higher than the compressible gels. This points out that the restricted productivities often associated with gel chromatography are the result of the predominant use of soft, compressible gels.

Table 4.14: Large Scale Gel Chromatography Application (121)

Protein	Separation Medium	Bed Demensions (D x L)	Productivity	Reference
		(cm x cm)	(l feed/l gel/ day)	
E. coli Iso-leucyl-t-RNA transferase	Sephadex G-200	21.5 x 80	0.026	122
E. coli Methionyl-t-RNA transferase	Sephadex G-200	14 x 180	0.0016	123
Citrobacter L-asparaginase	Sephadex G-200	4.2 L	0.007	124
Rape Seed Proteins	Sephadex G-200	45 x 85	0.025	125
Transferrin from Cohn Fraction IV	Sephadex G-200	45 x 75	0.019	126
Plasma Alkaline Phosphatase	Ultrogel AcA 34	16 x 100	0.045	127
Plasma Cholinesterase	Ultrogel AcA 34	9.3 x 90	0.026	127
Thyroglobulins	Sepharose 4B	37 x 45	0.125	128
Insulin	Sephadex G-50	37 x 90	0.071	129
Whey Protein Concentrate	Sephadex G-75	37 x 15	0.128	130
Continuous, annular Chromatograph	Sephadex G-75		0.37	131

The chromatographic purification of recombinant protein has been achieved with minimal effect on its tertiary structure and no effect on its biological activity using peptide fusion (132). Sassenfeld and Brewer used rDNA technology to produce human urogastrone fused to a C-terminal polyarginine. This fused polypeptide allowed a two-step chromatographic purification using SP-Sephadex C-25.

In the first step, a substantial purification was obtained due to the high basicity of the polyarginine fused protein. The polyarginine was then removed using carboxypeptidase. The second pass of the digested material through an SP-Sephadex C-25 column gave urogastrone at 95% purity with a yield of 44% in a pilot plant (32 liter feedstream) purification. The resolution from the previously co-chromatographed contaminants is shown in Figure 4.36.

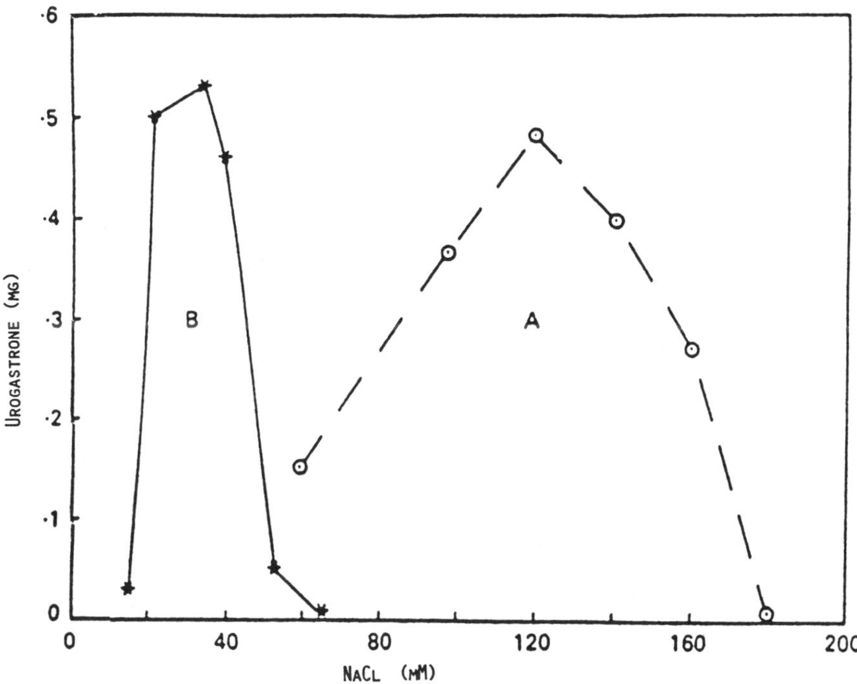

Figure 4.36. Purification of urogastrone on SP-Sephadex. Peak A is polyarginylurogastrone and Peak B is peak A after digestion with carboxypeptidase B and rechromatographing (Reference 132).

The use of non-functionalized resins (133) such as XAD-4 has been reported for the separation of amino acids and polypeptides. The authors were able to identify specific acid and base eluents that allowed di- and tripeptide separation at high purity and in good yield. The separation of a mixture of enkaphalin peptides is shown in Figure 4.37. While this work is still at the preparatory scale, it is anticipated that pharmaceutical companies will be utilizing this technique and other laboratory chromatographic techniques to obtain pure polypeptides for medicinal purposes.

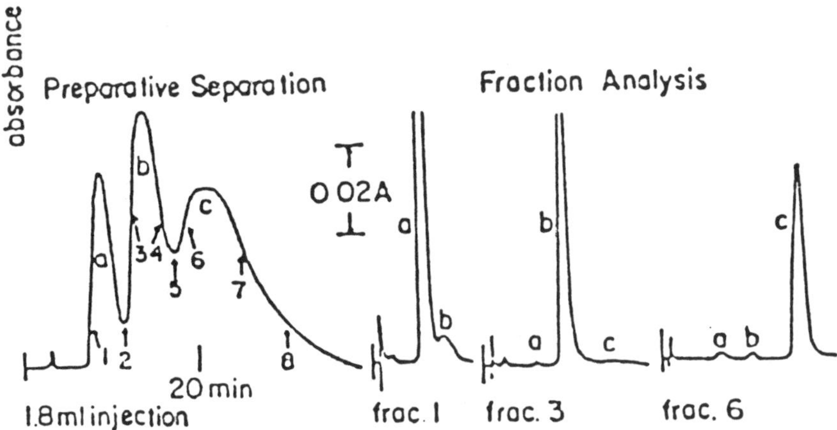

Figure 4.37. Analysis of feed and enkaphalin peptide fractions separated on XAD-4 (Reference 133).

Kato and coworkers (134) have examined the effect of eluent gradient, column length and sample loading on the retention and resolution of the proteins in commercial ovalbumin. The study used TSK-GEL DEAE ion exchange resins with a 10 micron diameter. While the two column lengths studied had little influence on column performance, the effect of eluent gradient was higher resolution as the gradient time became longer and lower retention as the gradient became less steep (Figure 4.38). The sample loading needs to be less than 1 to 2 mg per unit column cross section (cm^2) in high performance ion exchange chromatography. As the sample loading increased, the peak width became wider, elution was faster and the peak shape began tailing.

Dizdaroglu and coworkers (135) used a silica-based bonded-phase weak anion exchanger to isolate peptides with up to 30 residues. Recoveries of over 80% were obtained because of the volatility of the eluent (gradient of increasing 0.01M triethylammonium acetate buffer at pH 6.0 into acetonitrile) employed. A comparison of this approach with reverse phase HPLC showed the differences in the selectivities for each method (136). The combined application of these two different separation principles provides a greater opportunity for the complete separation of a given mixture of peptides into individual components.

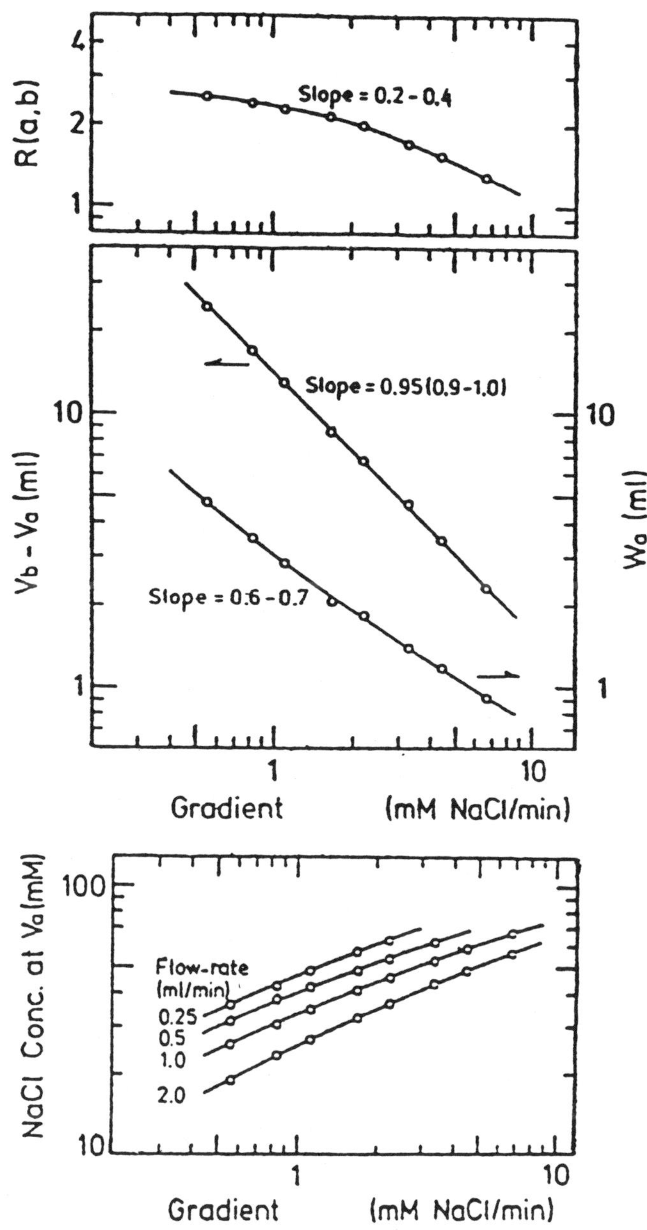

Figure 4.38. Dependence of resolution ($R_{a,b}$), peak interval (V_b - V_a) and peak width (W_a) on eluent concentration gradient change rate (Reference 134).

The recently developed Mono Q and Mono S resins from Pharmacia have been used in the recovery of several enzymes, while maintaining the enzyme activity. This is shown in Table 4.15 (137). That same article describes the use of eluent selection which can facilitate the use of these resins to purify and recovery proteins. The Mono S resin has been reported to isolate the five separate components of purified human bronchial proteinase inhibitor (138).

Table 4.15: Enzyme Activity Recoveries from Mono Q and Mono S Resins (137)

Protein	Mono Q Recovery, % Relative Activity	Mono S Recovery, % Relative Activity
β - Glucuronidase	106%	----
β - Glucosidase	----	93%
Phosphodiesterase	80%	----
Creatine kinase	90%	----
Enolase	----	95%
Lactate dehydrogenase	----	102%
Aldolase	----	94%

Carboxylic ion exchange resins have been used in the separation of proteinaceous materials using the resin's capability of dissociating only in certain pH ranges (139). At acidic conditions below pH 4.5, ion exchange resins based on methacrylate or acrylate polymers are undissociated. Thus, proteins may be adsorbed on the resins at pH 4.5. Following adsorption of the protein mixture, individual protein fractions may be separately removed from the column by the controlled increase of the pH of the eluent (Figure 4.39).

Adenosine, obtained by the fermentation of a *Bacillus* strain, ATCC No. 21616, was purified by column chromatography (140). The supernatant from the fermentation broth was adjusted to pH 4.0, then passed through a column of activated carbon for adsorption of adenosine. The column was washed with water; then the adenosine was eluted with a solution of methanol/isooctanol/ammonia/water (50:1:2:47). The eluate is concentrated and then separated by chromatography with an anion exchange resin (Dowex 1, chloride form) to get a pure adenosine fraction. That fraction is concentrated and then crystallized from ethanol.

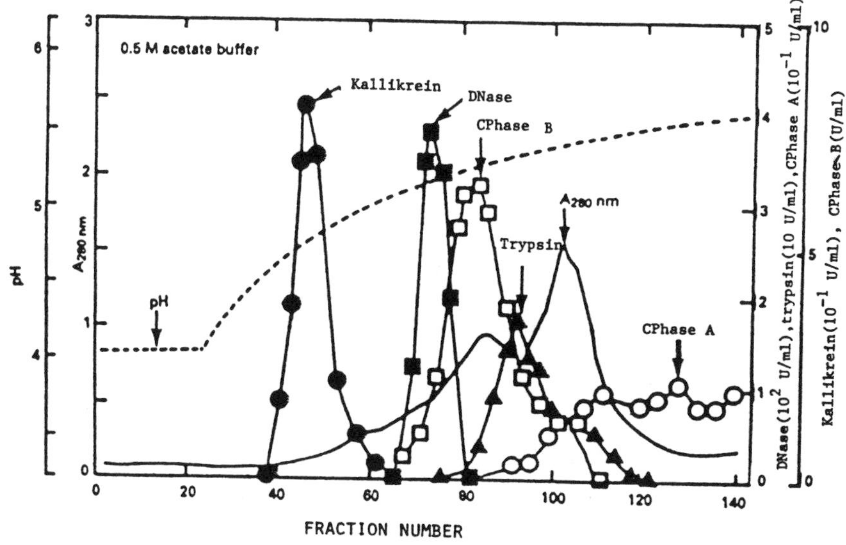

Figure 4.39. Example of the use of carboxylic acid resins in the purification of protein fractions (Reference 139).

There are few large scale applications in protein purification which utilize classical adsorption chromatography with alumina, silica or charcoal as the separation media. One example has been reported for the large scale purification with celite of β-lactamases from *Bacillus cerus* culture supernatant (141). A 20-fold purification and a 10-fold concentration was possible with this procedure.

Marshall (142) used a column of DEAE cellulose, followed by one of Ultragel AcA34, one of Sephadex G-100 and finally one of DEAE-Sepharose in his industrial scale separation of a specific glucoamylase-type of enzyme to be used in the production of low alcohol beverages. This four column procedure allowed the isolation of α-amylase, α-glucosidase, glucoamylase and a glucoamylase-type of enzyme, exopullulanase, which has the ability to degrade pullulan at a rapid rate.

Phosphocellulose and Sephadex C-25 have been used to isolate micrococcal nuclease from a crude culture of *Staphylacoccus aureus* after 16 to 18 hours of growth. The cation Sephadex showed significantly higher levels of recovery compared to the phosphocellulose as shown by the elution results in Table 4.16 (143).

Table 4.16: Recovery of Micrococcal Nuclease from SP-Sephadex C25 and Phosphocellulose (143)

Resin	pH	% Adsorption	% Elution
SP-Sephadex C25	5.0	98	82
SP-Sephadex C25	5.8	97	80
Phosphocellulose	5.0	95	30
Phosphocellulose	5.8	90	31

Large scale purifications of *Trichoderma reesei* strain L27 cultures have been routinely performed by applying the filtrate to a column of DEAE-Sepharose CL-6B and eluting the fractions with a NaCl gradient (144). A second DEAE-Sepharose CL-6B column was used to isolate exocellobiohydrolase I and endoglucanase. A cation SP-Sephadex C-50 column was used to purify β-glucosidase further. The recovery of these proteins was 83%.

Additonal examples, mostly of laboratory studies, are available in books specifically on the HPLC of peptides and proteins (145,146). "LC GC" and "Chromatography" are two periodicals with helpful HPLC operational suggestions.

4.8.7 Polyhydric Alcohol Separation

There have also been industrial chromatographic developments in the separation and purification of monosaccharides and polyhydric alcohols from lignocellulosic materials such as wood (147,148). These authors were able to separate xylitol and sorbitol from galactitol, mannitol and arabinitol using a resin in the calcium form. The aluminum form of the resin could be then used to separate xylitol from sorbitol. As a further refinement of this double fractionation process, the saccharide fraction was removed to allow the purification of xylose from mannose, glucose and galactose. Other variations on these process schemes may be utilizied to recover other polyhydric alcohols or monosaccharides.

4.8.8 Regenerant Recovery

The regeneration of ion exchange resins results in solutions which contain varying concentrations of the waste salts stripped from the resin and of the acid or base used for resin regeneration. Ion retardation can be used to separate these waste salts from the regenerant acids and bases. Special resins have been developed (149) which facilitate this separation. Figure 4.40 shows the effect of different resins in separating acids and salts. This method of regenerant recovery entails such capital expenditure for columns, controls and resin that only very large resin installations, producing large volumes of regenerant-salt waste stream, are economically justifiable. The high fructose corn syrup industry and large power installations are such applications.

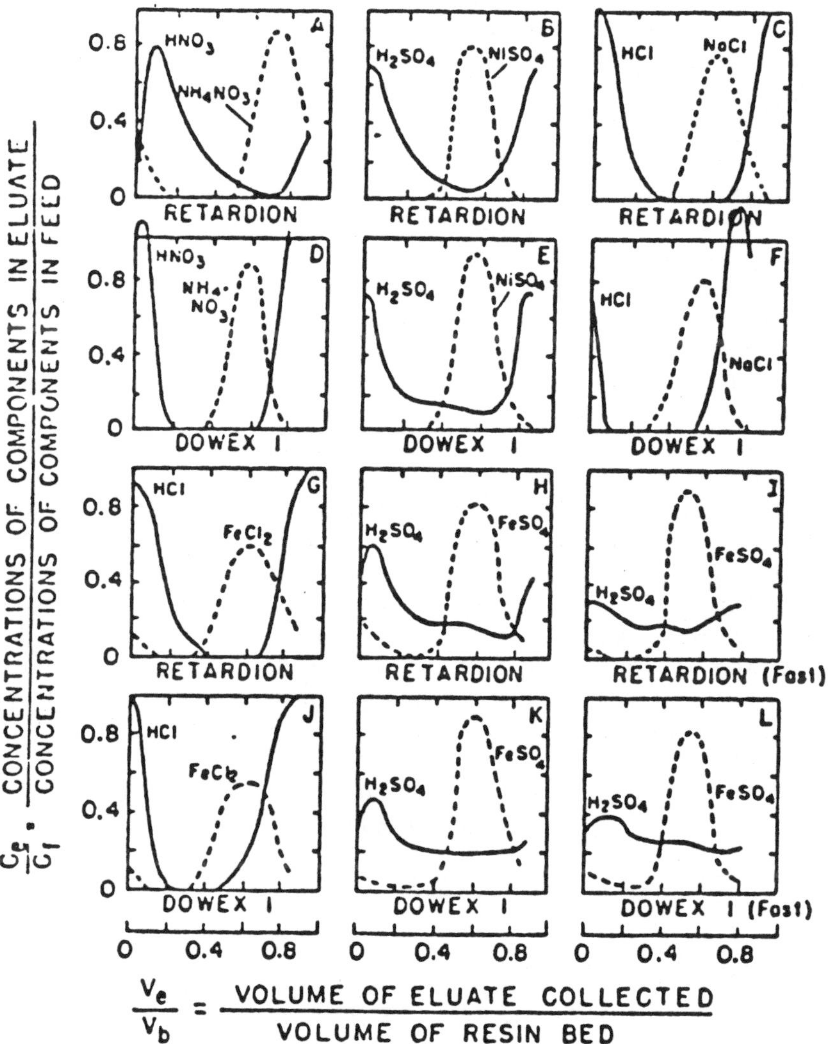

Figure 4.40. Effect of different resins on the elution profiles for various acids and salts (Reference 149).

There are two other situations which would warrant ion retardation systems: 1) when valuable salts must be separated from the eluent which stripped the salts from a concentrating ion exchange column or from a solvent extraction solution and 2) when ecological coinsiderations limit the acid, base or salt waste streams which may otherwise be discharged into the environment.

4.8.9 Sugar From Molasses

Molasses is a by-product in normal extraction and crystallization processing of cane sugar and beet sugar. The sugar content in molasses is usually between 30 and 60% as a percentage of the total dissolved solids. While it is possible to decrease this sucrose loss to molasses by typical ion exchange treatment to remove the salts prior to crystallization, that process leads to substantial waste loads (150).

When the molasses solution is added to the chromatographic column, the highly ionized ash is excluded from the resin beads while the non-ionic sugars are selectively adsorbed on the active sites of the resin beads. The distribution constants are given in Table 4.17 for K and Ca forms of the resin for various salts and sugars of molasses. In addition to the non-sugar inorganics, there are several non-sugar organics in the molasses. These non-sugar organics are different depending upon whether the source of the molasses is cane sugar or beet sugar. The non-sugar organics from cane molasses have a smaller amount of amino acids and a higher content of colored components resulting from the degradation of reducing sugars (glucose, glactose) compared to beet molasses.

Table 4.17: Distribution Coefficients for Dowex 50 WX4, 50-100 Mesh (150)

	K^+ FORM	Ca^{++} FORM
KCl	0.454	0.482
$CaCl_2$	0.466	0.530
Sucrose	0.623	0.623
Glucose	0.955	0.690
Fructose	1.07	1.26

Sargent (151) studied the variables of column shape, density gradients and feed concentrations for ion exclusion purifications of molasses. Stark (152) showed that both beet and cane molasses could be purified using Dowex 50WX4 in the K ion form. Fifty percent of the sucrose in beet molasses was recovered into a fraction with about 80% purity. Over 65% of the sucrose in cane refinery molasses was recovered with over 68% purity. At these levels of purity, the sucrose fractions could be returned to intermediate pans for sucrose recovery by crystallization. Takahashi and Takikawa (153) showed that ion exclusion combined with a recycling technique and under optimal conditions would yield an apparent purity of 90% and a sucrose recovery of 90% for beet molasses. Houssiau (154), using an ion exchange resin in the K form on a second strike molasses stream, was able to upgrade a sucrose stream at 72% purity to 95% purity at minimum loading, but only up to 80% purity at maximum column loading. The dissolved solids level of the recovered sucrose fraction was reduced to 6 - 18°Brix when the molasses feed stream was at 40°Brix. An engineering analysis (155) of the ion exclusion process for recovering sucrose from beet molasses showed that, under certain cost projections for molasses and sucrose, this process would be economically justified. Gross (50) reviewed many of these studies along with early patents on chromatographic separations of molasses.

Ito (156) and Hongisto (157) each have examined the two alternatives of separating molasses into a non-sugar fraction and a total sugar fraction (mixture of sucrose, glucose and fructose) or of inverting the sucrose of the molasses into glucose and fructose and then separating the inverted molasses into a non-sugar fraction and a glucose-fructose fraction.

In Ito's studies, the K form of Dowex 50WX4 was used to separate molasses at 44°Brix with a 42% sugar concentration into a non-sugar fraction, a sucrose fraction, and a glucose-fructose fraction. The sucrose fraction and the glucose-fructose fraction together recovered 77% of the sucrose and 96% of the glucose-fructose with a purity of 82 to 89%. When the inverted molasses was separated by a K form resin, 76.7% of the glucose-fructose was recovered at a 93.2% purity. When the Ca form of the same resin was used, the separations were not as good.

Hongisto's studies showed that inverting the molasses prior to separation will increase the capacity of a chromatographic column without a decrease in product purity because of the sharp separation of non-sugar from monosaccharides on a Na form resin. The purity obtained by Hongisto's process was between 92 and 96% with a recovery between 85 and 92%.

Many studies have been undertaken recently to examine the potential of isolating other non-sucrose organics from molasses by chromatography. For a Na form resin, the raffinose peak occurs before, while glucose and fructose peaks occur after, the sucrose peak. Of the main non-sugar components, betaine is eluted last.

Sayama and his coworkers (32) used a Ca form of Dowex 50WX4 to study the chromatographic separation of betaine from beet molasses. The effects of solids level, flow rate, column temperature and ionic composition of the resin on peak resolution were examined. For this type of separation, resolution increased with increasing temperature. The effects of other variables on resolution were as expected: decreasing concentration, decreasing flow rate and increasing percentage Ca form resin increased betaine resolution.

To separate sucrose from molasses, a recent study (158) that examined resin crosslinkage, ionic form, flow rate and feed load determined that optimum conditions were the Na form of Dowex 50MW4, 50-100 mesh, with a flow rate of 1.3 space velocity and a feed load of 0.125 liters of 30°Brix molasses per liter of resin per cycle. The chromatographic system could treat 2 tons of molasses per cubic meter of resin per day with a recovery of 85% of the sucrose at a purity of 87.5%.

Sayama and his coworkers also worked out procedures for recovering inositol (159), raffinose (160), and adenosine (161) from molasses. Over 70% of the inositol was recovered using a Ca form resin. For raffinose recovery, the Na form of the resin was superior to the Ca form. Depending on the sucrose to raffinose ratio, from 66% to 86% of the raffinose could be recovered at a purity of 91 to 93%. The adenosine recovered on a pilot scale chromatographic unit was 30% when betaine was recovered in a separate fraction. The Ca form resin could be loaded with 0.250 m^3 of 40°Brix molasses per m^3 of resin per cycle during the recovery of betaine and adenosine.

One of the most recent developments in the area of chromatographic treatment of molasses is the work of Neuzil and Fergin (162) who have applied the UOP continuous chromatography technique to recover sucrose from molasses. In this patent, a non-functionalized adsorbent and an alcohol containing eluent are utilized. From a model feed stream with 30% sucrose, 10% KCl and 10% betaine, a sucrose recovery of 90% with a purity of 99% was obtained.

4.9 REFERENCES

1. Tswett, M., Ber Deut Botan Ges 24:316 (1906)

2. Tswett, M., Biochem Z 5:6.

3. Seaborg, G.T., Science 104(2704):379 (Oct. 25, 1946)

4. Spedding, F.H., Fulmer, E.I., Butler, T.A., et al., J Am Chem Soc 69:2812 (1947)

5. Spedding, F.H., Powell, J., J Am Chem Soc 76:2545 (1954)

6. Spedding, F.H., Powell, J., Wheelwright, E., J Am Chem Soc 76:2557 (1954)

7. Sitrin, R.D., Chan, G., DePhillips, P., et al., In: Purification of Fermentation Products, (LeRoith, D., Shiloach, J., Leahy, T.J., eds.), American Chemical Society, Washington, DC, p. 71 (1985)

8. Porath, J., Flodin, P., Nature 183:1657 (1959)

9. Porath, J., Lab Pract 16:838 (1967)

10. Moore, J.C., J Polym Sci, Part A 2:835 (1964)

11. Cuatrecasas, P., Wilcheck, M., Anfinsen, C.B., Proc Nat Acad Sci 61:636 (1968)

12. Wheaton, R.M., Bauman, W.C., Ann NY Acad Sci 57:159 (1953)

13. Dean, J.A., Chemical Separation Methods, D. Van Nostrand, New York p 295 (1969)

14. Eisenbraun, A.A., U.S. Patent No. 3,045,026 (July 17, 1962)

15. Dean, J.A. Chemical Separation Methods, D. Van Nostrand Co., New York, p 60 (1969)

16. Janson, J.C., presented at BIOTECH 84 Europe Conference, London (1984)

17. Roy, A.K., Roy, S., Liquid Chromatography 1(3):181 (1983)

18. Schomberg, G., LC·GC 6(1):36 (1988)

19. Wilson, J.N., J Am Chem Soc 62:1583 (1940)

20. DeVault, D., J Am Chem Soc 65:532 (1943)

21. Glueckauf, E., J Chem Soc 69:1321 (1947)

22. Glueckauf, E., Discussions Faraday Society 7:42 (1949)

23. Martin, A.J.P., Synge, R.L.M., Biochem J 35:1358 (1941)

24. Mayer, S.W., Thompkins, E.R., J Am Chem Soc 69:2866 (1947)

25. Simpson, D.W., Wheaton, R.M., Chem Eng Prog 50(1):45 (1954)

26. Glueckauf, E., Trans Faraday Soc 51:34 (1955)

27. Beukenkamp, J., Reiman, W. III, Lindenbaum, S., Anal Chem 26:505 (1954)

28. Simpson, D.W., Bauman, W.C., Ind Eng Chem 46:1958 (1954)

29. Karger, B.L., J Chem Educ 43:47 (1966)

30. Hongisto, H.J., Int Sugar J 79(940):100 (1977)

31. Martinola, F., Siegers, G., Verfahrenstechnik (Mainz), 13(1):32 (1979)

32. Sayama, K., Senba, Y., Kawamoto, T., Proc Res Soc Japan Sugar Refin Tech 29:10 (1980)

33. Warren, W., Dwyer, M., Merion, M., J Analysis and Purif, 2(2):10 (1987)

34. Semenza, G., Chimica 14:325 (1960)

35. Mikes, O., Strop, P., Hosomska, Z., et al., J Chromatog, 261:363 (1983)

36. Sofer, G., Mason, C., Biotechnology 5(3):239 (1987)

37. Voser, W., Walliser, H.P., in Discovery and Isolation of Microbial Products, (Verrall, M.S., ed) Ellis Horwood Ltd., Chichester, UK, p 116 (1985)

38. Charm, S.E., Matteo, C.C., in Methods in Enzymology, (Jacoby, W.B., ed) Academic Press, New York, vol 22, p 476 (1971)

39. Laboratory Methods of Column Packing, Whatman Data Sheet, Reeve-Angel Co., Clifton, New Jersey (1968)

40. Stacey, C., Brooks, R., Merion, M., J of Analysis and Purif, 2(1):52 (1987)

41. Charm, S.E., Matteo, C.C., Carlson, R., Anal Biochem, 30:1 (1969)

42. Rudge, S.R., Ladisch, M.R., in Separation, Recovery and Purification in Biotechnology, (Asenjo, J.A., Hong, J., eds.) American Chemical Society, Washington, DC, p 122 (1986)

43. Sofer, G., Mason, C., Biotechnology, 5(3):239 (1987)

44. Heckendorf, A.H., Chem Proc, p. 33 (August 1987)

45. Heckendorf, A.H., Ashare, E., Rausch, C., In: Purification of Fermentation Products, (LeRoith, D., Shiloach, J., T.J., Leahy, eds.) American Chemical Society, Washington, DC, p 91 (1985)

46. Janson, J.C., Dunnill, P., In: Industrial Aspects of Biochemistry, American Elsevier, New York, p 81 (1974)

47. Strobel, G.J., Chemie Technik, 11:1354 (1982)

48. Rexova-Benkova, L., Omelkova, J., Mikes, O., et al., J Chromatography, 238:183 (1982)

49. Frolik, C.A., Dart, L.L., Sporn, M.B., Anal Biochem, 125:203 (1982)

50. Gross, D., Proc 14th Gen Assembly CITS, p 445 (1971)

51. Sutthoff, R.F., Nelson, W.J., U.S. Patent No. 4,022,637 (May 10, 1977)

52. Keller, H.W., Reents, A.C., Laraway, J.W., Starch/Staerke, 33:55 (1981)

53. Norman, L., Rorabaugh, G., Keller, H., J Am Soc Sugar Beet Tech, 12(5):363 (1963)

54. Broughton, D.B., Gerhold, C.G., U.S. Patent No. 2,985,589 (May 23, 1961)

55. For more information contact K. Schoenrock, Amalgamated Sugar, Ogden, Utah

56. Burke, D.J., presented at A.I.Ch.E. 23rd Annual Symposium (May 12, 1983)

57. Ishikawa, H., Tanabe, H., Usui, K., Japanese Patent No 102,288 (August 11, 1979)

58. Broughton, D.B., Gembicki, S.A., In: Fundamentals of Adsorption, (Myers, A.L., Belfort, G., eds.) Engineering Foundation, New York, p 115 (1984)

59. Barker, P., Thawait, S., Chem & Ind, 817 (Nov 7, 1983)

60. Wankat, P.C., Ind Eng Chem Fundam, 16:468 (1977)

61. Wankat, P.C., Ortiz, P.M., Ind Eng Chem Process Des Dev, 21:416 (1982)

62. McGary, R.S., Wankat, P.C., Ind Eng Chem Fund, 22:10 (1983)

63. Wankat, P.C., Ind Eng Chem Fund, 23:256 (1984)

64. Novak, L.J., Bowdle, P.H., U.S. Patent No. 4,155,846 (May 22, 1979).

65. Begovich, J.M., Byers, C.H., Sisson, W.G., Sep Sci Technol, 18(12 & 13):1167 (1983)

66. Begovich, J.M., Sisson, W.G., A.I.Ch.E.J., 30:705 (1984)

67. Hollein, H.C., Ma, H., Huang, C., et al., Ind Eng Chem Fundam, 21:205 (1982)

68. Moore, S., Stein, W.H., J Biol Chem, 211:893 (1954)

69. Doury-Berthod, M., Poitrenaud, C., Tremillon, B., J Chromatogr, 179:37 (1979)

70. Szczepaniak, W., Ciszewsha, W., Chromatographia, 15(1):38 (1982)

71. Hamilton, P.B., Anal Chem, 35:2055 (1963)

72. Dutcher, J.D., Hosansky, N., Donin, M.H., et al., J Am Chem Soc, 73:1384 (1951)

73. Weston, R.G., Putter, I., U.K. Patent No. 1,345,729 (Feb 6, 1974)

74. Yamashita, T., Naoi, N., Hidaka, T., et al., J Antibiotics (Tokyo), 32(4):330 (1979)

75. Johnson, D.L., Doyle, T.W., U.S. Patent No. 4,189,568 (Feb 19, 1980)

76. Brown, D., Giles, A.F., Noble, H.M., et al., U.S. Patent No. 4,379,920 (April 12, 1983)

77. Sakuma, S., Motomura, H., J Chromatogr, 400:293 (1987)

78. LeFevre, L.J., U.S. Patent No. 3,044,905 (July 17, 1962)

79. LeFevre, L.J., U.S. Patent No. 3,044,906 (July 17, 1962)

80. Serbia, G.R., U.S. Patent No. 3,044,904 (July 17, 1962)

81. Ghim, Y.S., Chang, H.N., Ind Eng Chem Fundam, 21:369 (1982)

82. Ho, C., Ching, C.B., Ruthven, D.M., Ind Eng Chem Res, 26:1407 (1987)

83. Bieser, H.J., deRosset, A.J., Starch/Staerke, 29:392 (1977)

84. Miyahara, S., Sakai, S., Matsuda, F., et al., Japanese Patent No. 118,400 (Sept 11, 1980)

85. Asher, D.R., Simpson, D.W., J Phys Chem, 60:518 (1956)

86. Sargent, R.N., Rieman, C.W. III, U.S. Patent No. 3,134,814 (May 26, 1964)

87. Hirota, T., Shioda, K., Japanese Patent No. 48,400 (April 7, 1980)

88. Regnier, F.E., In: High Perform Liq Chromatogr Proteins Pept, Proc Int Symp, 1st 1981, Academic Publications, New York, NY (Pub. 1983)

89. Chang, S.H., Gooding, K.M., Regnier, F.E., J Chromatogr, 125:103 (1976)

90. Chang, S.H., Noel, R.N., Regnier, F.E., Anal Chem, 48:1839 (1979)

91. Schlabach, T.D., Alpert, A.J., Regnier, F.E., Clin Chem 24:1351 (1978)

92. Schlabach, T.D., Fulton, J.A., Mockridge, P.B., et al., Clin Chem, 25:1600 (1979)

93. Schlabach, T.D., Fulton, J.A., Mockridge, P.B., et al., Anal Chem, 52:729 (1980)

94. Fulton, J.A., Schlabach, T.D., Kerl, J.E., et al., J Chromatogr, 175:269 (1979)

95. Denton, M.S., Bostick, W.D., Dinsmore, S.R., et al., Clin Chem, 24:1408 (1978)

96. Bostick, W.D., Denton, M.S., Dinsmore, S.R., Clin Chem 26:712 (1980)

97. Schlabach, T.D., Chang, S.H., Gooding, K.M., et al., J Chromatogr, 134:91 (1977)

98. Alpert, A.J., Ph.D., Thesis, Purdue University (1979)

99. Bostick, W.D., Dinsmore, S.R., Mrochek, J.R., et al., Clin Chem, 24:1305 (1978)

100. Gooding, K.M., Lu, L.C., Regnier, F.E., J Chromatogr, 164:506 (1979)

101. Hanash, S.M., Shapiro, D.N., Hemoglobin, 5:165 (1980)

102. Hanash, S.M., Kavadella, K., Amanulla, A., et al., In: Advances in Hemoglobin Analysis, (Hanash, S.M., Brewer, G.J., eds) A.R. Liss, New York, NY, p 53 (1981)

103. Gardiner, M.B., Wilson, J.B., Carver, J., et al., International Symosium on HPLC of Proteins and Peptides, Paper No. 203, Washington, D.C. (1981)

104. Umino, M., Watanabe, H., Komiya, K., et al., ibid., Paper No. 208, Washington, D.C. (1981)

105. Abraham, E.C., Cope, N.D., Braziel, N.N., et al., Biochem Biophys Acta, 577:159 (1979)

106. Gupta, S.P., Pfannkoch, E., Regier, F.E., Anal Biochem 128(1):196 (1983)

107. Umino, M., Watanabe, H., Komiya, K., Ref. 103, Paper No. 204

108. Radhakrishnan, A.N., Stein, S., Licht, A., et al., J Chromatogr, 132:552 (1977)

109. Vanecek, G., Regnier, F.E., Anal Biochem, 121(1):156 (1982)

110. Lu, K.C., Gooding, K.M., Regnier, F.E., Clin Chem, 25:1608 (1979)

111. Vanecek, G., Regnier, F.E., Anal Biochem, 109:345 (1980)

112. Kato, Y., Komiya, K., Hashimoto, T., Ref. 103, Paper No. 214

113. Alpert, A.J., Beaudet, A.L., Ref. 103, Paper No. 210

114. Ott, G.S., Shore, V.G., Ref. 103, Paper No. 201

115. Rudolph, F.B., Clark, S.W., Ref. 103, Paper No. 202

116. Frolick, C.A., Dart, L.L., Sporn, M.B., Ref 103, Paper No. 205

117. Ansari, G.A.S., Patel, N.T., Fritz, R.R., et al., Ref. 103, Paper No. 211.

118. Weaver, V.C., Chromagraphia, 2:555 (1969)

119. Morris, C.J.O.R., Morris, P., Separation Methods In Biochemistry, Interscience Publ, New York, NY (1963)

120. Gordon, A.H., In: Laboratory Techniques in Biochemistry and Molecular Biology, Vol 1, Part 1, (Work, T.S., Work, E., eds.) North Holland Publ Co, Amsterdam - London (1969)

121. Kelley, J.J., Wang, G.Y., Wang, H.Y., In: Separation, Recovery and Purification in Biotechnology, (Asenjo, J.A., Hong, J., eds.) American Chemical Society, Washington, D.C., p 193 (1986)

122. Durekovic, A., Flosdorf, J., Kula, M.R., Eur J Biochem, 36:528 (1973)

123. Bruton, C., Jakes, R., Atkinson, T., Eur J Biochem, 59:327 (1975)

124. Bascomb, S., Banks, G.T., Skorstedt, M.T., et al., J Gen Microbiol, 91:1 (1975)

125. Janson, J.C., J Agr Food Chem, 19:581 (1971)

126. Curling, J.M., Amer Lab, 8:26 (1976)

127. Hanford, R., Maycock, W., Vallet, L., In: Chromatography of Synethic and Biological Polymers, Vol 2, (Epton, R., ed) Ellis Harwood, Chichester, U.K., p 111 (1978)

128. Horton, T., Amer Lab, 4:83 (1972)

129. The Large-Scale Purification of Insulin by Gel Filtration Chromatography, Pharmacia, Uppsala (1983)

130. Donnelly, E.B., Delaney, R.A.M., J Food Technol, 12:493 (1977)

131. Nichols, R.A., Fox, J.B., J Chromatogr, 43:61 (1969)

132. Sassenfeld, H.M., Brewer, S.J., Biotechnology, 2(1):76 (1984)

133. Pietrzyk, D.J., Cahill, W.J., Jr., Stodola, J.D., J Liquid Chrom, 5(3):443 (1982)

134. Kato, Y., Komiya, K. Hishimoto, T., J Chromatogr, 246(1):13 (1982)

135. Dizdaroglu, M., Krutzsch, H.C., Simic, M.G., J Chromatogr, 237:417 (1982)

136. Dizdaroglu, M., Krutzsch, H.C., J Chromatogr, 264:223 (1983)

137. Richey, J.R., Amer Lab, 14:104 (1982)

138. Boudier, C., Carvallo, D., Roitsch, C., et al., Arch Biochem Biophys, 253(2):439 (1987)

139. Sasaki, I., Gotoh, H., Yamamoto, R., et al., J Biochem (Tokyo), 86(5):1537 (1979)

140. Komatsu, K., Saijo, A., Haneda, K., et al., U.S. Patent No. 3,730,836 (May 1, 1973)

141. Davis, R.B., Abraham, E.P., Melling, J., Biochem J, 143:115 (1974)

142. Marshall, J.J., U.S. Patent No. 4,318,927 (March 9, 1982)

143. Darbyshire, J., In: Topics in Enzyme and Fermentation Biotechnology 5, (Wiseman, A., ed) Ellis Harwood, Chichester, U.K., p 156 (1981)

144. Shoemaker, S., Watt, K., Tsitovsky, G., et al., Biotechnology, 1(8):687 (1983)

145. Horvath, C., ed., High Performance Liquid Chromatography, Vol 3, Academic Press, New York (1983)

146. Hamilton, R.J., Sewell, P.A., Introduction to High Performance Liquid Chromatography, 2nd Ed., Chapman and Hall, London (1982)

147. Melaja, A.J., Hamalainen, L., U.S. Patent No. 4,008,285 (Feb 15, 1977)

148. Melaja, A.J., Hamalainen, L., U.S. Patent No. 4,075,406 (Feb 21, 1978)

149. Hatch, M.J., Dillon, J.A., I&EC Proc Des Dev, 2:253 (1963)

150. Rousseau, G., Lancrenon, X., Sugar y Azucar, 27 (1984)

151. Sargent, R.N., Ind Eng Chem Proc Des Dev, 2(2):89 (1963)

152. Stark, J.B., J Am Soc Sugar Beet Technol, 13:492 (1965)

153. Takahashi, K., Takikawa, T., Proc Res Soc Japan Sugar Refin Tech, 16:51 (1965)

154. Houssiau, J., Sucrerie Belge, 87:423 (1968)

155. Schultz, W.G., Stark, J.B., Lowe, E., Int Sugar J, 69:35 (1967)

156. Ito, Y., Proc Res Soc Japan Sugar Refin Tech, 22:1 (1970)

157. Hongisto, H.J., S.I.T. Paper 78, 624 (1978)

158. Sayama, K., Senba, Y., Kawamoto, T., Proc Res Japan Sugar Refin Tech, 29:1 (1980)

159. Sayama, K., Senba, Y., Kawamoto, T., et al., Proc Res Japan Sugar Refin Tech, 29:20 (1980)

160. Sayama, K., Senba, Y., Kawamoto, T., et al., Proc Res Japan Sugar Refin Tech, 30:72 (1981)

161. Oikawa, S., Sayama, K., Senba, Y., et al., Proc Res Japan Sugar Refin Tech, 33:55 (1982)

162. Neuzil, R.W., Fergin, R.L., U.S. Patent No. 4,426,232 (Jan 17, 1984)

5

Affinity Chromatography

5.1 INTRODUCTION

Affinity chromatography was developed by Cuatrecasas, Wilchek and Anfinsen (1) in 1968 as a laboratory protein purification method that is based on the biospecific recognition between a solute molecule and a ligand immobilized onto a support material. As shown in Figure 5.1 (2), the affinity purification process can be idealized into the selective adsorption of enzyme A from a mixture of proteins. After the unadsorbed contaminating proteins are washed from the column, enzyme A can be eluted from the ligand and recovered. This technique has been used successfully to purify enzymes, proteins, antibodies, oligonucleotides and nucleic acids. Several recent review articles and books are available which cover these applications (2-7).

Generalized protein purification schemes are shown in Figure 5.2 (8) where conventional ion exchange and chromatography techniques are compared to affinity chromatography. The conventional techniques separate the protein on the basis of size, electrical charge or isoelectric point. Since these differences are often not sufficient to allow effective separation, many additional steps are required. This greatly reduces the overall yield of purified protein compared to the affinity chromatography technique.

The "chromatography" portion of the term affinity chromatography does not represent a true description of the process as actually practiced. More correct descriptive terms would be "affinity purification" or "affinity separation". However, for the sake of consistency with past authors on this subject, this text will continue to use "affinity chromatography."

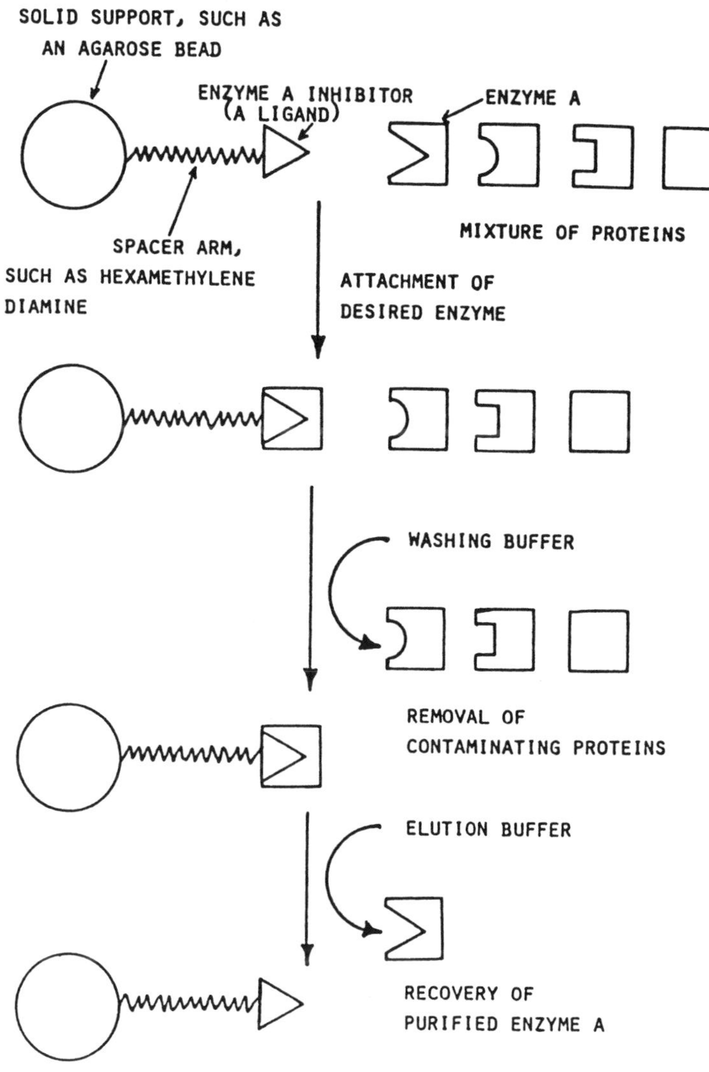

Figure 5.1. A schematic of the selective, reversible attachment of a protein to an immobilized ligand (Reference 2).

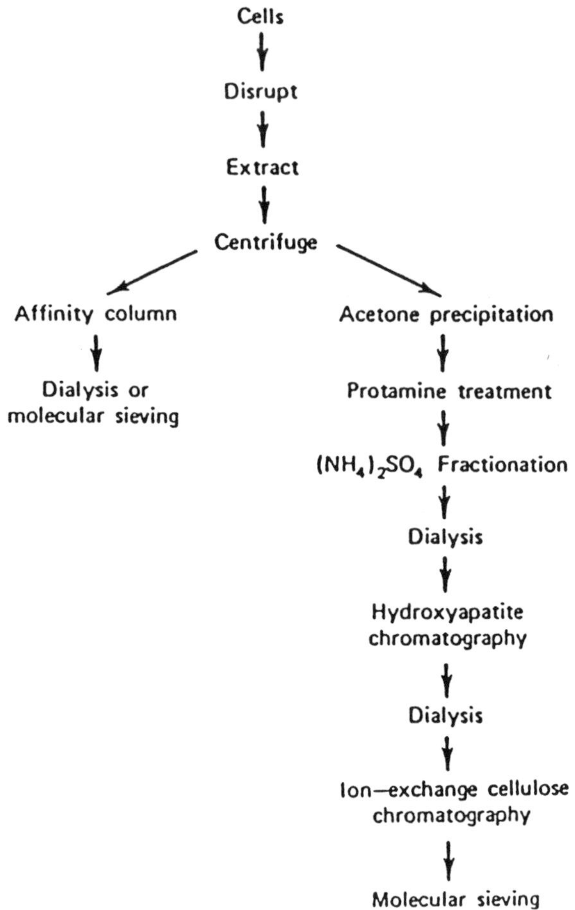

Figure 5.2. Comparison of conventional and affinity chromatography techniques (Reference 8).

5.2 THEORY OF AFFINITY CHROMATOGRAPHY

When compared to the vast literature available on empirical affinity chromatography studies, there is a paucity of theoretical work that has been reported explaining affinity chromatography. One reason is that the rigorous models developed for analyzing chromatographic systems require linear adsorption-desorption kinetics which do not occur for the affinity purification systems. Another complicating factor is the interference caused by non-specific adsorption which contributes complexity and oftentimes irreproducibility to many affinity chromatography systems.

Despite these difficulties, attempts to describe binding selectivity, equilibrium constants and column kinetics have provided useful insights into affinity purification processes.

5.2.1 Equilibrium and Binding Selectivity

The ideal ligand for affinity chromatography would possess the ability to bind strongly the solute to be isolated and the binding should be highly specific so that impurities with similar structures are not bound. The binding should also be reversible under readily obtained conditions so that the isolated solute can be easily eluted from the ligand. Thus, the binding strength, the selectivity and the elution behavior constitute the controlling factors in a quantitative description of affinity chromatography.

The binding strength and the selectivity are termed the avidity of the ligand for the solute (9). The solute becomes the ligate when it is bound to the ligand. The avidity is described mathematically by the association constant, K_{as}:

$$K_{as} = \frac{k_a}{k_b} = \frac{[EL]}{[E]\,[L]} \tag{5.1}$$

for the interaction of the ligand [L] and the ligate [E]:

$$E + L \underset{k_b}{\overset{k_a}{\rightleftharpoons}} EL \tag{5.2}$$

The adsorption of the ligate by the ligand can be represented by the Langmuir-type isotherm (10):

$$[EL] = \frac{K_{as}\,[E]\,[L_o]}{1 + K_{as}\,[E]} \tag{5.3}$$

where

$$[L_o] = [L] + [EL] \tag{5.4}$$

High values of K_{as} represent an extremely stable complex. Typical values are 10^{15} M for the avidin-biotin interaction, 10^6 to 10^{12} for antigen-antibody interaction and 10^4 to 10^6 for lectin-carbohydrate and enzyme-substrate interactions (11).

In most cases, there are additional materials which will compete with the desired ligate for ligand sites. In all cases, such competition occurs when it becomes necessary to elute the desired ligate from the ligand. These aspects of affinity chromatography theory have been developed by Bottomley and coworkers (12). When the competing ligate [D] is present in the solution, the following equation can be used to describe the relevant equilibria:

$$\frac{[E]}{[EL]} = \frac{1}{K_{as}[L_o]} \quad 1 + \frac{[D]}{r\,K_{ED}} + \frac{[E]}{[L_o]} \qquad (5.5)$$

In this equation, K_{ED} represents the dissociation constant for a complex formed between the two potential ligates in solution:

$$K_{ED} = \frac{[E]\,[D]}{[ED]} \qquad (5.6)$$

The parameter r represents the ratio of the partition coefficients α_{eq} and α_D. The partition coefficient accounts for the possibility that there are sections of the column which are accessible to the solvent but not to the large protein molecule, E.

Where for [E] and [EL] are obtained from the elution curve (Figure 5.3). The shaded area in this figure represents the amount of enzyme E bound to the ligand L. The values for [E] and [EL] from experiments with different concentrations of E at constant concentrations of D can be plotted as in Figure 5.4 to give straight lines whose abscissa values can be used to obtain values for K_{as} and K_{ED}.

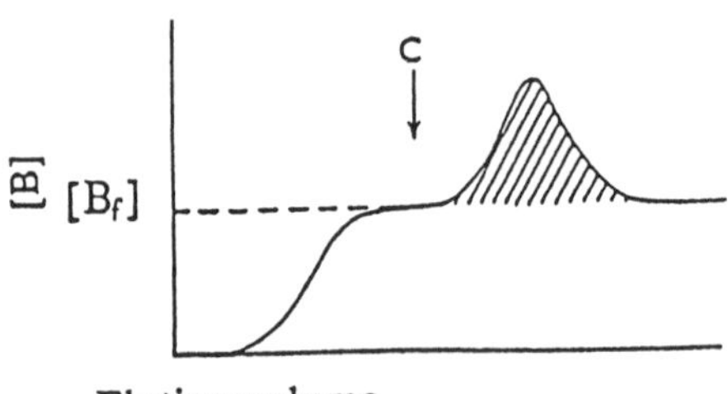

Figure 5.3. Elution profile of enzyme B applied to columns of immobilized A. The hatched area represents the amount of B bound to A. This is displaced from the column by some competing ligand C at concentration D (Reference 12).

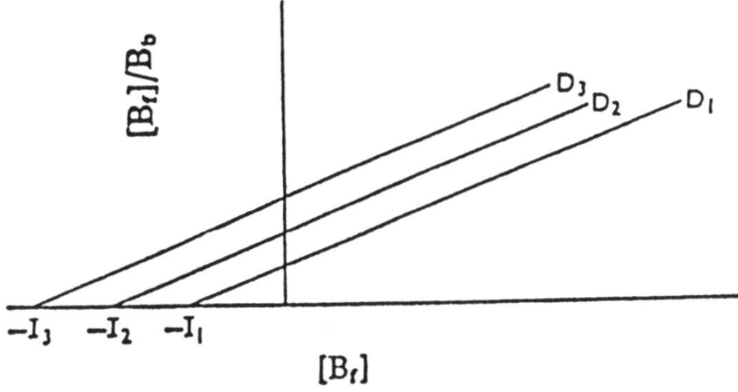

Figure 5.4. Primary plot for the three component system. Values of B_f and B_b are obtained from the chromatography experiments of Figure 5.3 at different concentrations (D_1, D_2, D_3, etc.). The slopes of all the lines should be the same (Reference 12).

Another useful quantity is the fraction of the total enzyme which will be bound to the ligand at equilibrium. This is given by the equation (9):

$$\frac{\text{Bound Enzyme}}{\text{Total Enzyme}} = \frac{[EL]}{[E]\,V} = \frac{[L_o]\,V}{K_i(V + v) + [L_o]v + [E]\,V} \tag{5.7}$$

where $K_i = 1/K_{as}$, V is the volume of enzyme solution added and v is the volume of solution entrapped within the affinity support particles. As is seen from Figure 5.5, the resulting curve for any value of K_i is a rectangular hyperbola similar to the Langmuir adsorption isotherm or the Michaelis-Menten equation. Typical ligand concentrations used in affinity chromatography experiments are 10mM. At this ligand concentration, K_i values of 10^{-3}M or less are necessary for effective enzyme binding; higher values would result in ineffective retention of the enzyme.

Figure 5.5 can also be used to evaluate the fractional saturation for a system by substituting [E]V for the quantity $[L_o]v$ on the abscissa and using K (V + v) + $[L_o]v$ instead of K (V + v) + [E]V as the variable parameter. The fraction $[EL]/[L_o]$ then is simply read on the ordinate scale. These values are given in Table 5.1 as percentages for several representative values of K_i, [E] and ω, where ω is the ratio of V to v.

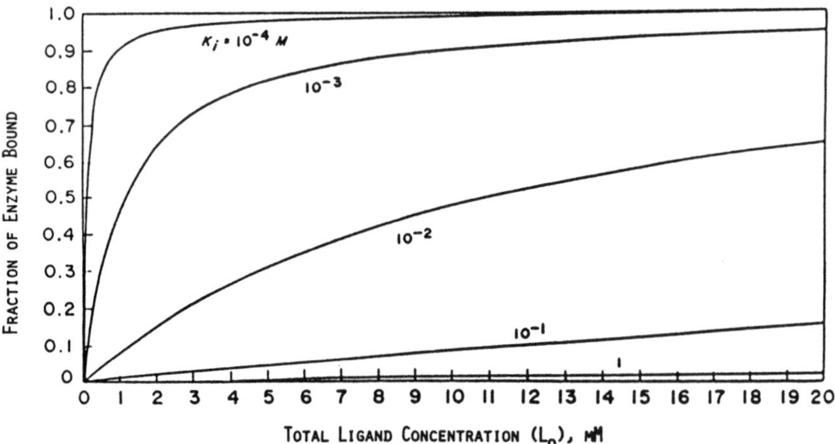

Figure 5.5. Fractional enzyme binding for low enzyme concentration (Reference 9).

Table 5.1: Percentage Ligand Saturation Under Various Conditions

(For $[L_o]$ = 0.01 M)

$[E_o]$ (M)	K_i (M)	$[EL]/[L_o]$ (%) ω = 1	ω = 10	ω = 100
10^{-4}	10^{-3}	0.8264	4.5455	8.2645
	10^{-4}	0.9709	8.2645	33.2226
	10^{-5}	0.9881	9.0009	47.5964
	10^{-6}	0.9899	9.0818	49.7488
10^{-5}	10^{-3}	0.0833	0.4739	0.8929
	10^{-4}	0.0979	0.8929	4.7393
	10^{-5}	0.0997	0.9794	8.3264
	10^{-6}	0.0999	0.9890	9.0082
10^{-6}	10^{-3}	0.0083	0.0476	0.0900
	10^{-4}	0.0098	0.0900	0.4950
	10^{-5}	0.0100	0.0988	0.9001
	10^{-6}	0.0100	0.0998	0.9803
10^{-7}	10^{-3}	0.0008	0.0048	0.0090
	10^{-4}	0.0010	0.0090	0.0497
	10^{-5}	0.0010	0.0099	0.0907
	10^{-6}	0.0010	0.0100	0.0989

Example 5.1

From Table 5.1, it is seen that when [E] = 10^{-6} M, K_i = 10^{-3} M and V = v, only about 83% of the enzyme in solution will bind to the ligand even though less than 0.1% of the ligand capacity is exhausted. The lower experimental capacity of affinity columns compared to titratable ligand has usually been attributed to steric exclusion or ionic repulsion. It is seen from this example that equilibrium effects alone may be a sufficient explanation.

Washing and elution by changing K_i to new values were also studied by Graves and Wu (9). K_i values of 10^{-3} M looked much less attractive than when binding alone was considered because much of the enzyme was lost by washing. If K_i was 10^{-5} M or less, batch adsorption and elution were as good or better than column chromatography. When K_i was changed to a new value during the elution step, the increase had to be at least 10^{-2} M in order to remove enzyme effectively from the ligand. Achieving such a high value can be a serious limitation if the initial K_i is very low.

The number of washing steps and the volume of each wash have as their primary effect an influence on the removal of contaminant. The degree of purification was strongly influenced by the thoroughness of washing. If K_i was less than 10^{-4} M, very little enzyme was lost even with a large amount of washing. This is illustrated in Figure 5.6 for simulated column chromatography.

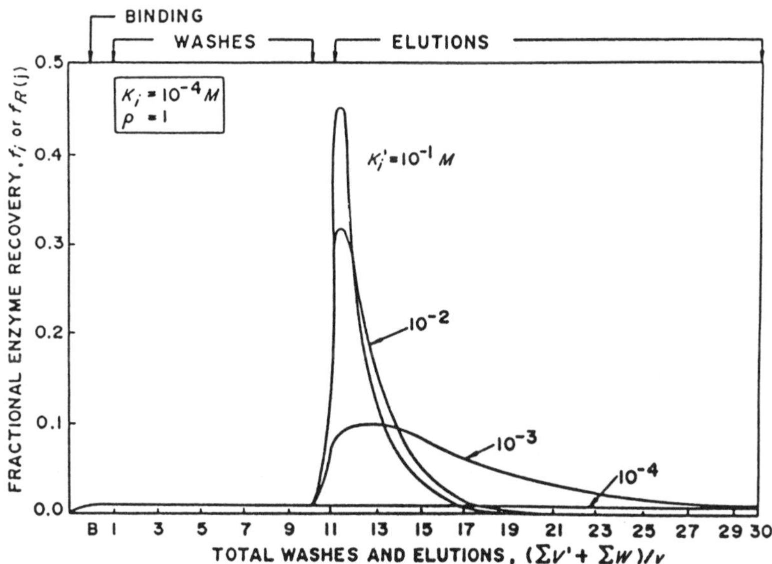

Figure 5.6. Simulated column results for moderately good binding (K_i = 10^{-4}M) (Reference 9).

5.2.2 Kinetics of Affinity Purification

In Chapter 4, the plate theory was used to model the column chromatography behavior. However, that theory is only applicable where a linear isotherm can be assumed because the plate height is a function of the fixed linear equilibrium constant or partition ratio (13). This is not the situation for affinity chromatography systems since equilibrium isotherms of such systems are usually highly curved. This points out one of the dangers of using the term affinity chromatography rather than affinity purification to describe these systems since it has led to the inappropriate use of plate theory with these systems. As Yang and Tsao (14) have pointed out, it is necessary to choose a model with appropriate assumptions for the affinity system studied. The porous particle, constant pattern rate theory (15) and the statistical moment theory (16), although more mathematically complicated, are more appropriate for describing affinity purification systems.

The mechanism of mass transfer in affinity chromatography is similar to what was observed with column adsorption in Chapter 2. It can be divided into three successive steps: the diffusion of the solute to the particle, its diffusion inside this particle and the reaction between the ligand and the ligate. Since the ligand-ligate reaction occurs rapidly, the two diffusion steps are considered to control the rate of mass transfer.

The mathematical description of the column kinetics for affinity chromatography follows the same type of material balance as was seen for volume elements in fixed bed adsorption. With the assumption of uniform velocity distribution over the cross section of the column and an isothermal operation, the basic equation is written as (17):

$$u \frac{\partial c}{\partial z} + \epsilon \frac{\partial c}{\partial t} + \rho_b \frac{\partial \bar{q}}{\partial t} - D_z \frac{\partial^2 c}{\partial z^2} = 0 \qquad (5.8)$$

The axial diffusion coefficient, D_z, is essentially the eddy diffusion since the effect of molecular diffusion can be neglected in affinity chromatography. The value of $\epsilon d_p u/D_z$ is constant when the Reynolds number is between 10^{-3} and 10. This would give:

$$D_z = \lambda d_p u \qquad (5.9)$$

where λ is a constant.

The adsorption rate can be expressed in terms of the volumetric coefficient of the overall mass transfer based on the fluid phase concentration, $K_f a$:

$$\rho_b \frac{\partial \bar{q}}{\partial t} = K_f a (c - c^*) \qquad (5.10)$$

With the assumption of a constant pattern adsorption zone, $\bar{q}$ can be set equal to $(\bar{q}_o/c_o)c$ so that Equation 5.10 can be integrated to give:

$$t_E - t_B = \frac{\rho_b \bar{q}_o}{\overline{K_f a}\, c_o} \int_{c_B}^{c_E} \frac{dc}{c - c^*} \qquad (5.11)$$

where $\overline{K_f a}$ is the average volumetric coefficient of overall mass transfer over the range between c_B and c_E. If the Freundlich adsorption isotherm, $\bar{q} = Kc^\beta$, adequately describes the system, an analytical integration of Equation 5.11 yields (18):

$$t_E - t_B = \frac{\rho_b \bar{q}_o}{\overline{K_f a}\, c_o} \left(\ln \frac{x_E}{x_B} + \frac{\beta}{1 - \beta} \ln \frac{1 - x_B^{(1-\beta)/\beta}}{1 - x_E^{(1-\beta)/\beta}} \right) \qquad (5.12)$$

Example 5.2

The breakthrough curve data for trypsin on Sepharose 4B columns with soybean trypsin inhibitor ligand (Figure 5.7) and arginine peptide ligand (Figure 5.8) have been used with Equation 5.12 to calculate the average values of the volumetric coefficients of the overall trypsin transfer. These values are shown in Table 5.2. The fact that changes in trypsin concentration or in ligand caused little change in $\overline{K_f a}$ indicates that the reaction rate of trypsin with the ligand was not rate limiting. The finite difference method of Crank and Nicolson (19) was used to generate the calculated breakthrough curves. Note that the measured breakthrough curves are somewhat steeper than the calculated curves at the beginning of breakthrough. This has been attributed to the variation of the intraparticle mass transfer coefficient with the progress of adsorption.

The very tight binding of the ligate during the adsorption step in affinity chromatography means that the

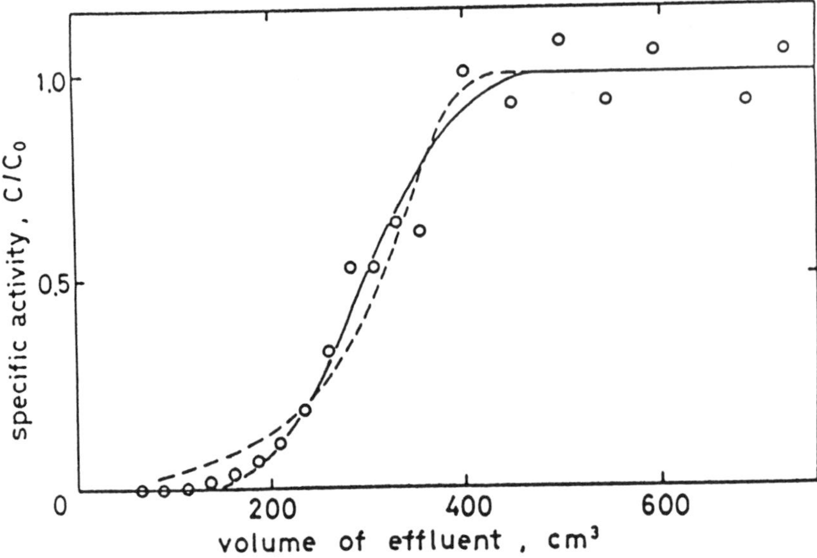

Figure 5.7. Breakthrough curve for trypsin on Sepharose 4B columns with soybean trypsin inhibitor ligand (C_o = 0.190 mg/cm^3, liquid flow rate is 0.109 cm^3/sec, column diameter is 2 cm) (Reference 17).

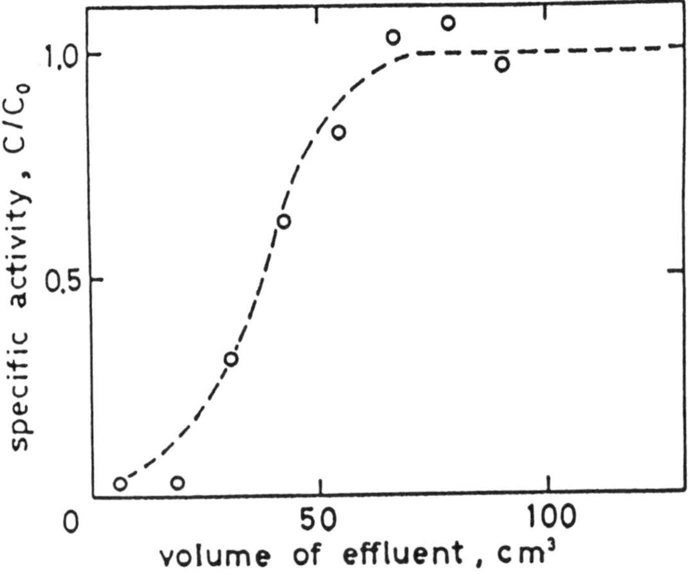

Figure 5.8. Breakthrough curve for trypsin on Sepharose 4B columns with arginine peptide ligand (C_o = 0.116 mg/cm^3, liquid flow rate is 0.056 cm^3/sec, column diameter is 2 cm) (Reference 17).

Table 5.2: Volumetric Coefficients ($\overline{K_f a}$) for Sepharose 4B-Soybean Trypsin Inhibitor and Sepharose 4B-Arginine Peptides Systems (17)

Ligand	Flow Rate (cm^2/sec)	Trypsin Concentration (mg/cm^3)	$\overline{K_f a}$ (sec^{-1})
Soybean trypsin inhibitor	0.056	0.078	0.0157
	0.056	0.176	0.0138
	0.109	0.083	0.0205
	0.109	0.190	0.0189
Arginine peptides	0.056	0.009	0.0162
	0.056	0.011	0.0148
	0.056	0.019	0.0134
	0.056	0.074	0.0168
	0.056	0.116	0.0150
	0.135	0.041	0.0210
	0.135	0.048	0.0177
	0.135	0.065	0.0146

shape of the breakthrough curve depends only on the rate limiting mass transfer (or reaction) mechanism. Arnold and coworkers (20) have presented analytical expressions for the breakthrough curves for four cases in terms of the number of transfer units (N) and a dimensionless throughput parameter (T) when the particles had a 0.01 cm diameter, the superficial velocity was 0.01 cm/sec and the solute bulk diffusivity was 7×10^{-7} cm^2/sec.

The breakthrough curve for the irreversible case with pore diffusion is:

$$N_{pore} (T - 1) = 2.44 - 0.273 \sqrt{1 - X} \qquad (5.13)$$

where X is the dimensionless liquid phase concentration (c/c_{FEED}),

$$N_{pore} \cong 27 (D_{EFF}/D_{BULK}) L \qquad (5.14)$$

The effective diffusivity (D_{EFF}) is usually 1/100 of the value of the bulk diffusivity (D_{BULK}) (21). L is the column bed length and

$$T = (V - \epsilon v) / \Gamma v \qquad (5.15)$$

with V the throughput volume, v the column volume, ϵ the void fraction and Γ the dimensionless distribution parameter.

The breakthrough curve for the solid homogeneous diffusion case is:

$$N_p\ (T - 1) = -1.69 \left[\ln(1 - X)^2 + 0.61\right] \qquad (5.16)$$

where

$$N_p \cong 21\,\Gamma\,(D_p/D_{BULK})\ L \qquad (5.17)$$

with D_p the pore diffusivity.

The combination of pore and film diffusion results in a breakthrough curve given by:

$$(T - 1) = \left(\frac{1}{N_{pore}} + \frac{1}{N_f}\right) \left\{\frac{\phi(X) + \frac{N_{pore}}{N_f}(\ln X + 1)}{\frac{N_{pore}}{N_f} + 1}\right\} \qquad (5.18)$$

where

$$\phi(X) \cong 2.39 - 3.59\sqrt{1 - X} \qquad (5.19)$$

and

$$N_f \cong 21\ L \qquad (5.20)$$

The combination of solid homogeneous diffusion and film mass transfer has a breakthrough curve given by:

$$N_f\ (T - 1) = -m\,N_p\ (T - 1) = \begin{cases} \ln\left(\dfrac{X}{\beta}\right) + 1 + m; & 0 \leqslant X \leqslant \beta \qquad (5.21) \\ m\ln\dfrac{(1 - X)}{(1 - \beta)} + 1 + m; & \beta \leqslant X \leqslant 1 \qquad (5.22) \end{cases}$$

where

$$m = -N_f/N \qquad (5.23)$$

and

$$\beta = 1/(1 + N_f/N_p) \qquad (5.24)$$

Example 5.3

Typical breakthrough curves for the adsorption of bovine serum albumin conjugated to arsanilic acid are shown in Figure 5.9, on controlled pore glass and in Figure 5.10, on

Sepharose gel. The ligand for each column was mouse monoclonal anti-benzenarsonate IgG.

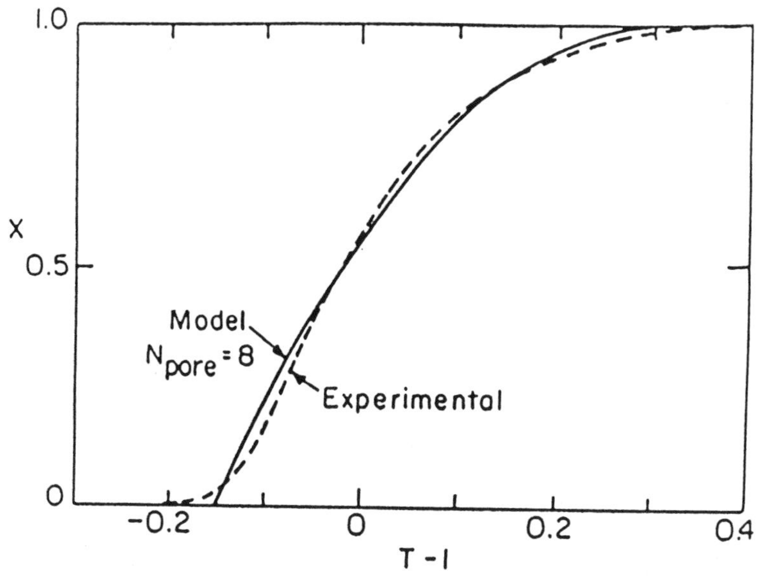

Figure 5.9. Experimental and calculated breakthrough curve for bovine serum albumin on 1.5 x 16.5 cm controlled pore glass-monoclonal antibody column (Reference 20).

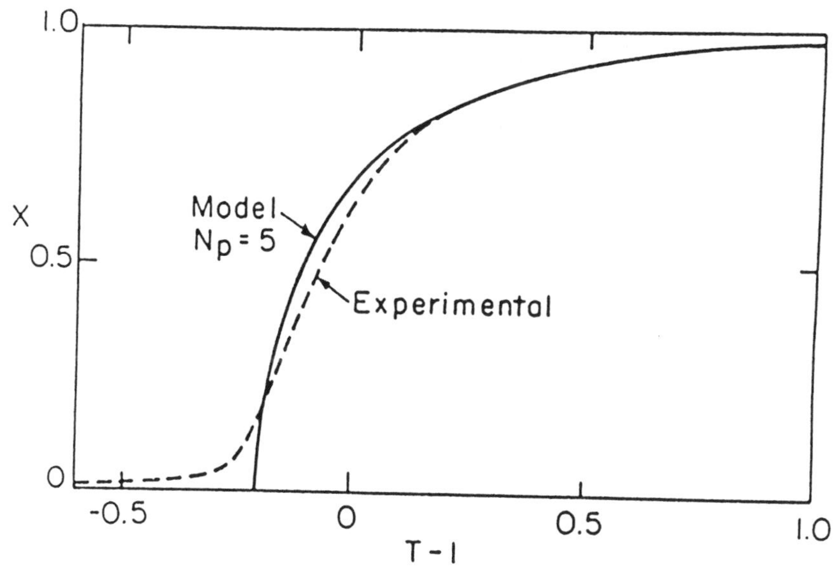

Figure 5.10. Experimental and calculated breakthrough curve for bovine serum albumin on 1.5 x 18.3 cm Sepharose 4B-monoclonal antibody column (Reference 20).

The experimental curve for the controlled pore glass columns was better modeled using the pore diffusion equation (5.13) with $N_{pore} = 8$. However, the breakthrough curve for the Sepharose column showed a greater tendency to tail off at the upper end and was better modeled using the solid diffusion equation (5.16) with $N_p = 5$.

5.3 AFFINITY CHROMATOGRAPHY MATERIALS

5.3.1 Supports

The earliest supports used in affinity chromatography were agarose gels, a purified form of the naturally occurring polysaccharide agar. Microporous glass beads, cellulose, polyacrylamide, crosslinked dextrans and polystyrenes have also been used as supports.

The proper selection of the support or matrix for the ligands is of critical importance for the success of an affinity chromatography system. An ideal support would possess the following properties (22):

(1) It must interact very weakly with proteins in general to minimize the non-specific adsorption of proteins.

(2) It should be an insoluble, rigid material which exhibits good flow properties which are retained after coupling.

(3) It must possess chemical groups which can be activated or modified, under conditions innocuous to the structure of the matrix, to allow the chemical linkage of the ligand.

(4) These chemical groups should be abundant in order to allow attainment of a high effective concentration of coupling sites, so that satisfactory adsorption can be obtained even with protein-ligand systems of low affinity.

(5) It must be mechanically and chemically stable to the conditions of coupling and to the varying conditions of pH, ionic strength, temperature and presence of denaturants (e.g., urea, guanidine

hydrochloride) which may be needed for adsorption or elution. These properties permit the repeated use of the support-ligand material.

(6) It should form a very loose, porous network which permits uniform and unimpaired entry and exit of large macromolecules throughout the entire matrix.

May and Landgraff (23) added three additonal requirements:

(7) The support should be resistent to microbial attack.

(8) The support should be highly "wettable" hydrophilic material when isolating soluble macromolecules.

(9) It should be available at a reasonable cost.

No support material meets all of the requirements of this ideal. The most common matrices used are the polysaccharide, agarose, polyacrylamide and controlled pore glass. Other less successful materials used include polystyrene, cellulose and dextrans. Hyrophobic supports, such as polystyrene, cause drastic alterations in the three-dimensional structure of proteins, resulting in substantial decreases in the proteins' activity (24).

Cellulose has a heterogeneous microstructure which strongly detracts from its suitability as a support and, in addition, has a significant amount of non-specific adsorption (25). Cellulose particles are formed with more difficulty compared to agarose gels. However, it has seen use as the support material in the affinity chromatographic isolation of immunoglobulins (26). The principal advantage of cellulose is that it is the cheapest of the support materials.

Crosslinked dextrans are polysaccharides like agarose and possess most of agarose's desirable features, except for its degree of porosity. The low porosity of dextrans make them ineffective as adsorbents for enzyme purification (22).

5.3.1.1 Agarose: Agarose is a linear polysaccharide

consisting of alternating β-D-galactose and 3,6-anhydro-α-L-galactose residues linked 1,3- and 1,4- respectively. Agarose is not chemically crosslinked and the size exclusion limit depends on the percentage of agarose in the gel. A crosslinked agarose (Figure 5.11) has been prepared with 2,3-dibromopropanol which substantially increases the chemical and mechanical stability of the support.

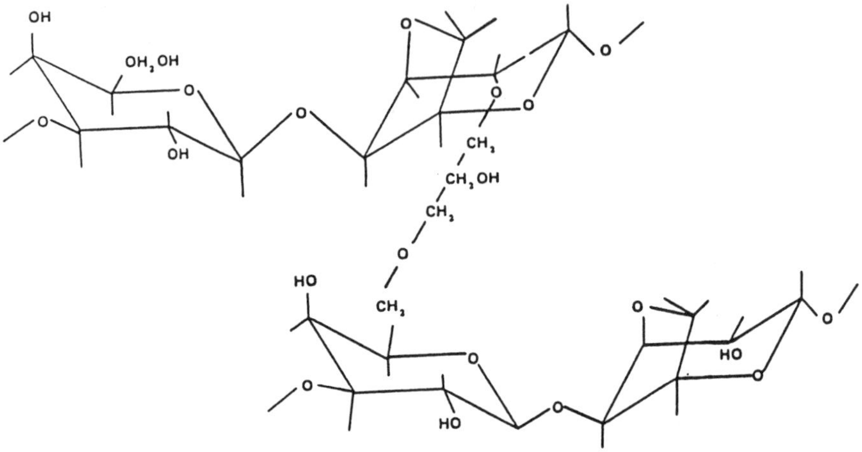

Figure 5.11. Chemical structure of crosslinked agarose (Reference 23).

Agarose and other polysaccharides are usually activated using the CNB method (27). This method may be carried out in alkaline solution in a single step. Various procedures involving addition of CNB solid or solution have been developed. Activation is usually complete in less than 30 minutes. The activated gel must then be rapidly washed with cold buffer and then the ligand is immediately attached. The chemical reactions proposed for the activation process are shown in Figure 5.12, along with the attachment of an amino-containing ligand.

Agarose can also be activated using other chemical reactions. The matrix can be activated by epichlorophydrins or diepoxides which introduce reactive epoxide functionalities. Figure 5.13 shows the epoxide formation, along with the sulfhydryl coupling and derivatization with aminoalkyl arms which allow spacing between the matrix and the ligand.

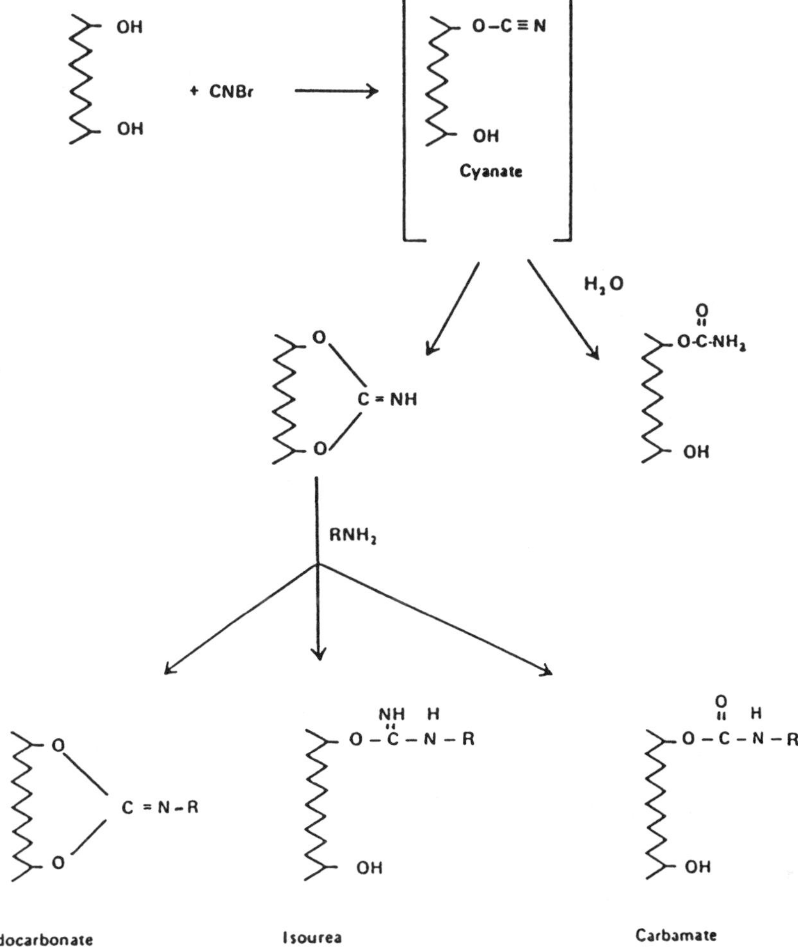

Figure 5.12. Activation of polysaccharides with CNBr and coupling of amino ligands (Reference 23).

Agarose continues to be one of the most popular supports used in affinity chromatography. The porosity of agarose results from the polymer's ability to form a highly hydrated open structure. Agarose is quite stable in aqueous solutions of intermediate pH and ionic strength. The disadvantages of agarose are its matrix compressibility, its susceptibility to microbial attack, its instability at temperatures above 40°C and its relatively high cost. It is possible, however to strengthen its matrix by crosslinking it with epichlorohydrin.

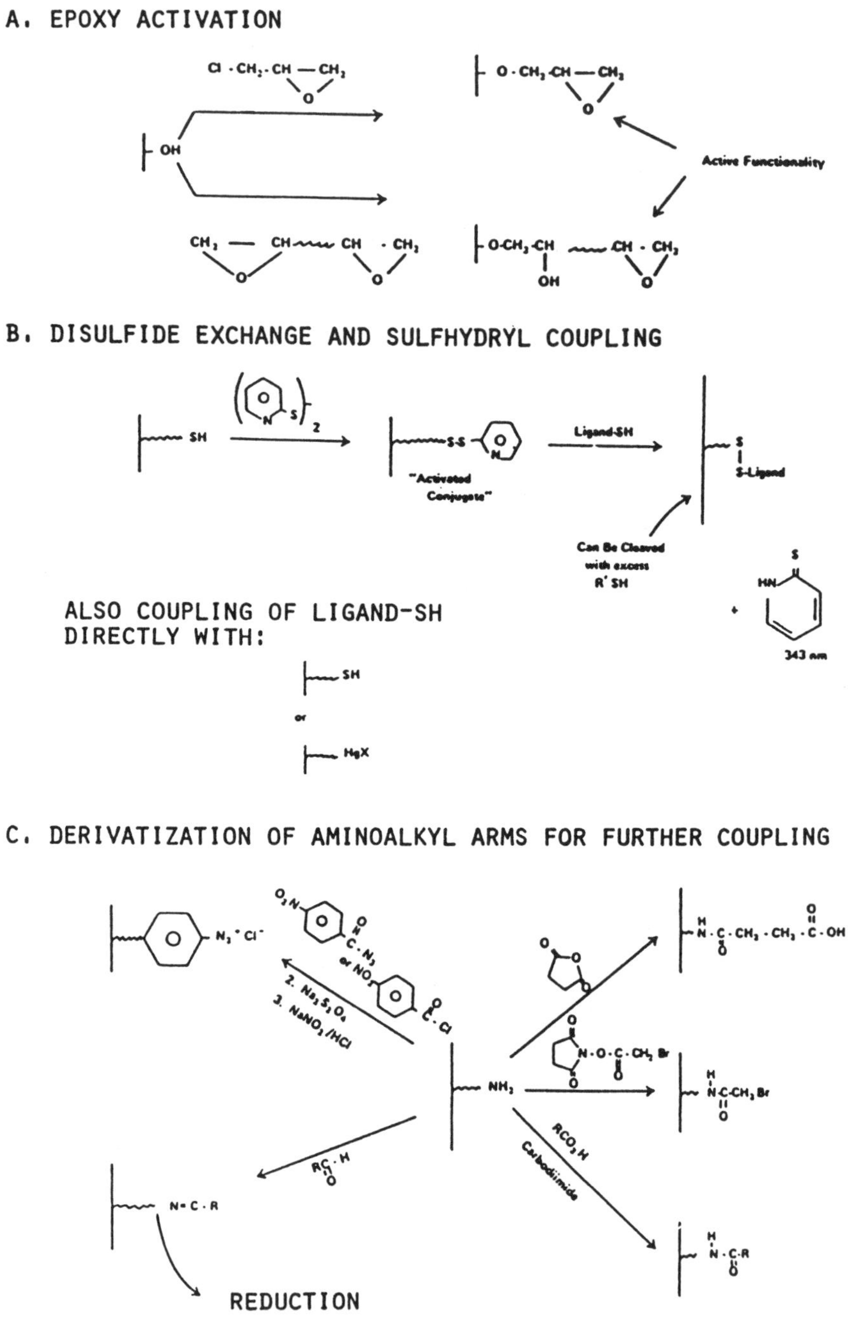

A. EPOXY ACTIVATION

Active Functionality

B. DISULFIDE EXCHANGE AND SULFHYDRYL COUPLING

"Activated Conjugate"

Ligand-SH

Can Be Cleaved with excess R' SH

343 nm

ALSO COUPLING OF LIGAND-SH
DIRECTLY WITH:

C. DERIVATIZATION OF AMINOALKYL ARMS FOR FURTHER COUPLING

REDUCTION

Figure 5.13. Formation of agarose derivatives by (A) epoxy activation, (B) disulfide exchange and sulfhydryl coupling and (C) derivatization of aminoalkyl arms for further coupling (Reference 23).

5.3.1.2 Polyacrylamide Gels: Polyacrylamide gels are synthetic, crosslinked polymers prepared by the copolymerization of acrylamide and N,N-methylene-bis-acrylamide (Figure 5.14). The chemistry of activation for the polyacrylamide support is based on the exchangeability of the carboxamide nitrogen with other amino-containing compounds, as shown in Figure 5.15. Further modification of these derivatives and the attachment of ligands is accomplished using chemistry similar to that used with the agarose activation (28).

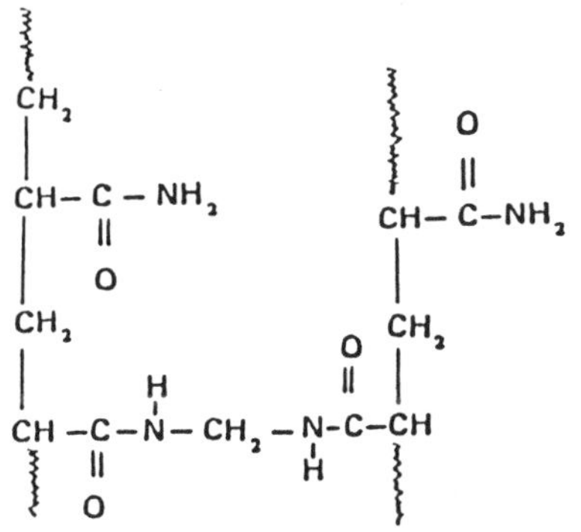

Figure 5.14. Chemical structure of acrylamide copolymerized with N,N-methylene-bis-acrylamide (Reference 23).

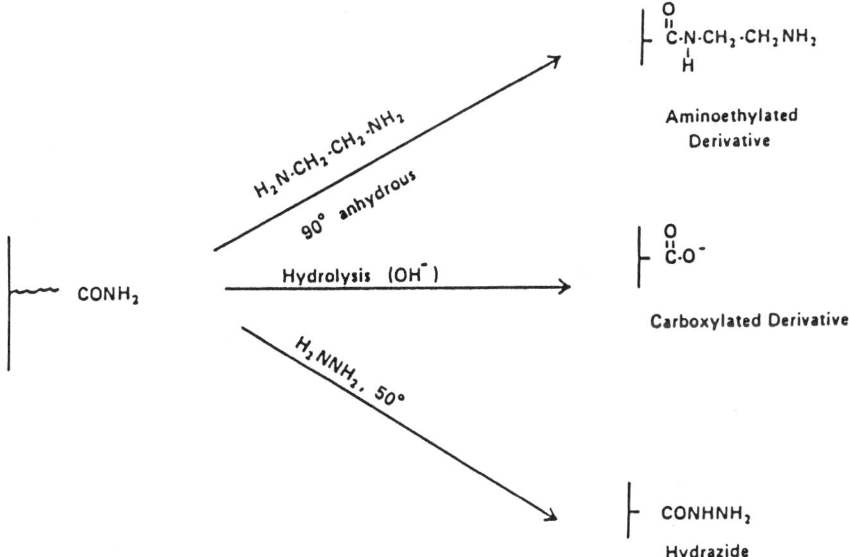

Figure 5.15. The chemistry of the activation of the polyacrylamide support (Reference 23).

The advantages of polyacryamide gels include the chemical stability due to its polyethylene backbone, a uniform physical matrix and porosity, low non-specific adsorption of proteins and a high concentration of carboxamide groups which can be used as sites for ligand attachment. The major drawback of this material is that its porosity is reduced after the derivatization process.

A material is commercially available that combines the three-dimensional polyacrylamide lattice with an interstitial agarose gel (29). This material has both the hydroxyl and carboxamide functionalities of the agarose and polyacrylamide, respectively, which can be activated independently and coupled to two separate ligands using independent reaction schemes. The flow difficulties noted with agarose are overcome by having the polyacrylamide lattice while the low porosity of derivatized polyacrylamide supports are eliminated by having the interstitial agarose. These combination materials are available in varying concentrations of each component.

5.3.1.3 Controlled Pore Glass: Ohlson (30) was the first to use microporous glass beads as a support material in affinity chromatography. Controlled pore glass has also seen wide use as a support in affinity chromatography.

The chemistry of derivatizing glass supports is shown in Figure 5.16. The glass beads must first be silanized before they can be activated. Silanization is the coupling of the silane to the glass bead support through the surface silanol, or of the oxide groups to the silylalkoxy groups (31). The reaction sequence as shown in Figure 5.16 used triethoxysilane. Specific silane reaction procedures have been described by Weetall for organic (32) and aqueous (33) solvents.

The silanized controlled pore glass beads can then be activated to form an arylamine, carboxyl or aldehyde derivative (34). The arylamine is prepared from the silanized support by reaction with p-nitrobenzoylchloride, followed by reduction with sodium dithionite. The reaction is shown with a γ-aminopropyltriethoxysilanized support in Figure 5.17. The carboxylated derivative is prepared by reaction of the alkylamine carrier with succinic anhydride, as shown in Figure 5.18. The aldehyde derivative is prepared by reaction with gluteraldehyde, which is illustrated in Figure 5.19.

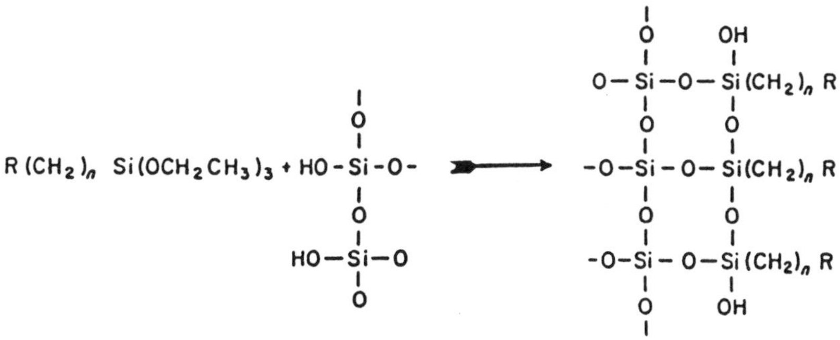

Figure 5.16. Silanization of the surface of porous glass. R represents an organic functional group (Reference 34).

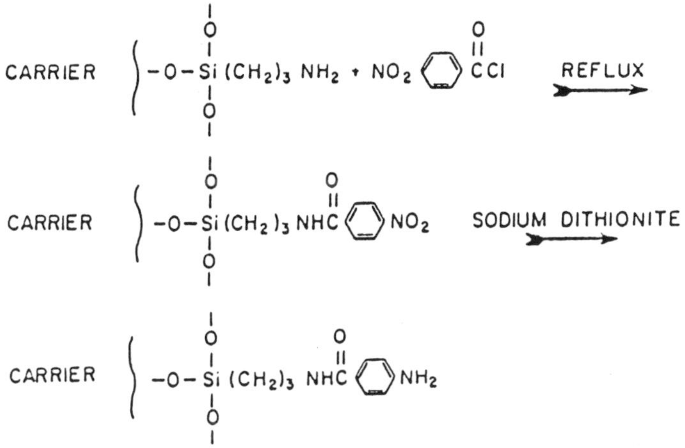

Figure 5.17. Preparation of the arylamine derivative from the alkylamine derivative (Reference 34).

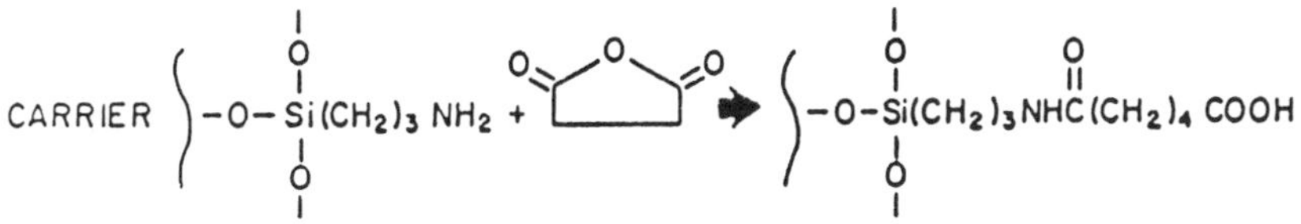

Figure 5.18. Preparation of the carboxyl derivative from the alkylamine derivative (Reference 34).

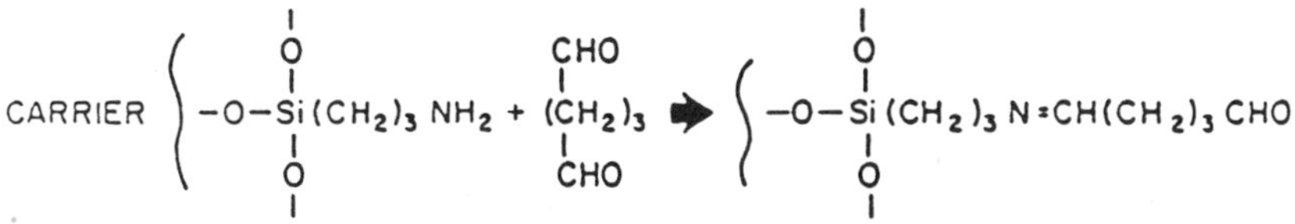

Figure 5.19. Preparation of the aldehyde derivative from the alkylamine derivative (Reference 34).

Controlled-pore glass beads are now available with surface coatings of glycerol and glycol-like structures to eliminate non-specific protein adsorption. The advantages of this type of support are its mechanical strength, its well-defined and accessible pore volume, its resistance to microbial attack and flexibility of chemical modification procedures for ligand coupling. The principal disadvantage is that silica is susceptible to dissolution in alkaline media which restricts its use to solution pH values of less than 7.5. This support is even more expensive than the agarose gel beads.

5.3.2 Spacers

Hydrocarbon spacers between ligand molecules and the support matrix are used extensively in affinity chromatography. The usefulness of these spacers has been attributed to the relief of steric restrictions imposed by the matrix backbone and to an increased flexibility and mobility of the ligand as it protrudes farther into the solvent (35). Shaltiel (36) has pointed out, however, that hydrocarbon extensions by themselves may bind proteins through mechanisms unrelated to specific recognition of the ligand. Such binding is more likely to occur when the affinity chromatography material is prepared by first coating the matrix with hydrocarbon arms and then attaching the specific ligand to those extensions. Those hydrocarbon spacers which do not have an attached ligand may bind other proteins through hydrophobic interactions, thereby reducing the specificity of the system. Therefore, when spacers are employed between the matrix and the ligand, it is advisable first to synthesize the ligand with the hydrocarbon attached and then to attach the elongated ligand to the support matrix. This procedure will minimize the possibility of hydrophobic interactions of the space interfering with the intended affinity chromatography.

Example 5.4

Figure 5.20 (37) shows the effect of hydrocarbon spacers on the reversible binding of glycogen phosphorylase b to alkyl agarose supports. The columns were equilibrated at 22°C with a buffer of 50 mM sodium β-glycerophosphate, 50 mM 2-mercaptoethanol and 1mM EDTA at pH 7.0 before applying the protein sample to the column. Nonadsorbed protein was washed off with the same buffer. Elution, initiated at the arrow in the figure, was carried out with a deforming buffer of 0.4M imidazole, 50mM 2-mercaptoethanol adjusted to pH 7.0 with citric acid. The binding on the C_4 support is quite specific, since 95% of the crude muscle extract, containing no phosphorylase b activity, is not adsorbed while the small amount of protein eluted by the deforming buffer has a very high phosporylase activity and yields over 95% of the enzyme activity. This is in contrast to the results with C_1, where phosphorylase b is not adsorbed, with C_3, where the enzyme is retarded, and with C_6, where the enzyme is adsorbed but could not be eluted.

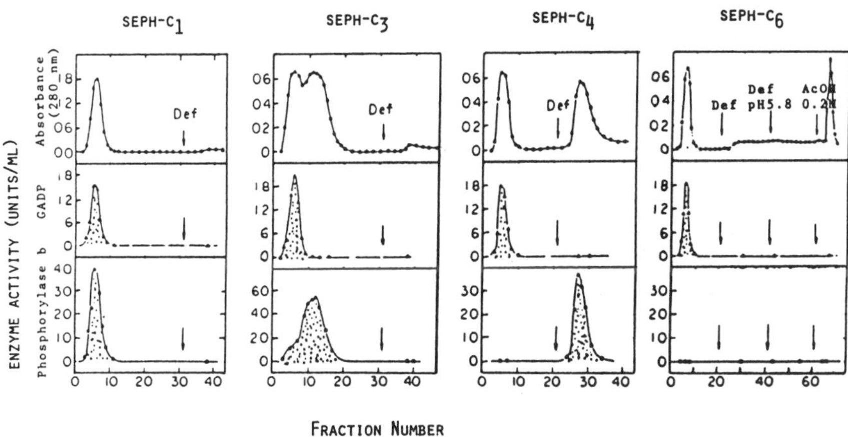

Figure 5.20. Preferential adsorption of glycogen phosphorylase b on hydrocarbon-coated agaroses depends on the length of their alkyl side chains (Reference 37).

5.3.3 Ligands

The ligand is usually a small molecule, covalently attached to the solid support, that displays a special and unique affinity for the molecule to be purified. This small molecule must possess chemical groups that can be modified for linkage to the solid support without destroying or seriously altering interactions with the designated protein. When the target protein has a high molecular weight, the ligand groups which interact with that protein must be sufficiently distant from the solid matrix to minimize steric interference. The use of "spacers" to create this distance has just been discussed.

Enzymes constitute the largest category of proteins purified by affinity chromatography. Purification of enzymes can be divided into two categories: (A) enzymes purified with the aid of specific inhibitors or substrates or (B) the purification of a wide range of enzymes using a general ligand.

A large number of different ligands are available for specific interactions with enzymes; these include competitive enzyme inhibitors, coenzymes, other substrates and cofactors. The columns prepared from such ligands can only be used to purify the specific enzymes for which the column was designed. At times, specific columns may also serve for the purification of enzymes with the identical catalytic activity but which are derived from different organisms. Table 5.3 (38) shows some of the ligand-enzyme combinations based on specific interactions.

Table 5.3: Ligand-Enzyme Combinations Based on Specific Interactions (38)

Enzyme	Ligand	Eluent
N - Acetylglucosamidase	Thio-N-acetylglucosamide	NaCl
Adenosine (phosphate) deaminase	Inosine	Adenosine
D-Alanine carboxypeptidase	Penicillins	NaCl, hydroxylamine
Carboxylpeptidase N	Aminobenzoyl arginine	Guanidinoethyl-mercapto succinate
Choline dehydrogenase	Choline	Dithiothreitol
Dipeptidyl peptidase	4-Phenylbutylglycylproline	Ethylene glycol, NaCl
Elastase	Elastin, $(Ala)_3$	Salt, Ala
α - L - Fucosidase	Fucosamine Thiofuco-pyranoside	Fucose
β - Galactosidase	Thiogalactoside	Thiogalactoside, ethylene glycol
Glyoxylase	Glutathione	pH
Histaminase	Cadaverine	Heparin
3 β - Hydroxysteroid oxidase	Cholesterol	Triton X - 100
Myosin kinase (I, II)	Calmodulin	EGTA
Phenylalanine hydroxylase	Pteridine derivative	Phenylalanine, pH
Prostaglandin cyclooxygenase	Flurbiprofen	Flufenamic acid, ethylene glycol
Renin	Pepstatin, hemoglobin octapeptide	pH
Urate oxidase	8 - Aminoxanthine	Urate

The general ligand is a ligand which may be used with a larger group of enzymes (39). Since about 30% of the known enzymes require a coenzyme for their activity, using this coenzyme as the ligand will allow purification of a relatively large range of proteins on the same column. The coenzymes most commonly used are NAD, NADP, ATP, CoA, and other derivatives of adenine nucleotides. Table 5.4 (38) shows an alphabetical list of some enzymes that have been purified with general ligands.

Table 5.4: Enzymes Purified Using General Ligands

Enzyme	Ligand	Eluent
Adenosine	ATP, AMP	Adenosine, pH
Alanine dehydrogenase	NADP	NaCl
L - Arabinose kinase	ATP	NaCl
Choline acetyltransferase	Coenzyme A	Citrate-phosphate EDTA, dithioerythritol
Cytochrome reductase	NADP, ADP	NADP
L - Fucose dehydrogenase	AMP	pH
Glycogen phosphorylase	AMP	AMP
15 - Hydroxyprostaglandin dehydrogenase	NAD	NADH
3 β - Hydroxysteroid dehydrogenase	NAD	P_i
Isocitrate dehydrogenase	AMP	NAD
Malate thiokinase	ADP	CoA, ATP, malate
Nucleotide pyrophosphatase	AMP	NaCl
Phosphoglucose isomerase	ATP	Glucose 6 - phosphate
Phosphorylase a	AMP	AMP
Pyridine dinucleotide transhydrogenase	NAD, 2', 5'-ADP	NADPH, 2'-AMP
Ribonuclease T_2	AMP	2' (3)-AMP, NaCl
Sorbitol dehydrogenase	NAD	NAD
Threonine dehydratase	AMP	AMP

The lectin, concanavalin A, has been used as a general ligand for polysaccharides and glycoproteins (40). Concanavalin A will selectively bind glucose and mannose residues in macromolecules. Therefore, when used as a ligand, it can be used to isolate a wide variety of biopolymers which contain one of the two sugar groups.

B-12 coenzyme and other cobalamins have been used as general ligands by coupling to substituted agarose either using the phosphate group or by amide formation at one of the side chains (41). Intrinsic factor, transcobalamins and B-12-dependent enzymes bind to these adsorbents.

The polyanionic aromatic dye chromophores mimic the overall shape, size and charge distribution of the naturally occurring biological heterocycles, such as the nucleotides and coenzymes. This is seen in the use of dyes, especially the reactive triazine-based textile dyes, as general ligands (42,43).

Cibacron Blue F3G-A has been shown (42,44) to be an especially effective adsorbent for the purification of pyridine nucleotide-dependent oxidoreductases, phosphokinases, coenzyme A-dependent enzymes, hydrolases, acetyl-, phosphoribosyl- and aminotransferases, RNA and DNA nucleases and polymerases, restriction endonucleases, synthetases, hydroxylases, glycolytic enzymes, phosphodiesterases, decarboxylases, sulfohydrolases and many other seemingly unrelated proteins including interferon and phytochrome.

Other triazine dyes, such as Procion Red H-E3B, Procion Red H-8BN, Procion Green H-4G and Procion Brown MX-5BR, have also been found (45) to be suitable ligands for the selective purification of individual pyridine nucleotide-dependent dehydrogenases, phosphokinases, plasminogen, carboxypeptidase G2, alkaline phosphatase and L-aminoacyl-tRNA synthetases.

These synthetic dyes offer five advantages as general ligands when compared to immobilized coenzymes or other biological ligands for use in large scale affinity chromatography:

1. The protein-binding capacities of immobilized dye adsorbents exceed those of biological origin by factors of 10 to 100.

2. The synthetic dyes may be easily coupled

to the matrix materials and are resistant to chemical and enzymatic degradation.

3. These dyes have general applicability and allow easy elution of bound proteins with good yields.

4. The characteristic spectral properties of the dyes permit monitoring of ligand concentrations and identification of column materials (Table 5.5).

5. These dyes are readily available at a reasonable cost.

Some of the commercially available affinity adsorbents are shown in Table 5.6.

Table 5.5: Properties of Some Triazine Dyes (45)

Reactive Dye	Mr (Na salt)	λmax (nm)	Molar absorption coefficient (M^{-1} cm^{-1})
Cibacron Blue F3G-A	773.5	610	13,600
Procion Brown MX-5BR	588.2	530	15,000
Procion Green H-4G	1760.1	675	57,400
Procion Red H-8BN	801.2	546	21,300
Procion Red H-E3B	-----	522	30,000

5.4 LABORATORY PRACTICES

5.4.1 Selecting a Ligand

The task of selecting a suitable ligand remains an empirical process involving the testing of a number of potential ligands for their ability to bind the desired protein.

As an example of a typical empirical approach, the relative suitability of several dyes as general ligands may be conveniently tested with a series of small columns, each containing 500 mg of agarose-bound dye equilibrated with an appropriate buffer in 0.8 x 4 cm disposable polypropylene columns (43). A small sample containing the desired protein is applied to each column and the flow is interrupted for 10 minutes to allow equilibration of the sample with the dye ligand. Nonadsorbed protein is purged with 5 to 10 mL of the equilibration buffer. The bound protein is then eluted with 1M KCl in the same buffer. Both the void and the

Table 5.6: Some Commercially Available Affinity Adsorbents

Tradename	Matrix	Ligand	Source
Dyematrex Blue A	Silica	Cibacron Blue F3GA	Amicon
Dyematrex Red A	Silica	Procion Red HE-3B	Amicon
Matrex Gel PBA	Agarose	Phenyl boronate	Amicon
Affi-Gel Blue	Agarose	Cibacron Blue F3GA	Bio-Rad
Affi-Gel Calmodulin	Agarose	Calmodulin	Bio-Rad
Affi-Gel Ovalbumin	Agarose	Ovalbumin	Bio-Rad
Affi-Gel Galactosamine	Agarose	N- acetyl galactosamine	Bio-Rad
Affi-Gel Phenothiazine	Agarose	Phenothiazine	Bio-Rad
Affi-Gel 731	Polyacrylamide	Polycations	Bio-Rad
7-Methyl-GTP Sepharose	Agarose	P- aminophenylester of 7- methylguanosine 5- triphosphate	Pharmacia
Oligo (dT)-Cellulose Type 7	Cellulose	Oligo (dT)	Pharmacia
DNA-Agarose	Agarose	DNA	Pharmacia
AG POLY (A) Type 6	Agarose	Polyriboadenylic acid	Pharmacia
AG POLY (C) Type 6	Agarose	Polyribocytidylic acid	Pharmacia
AG POLY (I) Type 6	Agarose	Polyriboinosinic acid	Pharmacia
Poly (U) Sepharose 4B	Agarose	Polyuridylic acid	Pharmacia
Progel-TSK Chelate - 5PW	Synthetic polymer	Iminodiacetic acid	Supelco/Toyo Soda
Progel-TSK Heparin - 5PW	Synthetic polymer	Heparin	Supelco/Toyo Soda
Progel-TSK Blue - 5PW	Synthetic polymer	Cibacron Blue F3GA	Supelco/Toyo Soda
Progel-TSK Boronate - 5PW	Synthetic polymer	M-Aminophenyl boronic acid	Supelco/Toyo Soda
Progel-TSK ABA - 5PW	Synthetic polymer	P-Aminobenzamidine	Supelco/Toyo Soda

eluate fractions are assayed for protein and activity to allow evaluation of the purification, recovery and capacity of the dye-protein combinations. Many other examples of ligand screening procedures are to be found in the literature (46-48).

The kinetic inhibition constants or dissociation constants of the protein-ligand interaction for a series of ligands in free solution represents an alternate method of screening potential ligands (49). For dye ligands, there is a general trend of protein-dye dissociation constants (K_d) across the color spectrum. The blue and green dyes tend to bind more tightly to pig heart lactate dehydrogenase (K_d < 8M) than yellow dyes (K_d > 36M), while orange and red dyes display intermediate affinities (10 M < K_d < 33 M).

After these screening tests, the choice of a ligand should be determined by two factors (50). First, it should have a combination of chemical functionalities that allow it to be attached to the activated support while retaining its selective binding ability. Second, the ligand should have an affinity for the desired product in the 10^{-4} to 10^{-8} M range in free solution to allow maximum binding without the risk of unacceptably difficult product elution.

Table 5.7 shows examples of many types of ligands which have been used with Sepharose derivatives in affinity purification applications. In general, enzymes have been used as ligands to isolate a substrate analogue, an inhibitor or a cofactor; an antibody may be used to isolate an antigen, a virus or a cell type; a lectin for a polysaccharide, a glycoprotein, a cell surface receptor or a cell type; a nucleic acid for a complementary base sequence, a histone, a nucleic acid polymerase or a binding protein; a hormone or a vitamin for a receptor or a carrier protein; and a specific cell for a cell surface specific protein or a lectin.

The capacity of affinity adsorbents has been shown to depend on the degree of ligand substitution (51). The greater the amount of ligand on the matrix, the lower the specificity of the protein adsorption. The larger amounts of ligand are likely to increase steric hindrance and non-specific binding and to make elution difficult. Ten milligrams of protein ligand per mL of support (2 μmols per mL) is a useful upper limit on substitution.

Table 5.7: Coupling Gels for Ligand Immobilization (50)

Type of Ligand	Derivative of Sepharose	Comments
Proteins, peptides, amino acids, nucleic acids (using $-NH_2$ group)	CNBr -activated Sepharose 4B	Method of choice for proteins and nucleic acids. Very well documented.
	CH-Sepharose 4B	Coupling via 6-carbon spacer arm. Organic solvents can be used for water insoluble ligands. Carbodiimide coupling method used.
	Activated CH-Sepharose 4B	Activated gel for spontaneous coupling via spacer arm. Especially useful for small, sensitive ligands.
	Epoxy-activated Sepharose 6B	Activated gel for coupling amino acids and peptides. Hydrophilic spacer arm. Organic solvents can be used.
Amino acids, keto acids, carboxylic acids (using -COOH group)	AH-Sepharose 4B	Coupling via 6-carbon spacer arm. Organic solvents can be used for water insoluble ligands. Carbodiimide coupling method is used.
Sugars, other hydroxyl compounds (using -OH group)	Epoxy-activated Sepharose 6B	Activated gel for coupling via extremely stable ether bond. Hydrophilic spacer arm. Organic solvents can be used.
Proteins (using -SH group)	Thiopropyl Sepharose 6-B Activated Thiol Sepharose 4B	For covalent chromatography of -SH containing substances. Coupling reactions are reversible.
Amino acids, other low MW compounds	Epoxy-activated Sepharose 6B	For a stable irreversibly coupled product.
Low MW thiol (using -SH group)	Thiopropyl Sepharose 6B	Ligand-matrix bond cleavable. (i.e. Coenzyme A)

5.4.2 Preparing the Support-Ligand

The ligand must be coupled to the solid support under mild conditions that are well tolerated by both ligand and support. The resulting support-ligand must be washed exhaustively to make certain that all material which is not covalently bound is removed. With highly aromatic compounds, such as estradiol, complete removal of adsorbed material is very difficult and will require many days of continuous washing. In extreme cases, such as for Congo red, it is necessary to wash the column with large quantities of the protein to be purified, followed by elution to regenerate the column.

A prerequisite to affinity chromatography experiments is an accurate method for determining the amount of material attached to the solid support. This is preferably determined by measuring the amount of ligand released from the support-ligand material after acid or alkaline hydrolysis.

Exhaustive digestion with pronase or carboxypeptidase has been used in some cases in which oligopeptides are attached to agarose. The degree of ligand substitution on the support is then expressed in terms of concentration, such as micromoles of ligand per milliliter of swollen support.

The quantity of ligand attached to the support can be controlled by varying several parameters. Most important is the amount of ligand added to the support. This is shown in Table 5.8 for the attachment of 3'-(4-aminophenylphosphoryl) deoxythymidine 5'-phosphate to agarose. When highly substituted derivatives are desired, the amont of ligand added should be 20-30 times higher than that which is desired in the final product.

Table 5.8: Efficiency of Coupling 3'-(4-Aminophenylphosphoryl) Deoxythymidine 5'-Phosphate to Agarose (1)

Experiment	μ Moles of inhibitor/ml agarose	
	Added	Coupled
A	4.1	2.3
B	2.5	1.5
C	1.5	1.0
D	0.5	0.3

The pH at which the attachment is performed will also determine the degree of attachment since it is the unprotonated form of the amino group which is reactive. Compounds containing an α-amino group, with a pK about 8, will react optimally at a pH of about 9.5 to 10.0. This is shown in Table 5.9 for the attachment of alanine to argarose (35).

Table 5.9: Effect of pH on the Coupling of Alanine to Activated Agarose (35)

Conditions for Coupling Reaction		Alanine Coupled
Buffer	pH	(moles per ml agarose)
Sodium citrate, 0.1 M	6.0	4.2
Sodium phosphate, 0.1 M	7.5	8.0
Sodium borate, 0.1 M	8.5	11.0
Sodium borate, 0.1 M	9.5	12.5
Sodium carbonate, 0.1 M	10.5	10.5
Sodium carbonate, 0.1 M	11.5	0.2

The time and temperature will also affect the quantity of ligand attached. Swollen polyacrylamide beads are reacted with 3 to 6M hydrazine hydrate in a constant temperature bath at temperatures between 45°C and 50°C. Figure 5.21 shows the time required to achieve certain degrees of derivatization (52). The amount of substitution varies linearly for at least an eight hour period.

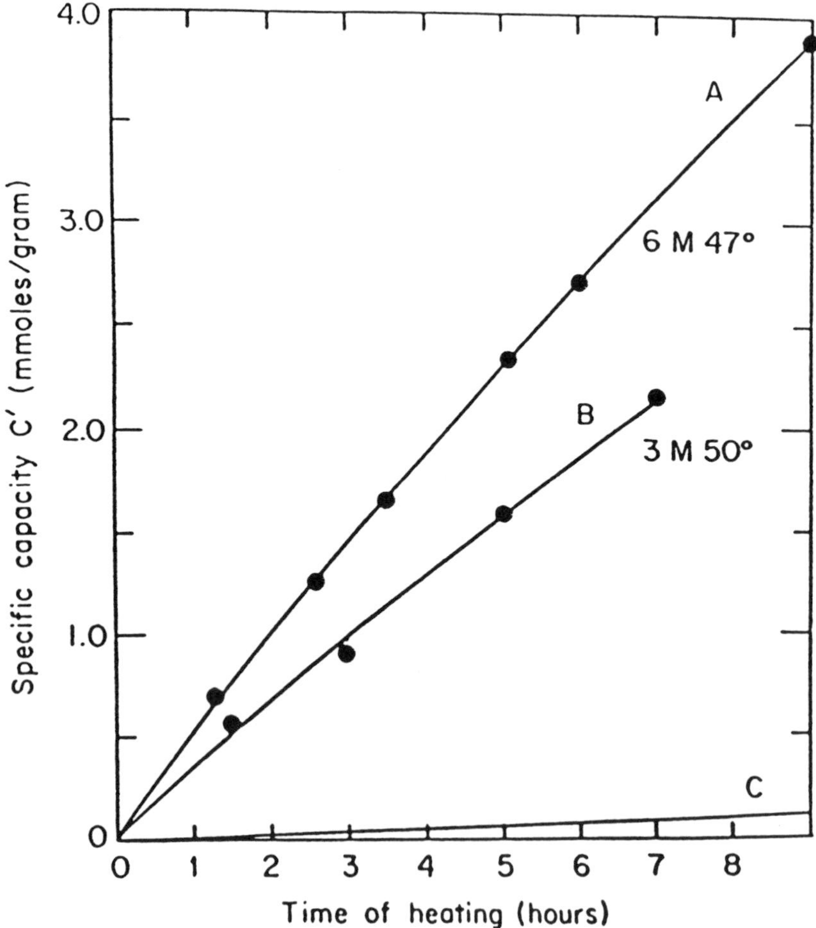

Figure 5.21. Time course of the reaction between polyacrylamide BIO-GEL P-60 (100-200 mesh) and aqueous hydrazine. C' refers to mmoles of hydrazine groups found on the amount of derivative produced from 1 gram of the original dry polyacrylamide (Reference 52).

5.4.3 Ligate Adsorption

The specific conditions for affinity chromatography adsorption are dictated by the specific properties of the ligate to be purified. When ligands with mixed electrostatic-hydrophobic characteristics are used, the pH, ionic strength and temperature conditions can often be adjusted to improve adsorption of the ligate. The optimal sample size, flow rate, buffer composition, pH, ionic strength and temperature will have to be defined for the specific ligand-ligate combination. Table 5.10 (45) lists a selection of conditions that have been used to adsorb proteins to immobilized dye ligands.

Table 5.10: Operational Parameters for the Adsorption of Proteins to Immobilized Dyes (45)

Condition	Data
Equilibration buffers	
Molarity	5 - 100 mM
pH	5.5 - 9.0
Composition	Tris, phosphate, acetate, triethanolamine, tricine, HEPES, MOPS, bicarbonate
Temperature	4 - 50°C
Additives	EDTA, KCl, NaCl, NH_4Cl, $(NH_4)_2SO_4$, 2-mercaptoethanol, dithiothreitol, cysteine, thioglycerol, phenylmethylsulfonyl fluoride, urea, sucrose, glycerol, ethylene glycol, detergents, chloramphenicol, amphotericin B
Metal Ions	Na, K, Mg, Ca, Sr, Ba, Zm, Cu, Co, Ni, Mn, Fe, Al, Cr.
Sample applied per bed volume	0.03 - 950 mg/ml

It is not necessary that the adsorption process take place in a column. It may be advisable to use batch purification, for example, when small amounts of protein are to be extracted from very crude protein mixtures with an adsorbent of very high affinity. The flow rate of such a mixture through a column would be very slow, decreasing with time for those cases when particulates are present in the protein mixture.

In some cases involving very high affinity complexes, such as occurs with certain antibody-antigen systems, it is preferable to adsorb the protein to the support-ligand in a column, to wash it there extensively and then to elute the protein by removing the adsorbent from the column and suspending it in a vessel with an appropriate solvent. This means of elution requires less drastic conditions and may give higher yields because of the thoroughness of mixing, higher dilution of the insoluble ligand and easier control of time and temperature which are possible.

The operational capacity for specific adsorption with a given affinity adsorbent can be determined by slowly passing an excess amount of enzyme or protein through a sample of the adsorbent packed in a column. After washing with buffer until negligible protein is present in the effluent, the adsorbed protein is eluted and quantified. The operational capacity is the amount eluted.

Another approach to determine operational capacity is to add successive, small samples of pure protein to the column until protein or enzymatic activity is noticed in the column effluent. The operational capacity is the total amount added until breakthrough occurs.

5.4.4 Ligate Elution

In most cases, the elution of the ligate is carried out by changing the pH, ionic strength or temperature of the buffer. Removal of proteins from very high affinity adsorbents may require protein denaturants, such as urea or guanidine hydrochloride. Ideal elution of a tightly bound ligate should utilize a solvent which causes sufficient alteration of the conformation of the protein to decrease significantly the affinity of the protein for the ligand but not so severely that the protein is completely denatured or unfolded. The eluted protein should be neutralized, diluted or dialyzed at once to allow prompt reconstitution of the native protein structure.

Another method for elution is to cleave the support-ligand bond selectively. This allows the removal of the intact ligand-protein complex. Excess ligand can then be removed by dialysis or by gel filtration. This method can be used for ligands attached to the support by azo linkages or by thiol or alcohol ester bonds.

There is no single method that serves as a general example for all affinity chromatography elutions. However, several elution methods for both specific and general elution are being used with success.

Elution of the adsorbed protein may also be carried out with a solution containing a specific inhibitor or substrate. The inhibitor can either be the one that is covalently attached to the matrix or an alternate, stronger competitive inhibitor. If the inhibitor is the same as attached to the support, it must be present at higher concentrations in the elution solution.

Specific elution is also called affinity elution chromatography since it utilizes competition for the ligand site by hapten, ligands or inhibitor in solution to elute the adsorbed species. This use of the bioaffinity of the adsorbed species for elution represents the best conditions for highly selective purification. On the other hand, non-specific methods can be applied when apparently specific eluants fail to elute the desired protein.

Non-specific methods include solvent or buffer changes, pH or ionic strength changes, temperature changes, reversible denaturation and the chemical cleavage of the ligand from the support (53).

A good strategy for designing an elution method is the initial examination of several means of non-specific elution (38). Non-specific elution can be considered as an extra step in washing the column if the desired compound is not eluted since it serves to remove contaminating macromolecules. The washing should be started with a high salt concentration in the buffer so that compounds which are bound solely due to their ion exchange properties will be eluted. Occasionally low salt concentrations may be used initially to elute compounds bound through hydrophobic interaction. When changes in ionic strength are ineffective, different buffers, such as a change from a Tris to a borate buffer, may be tried. The next step in this strategy is to change the pH of the eluant, beginning with small changes followed by the addition of 0.1N acetic acid or 0.5M NH_4OH. If elution of the desired protein does not occur, treatment with ethylene glycol, dioxane or propionic acid may result in elution.

Table 5.11 (45) lists a number of conditions which have been used to elute bound proteins from immobilized dye ligands. The two techniques used most often are pH and salt elution. The ionic strength of the eluant may be increased in steps, pulses or gradients (54).

Table 5.11: Reported Eluents of Proteins from Immobilized Dye Affinity Adsorbents (45)

Method	Data
Nonspecific	
Molarity	0.025 - 6 M
Salts	NaCl, KCl, $CaCl_2$, NH_4Cl, $(NH_4)_2SO_4$
Polyols	Glycerol, ethylene glycol
Chaotropes	Urea, NaSCN, KSCN
Chelating agents	EDTA, 2,2'-bypyridyl, pyridine dicarboxylic acid
Detergents	0.1% (w/v) SDS, Triton X - 100
Specific elutants	
Molarity	0.001 - 25 mM
Composition	Coenzymes, nucleotides, polynucleotides, ternary complexes, substrates, dyes

When the elution tests just described are not effective, urea or guanidine may be added to remove the desired protein through denaturation. The denaturation is time dependent and reversible in many cases. Therefore, it is necessary that the eluant with the denatured, desired protein be collected in a dilution buffer immediately after leaving the chromatography column to reverse the denaturation.

The binding of the desired protein to the affinity column may be so tight as to preclude recovery of the protein. Under such circumstances, the ligand should be attached by means of a hydrolyzable bond or through a destructible spacer arm. Esters (55) are readily hydrolyzed with mild bases; vicinal hydroxyl groups are removed with periodate; and diazo bonds are broken with dithionates (56).

In order to maintain specificity at the elution step, it is advisable to perform the elution with a substrate or inhibitor rather than other solvents. In either case, a control column with ligand-free spacers should be used in order to evaluate any contribution of hydrophobic interactions by the spacers to the retention capability of affinity chromatography columns.

In some instances, a suitable combination of substrates

and coenzymes may elute the desired protein with a much greater yield than any of the individual components. As an example, *E. coli* adenylosuccinate synthetase is eluted efficiently from immobilized Procion Blue H-B using quaternary complex formation with IMP, GTP and L-aspartate, while much lower yields are obtained when a single or pair of these substrates are employed (48).

Suzuki and Karube (57) have demonstrated the use of light to control the affinity adsorption or elution of enzymes to agarose modified with a spiropyran compounds, using the isomerization shown in Figure 5.22. Trypsin from bovine pancrease was bound in the dark on a spiropyran gel with a soybean trypsin inhibitor ligand and was released with visible light irradiation, as shown in Figure 5.23. Approximately 60 to 80% of bound trypsin was released with visible light irradiation. The activity of the released trypsin was the same as that of native trypsin. Approximately 21-fold purification of trypsin was obtained using this technique. A limitation of this method is the limited number of cycles the spiropyran gel may be used because the spiropyran derivative undergoes irreversible isomerization by the light irradiation.

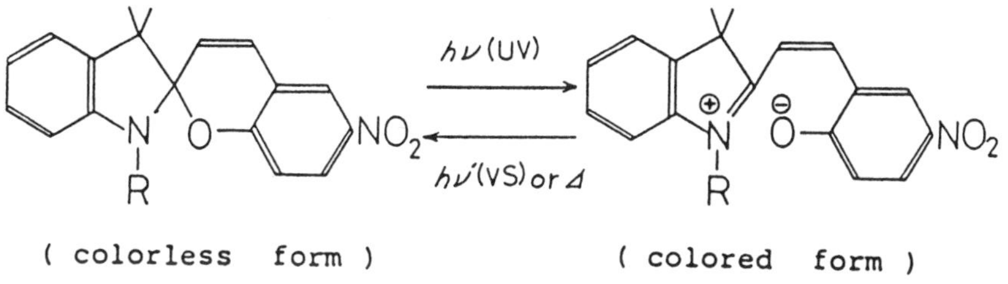

Figure 5.22. Photoreversible isomerization of spiropyran compounds (Reference 57).

5.5 SCALE-UP OF LABORATORY PROCEDURES

When the laboratory procedure has been carefully optimized, it is relatively easy to scale-up the process since it only requires extrapolation of the volumes of affinity adsorbent, washing buffer solution and elution solutions to quantities sufficient to process the required amount of feed stream.

As with other adsorption processes, the feedstream needs to be pretreated to give a clear, particle-free solution

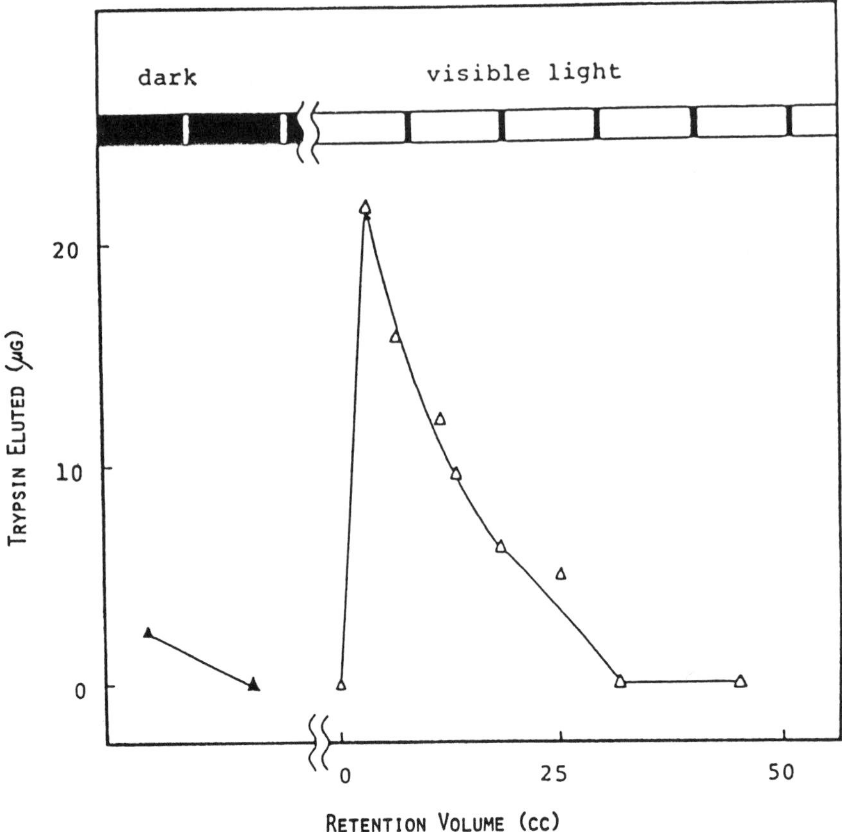

Figure 5.23. Elution pattern of trypsin from the column of light-activated ligand (Reference 57).

to prevent clogging if a column is used. It is desirable to pretreat the feedstream to reduce the amount of unwanted proteins to minimize any nonspecific binding to the affinity adsorbent.

Columns for process scale affinity chromatography are short and wide, having a width to height ratio of 1 to 2-4 to provide rapid throughput and reduced process time (Figure 5.24). When very large volumes (50 to 500 liters) of a feedstream are to be processed, it is convenient to carry out batch adsorption and elution in the same vessel. A very wide column is used, fitted with a slow-speed overhead stirrer. After batch mixing in the column, the affinity particles are allowed to settle for washing to remove the unbound material, followed by elution of the desired substances. The settled bed height is typically about 10 cm which allows high flow rates and rapid processing of large volumes of the feedstream.

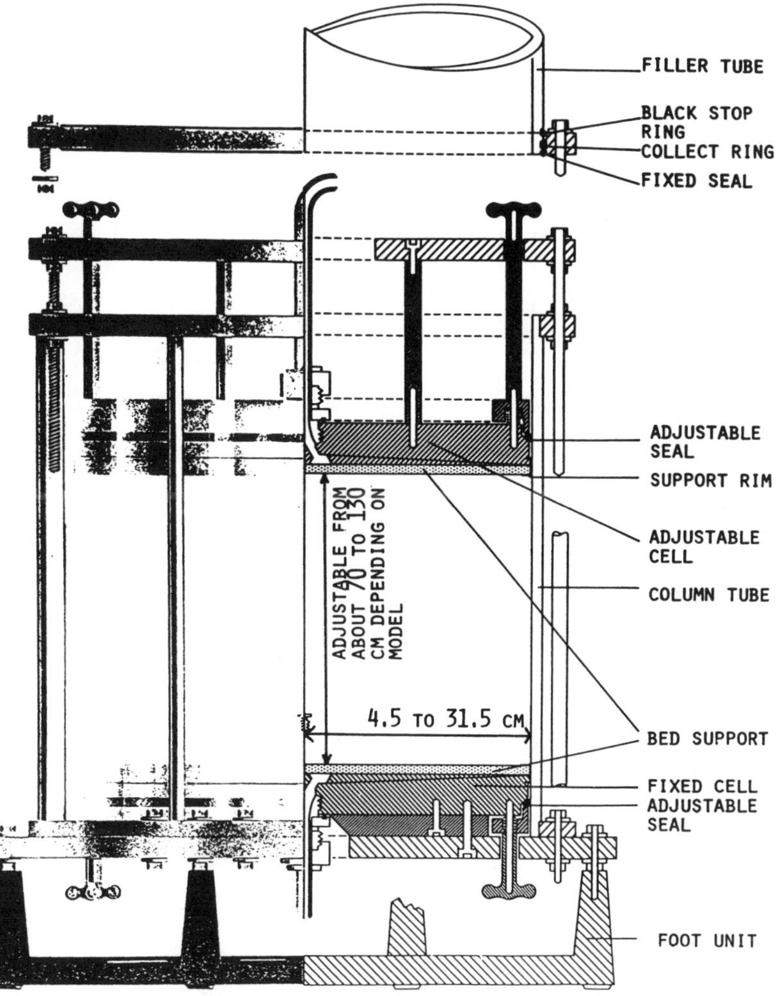

Figure 5.24. Process scale affinity column with adjustable height, manufactured by Amicon.

The purification scheme will usually require a subsequent step to remove the eluting agent from the purified product, particularly when partial denaturization of the product occurs during elution. Short connecting pipelines and the highest possible flow rates are parameters that must be designed into the process.

Automated instrumentation is usually employed to control the affinity purification process. An integrated automated system consists of a pump to maintain liquid flow, a UV monitor and recorder connected to the column outlet to follow the effluent composition and concentration, a programmable fraction collector and a system of solenoid valves to control feedstream and eluent application to the column, collection of fractions from the column and changes of eluents to clean and to regenerate the affinity adsorbent (Figure 5.25) (58).

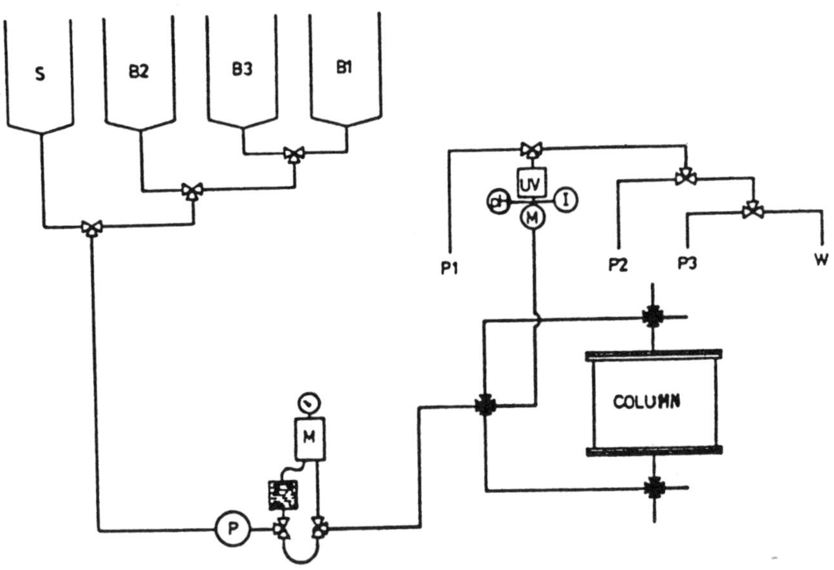

Figure 5.25. Process diagram for large-scale affinity chromatography. S: sample; B1 to B3: buffers 1 to 3; P1 to P3: products 1 to 3; W: waste (Reference 58).

5.6 AFFINITY CHROMATOGRAPHY OPERATIONS

5.6.1 Batch Operations

As with the other batch operations described in this book, the affinity particles are simply suspended in the sample fluid and agitated for the desired length of time, after which they are collected by filtration. The particles are then washed to remove non-bound substances, after which the particles are suspended in a solution which elutes the desired protein from the affinity particles. This scheme is shown in Figure 5.26. It is also possible to place the washed particles in a column for the elution of the desired protein.

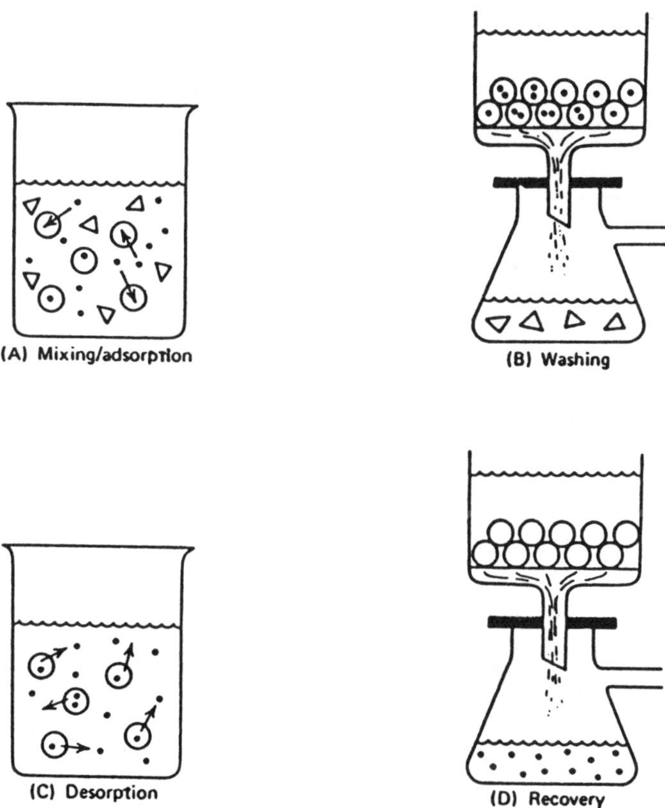

Figure 5.26. Batch processing of biopolymers on an affinity sorbent material. The open circles represent the affinity material, the dots are the ligate molecules and the triangles are the contaminants.

Sundberg and coworkers (59) have described a laboratory batch operation known as the "tea bag" method. With this method, shown in Figure 5.27, the affinity particles are placed inside a porous fabric bag which is immersed into the sample fluid. Several such bags, such containing affinity particles with different ligands, may be used to extract several proteins simultaneously from the same sample fluid. This simultaneous extraction offers the advantages of savings in time, of conserving precious starting materials and of decreasing losses of labile components.

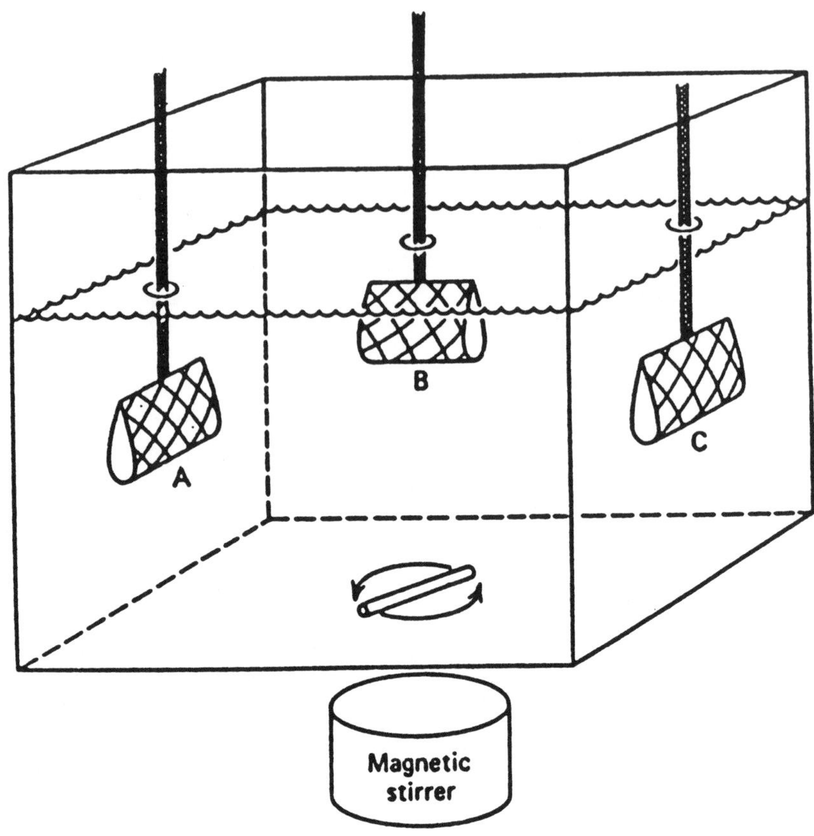

Figure 5.27. Tea-bag processing of different ligate molecules on three different affinity materials, A, B and C (Reference 59).

Wong and Charm (60) have taken the porous fabric bag approach and converted it into a continuous belt process, shown in Figure 5.28. The porous belt containing the affinity particles forms a continuous loop which moves through a series of tanks so that affinity adsorption, washing, elution and regeneration occur sequentially. This process requires that the ligate have a very high affinity for the sorbent material.

5.6.2 Column Operations

Affinity chromatography is most commonly carried out in a column operation. A typical arrangement of such an operation is shown in Figure 5.29. The feedstream is pumped onto the column, followed by a washing buffer, and finally the elution solution. The effluent from the column may be directed to the waste stream, to product collection or to further processing operations. A monitor on the effluent is needed to identify the process stage to facilitate product recovery.

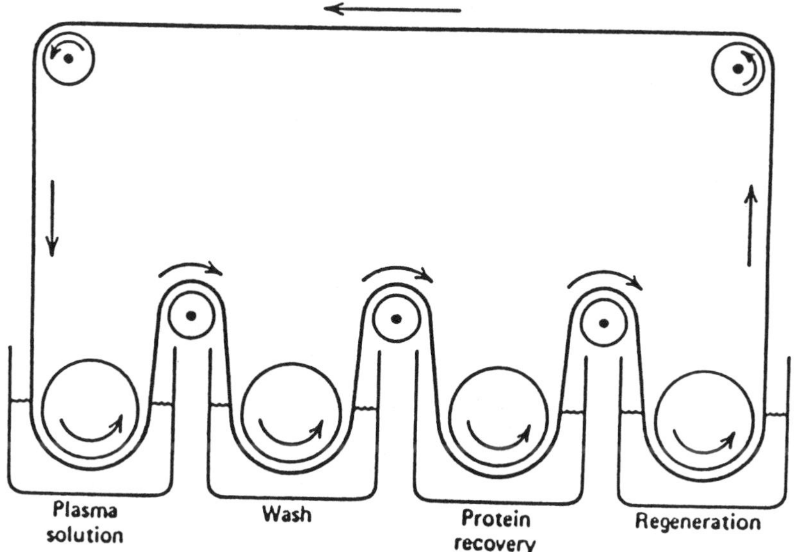

Figure 5.28. Continuous belt process for the affinity adsorption of proteins (Reference 60).

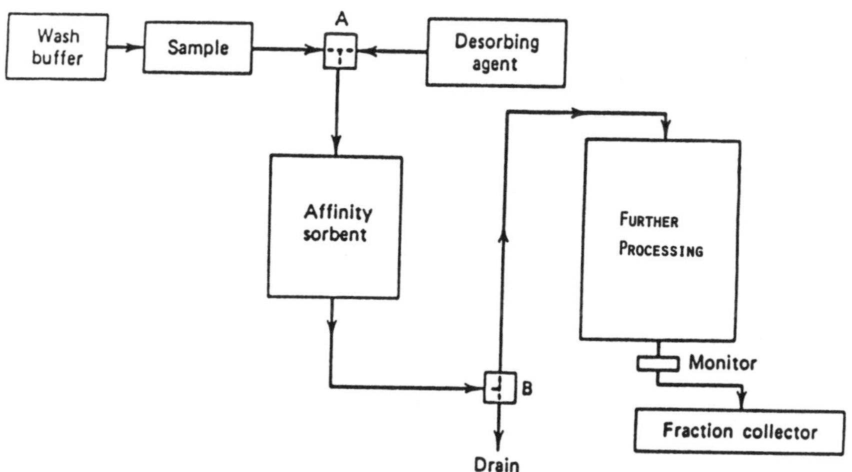

Figure 5.29. Affinity purification carried out in a column, where valve A selects either sample, wash buffer or desorbing solution. Valve B directs the column effluent to waste or to further processing.

The industrial use of column affinity chromatography requires the development of automatic control systems and monitoring in order to achieve reliable operation. A suitable control system consists of a set of valves to regulate the application of fluid streams to the column (feedstream, washes, eluants) and the collection of the effluent from the bed (product and waste streams), along with a set of instruments to provide a record of the separation. Figure 5.30 (61) shows the schematic of the microcomputer-controlled automatic system developed by Chase. The process includes an on-line gel filtration system for the rapid removal of eluent from the eluted protein. This is a desirable feature since eluents are often denaturants whose rapid removal is required.

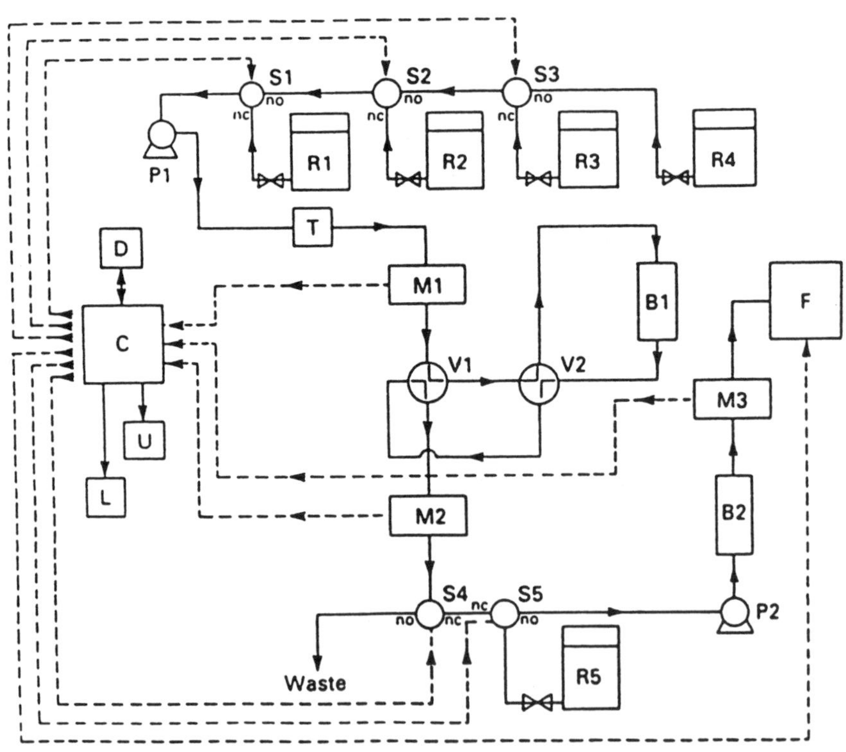

Figure 5.30. Microcomputer controlled system for the automatic control of fixed bed affinity separations. B1: immunosorbent bed; B2: gel filtration; C: computer; D: data storage; F: fraction collector; P: pumps; R: reservoirs; S: solenoid valve (Reference 61).

Several columns, each loaded with a different affinity sorbent, may be used in series to isolate several desired substances from a single feedstream. As an example (62), rat liver extract was passed through two columns in series which removed dihydrofolate reductase and guanine deaminase, respectively.

Begovich of the Oak Ridge National Laboratory (63) has developed a continuous column affinity chromatography procedure using a rotating chromatograph. Instead of changing feed, wash, elution and regeneration streams at preset times, as occurs with conventional columns, all four streams are fed continuously to separate segments of a rotating column bed, as shown in Figure 5.31. These streams are introduced at fixed points around the top of the column.

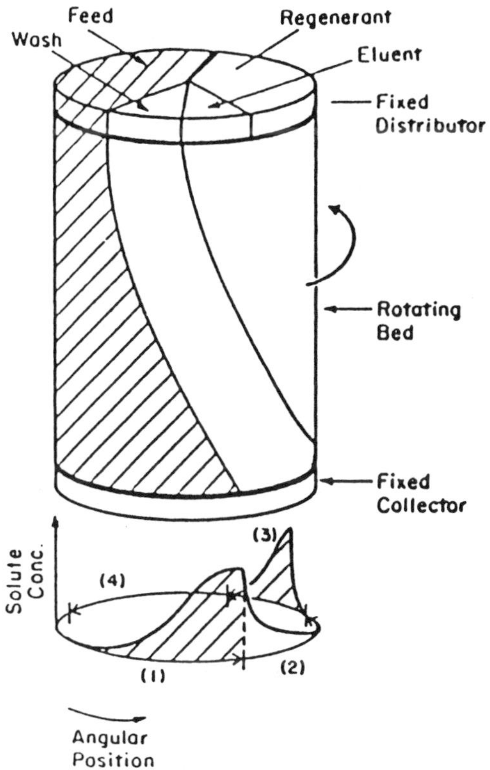

Figure 5.31. Continuous affinity chromatography on a rotating chromatograph. The effluent concentrations are a function of angular position rather than time: (1) breakthrough curve for adsorption; (2) wash profile; (3) elution profile; (4) regeneration (Reference 63).

The fraction of the column fed with each stream is equal to the fraction of the overall process time devoted to that operation in a conventional column, assuming all flow rates were equal. As the bed rotates, the adsorbent material is contacted with one or another of these streams. Purified product is withdrawn at a particular position as a continuous stream. The breakthrough, wash and elution profiles for this type of column are plots of effluent concentration at different angular positions around the column. While this technique offers no advantages over conventional column operations for small scale operations, it offers easier operation and more uniform product quality when used in preparative and manufacturing operations.

5.6.3 Affinity Purification with Membrane Operations

Affinity chromatography has been combined in a single step with cross-flow filtration to provide a high resolution and high recovery technique which is capable of treating unclarified and viscous feedstreams. This technique, illustrated in Figure 5.32 (64), has been used to purify concanavalin A from a crude extract of Jack beans by using *Saccharomyces cerevisiae* cells as the affinity ligand. Water-soluble macroligands are retained on one side of the ultrafiltration membrane. The feedstream is filtered in the presence of these ligands, the macroligand-product complex is retained while all impurities pass through the membrane. The product can then be eluted from the affinity ligand. A highly purified product at a yield of 70% was obtained using 0.8M D-glucose as the eluant.

Luong and coworkers (65) synthesized an acrylamide polymer with pendant aminobenzamidine groups to be used with this technique to purify trypsin. The trypsin-binding capacity of the water-soluble macroligands was close to the theoretical value 120:1 (weight of bound trypsin to weight of aminobenzamidine), which is considerably higher than that obtained with ligands on crosslinked gel supports. The eluant used was 0.5M arginine. This one-step procedure was able to purify trypsin from a trypsin-chymotrypsin mixture with 90% yield and 98% purity.

Mandaro and coworkers (66) have reported the use of an affinity filter matrix which had either Protein A or Protein G as ligands to purify immunoglobulins from various animal blood serum samples. The filter matrix was a composite of vinyl monomers grafted to a cellulose support which contained large pores. The pores allowed the migration of macromolecules through the matrix even though the matrix has the physical strength to be formed into a high-flow rate filter cartridge (Figure 5.33).

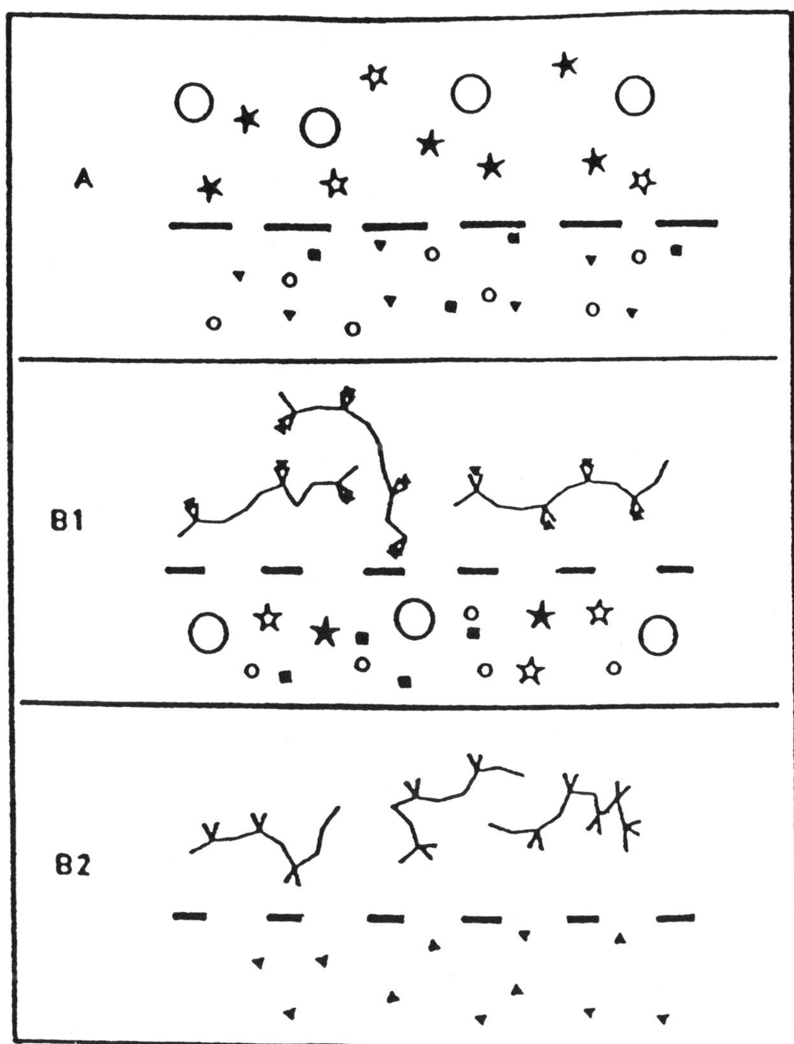

Figure 5.32. A: Cross-flow filtration, substances separated by sizes. B1: Binding and elimination in affinity cross-flow filtration. The binder is retained by specific attachment to macroligands. B2: Dissociation of the product from macroligands (Reference 64).

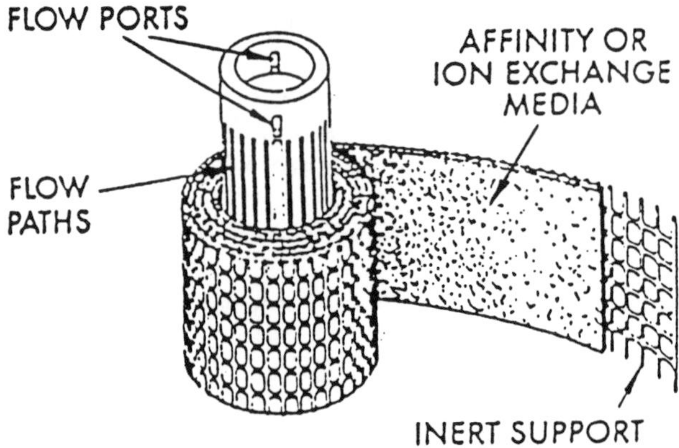

Figure 5.33. Filtration support for affinity separations (Reference 66).

These supports were activated with gluteraldehyde to provide an aldehyde group at the end of an 18-atom spacer arm. Depending on the ligand attachment process, between 4 and 5 mg of immobilized ligand is attached per gram of matrix. Table 5.12 shows the capacity of a filter matrix with 12.4 mg of immobilized Protein A or with 10.6 mg of immobilized Protein G. The flow rates of about 2 mL/min resulted in pressures less than 1 bar. Larger systems with ligand immobilization capacities of 80 mg and 3000 mg allowed flow rates of 5 and 125 mL/min.

Table 5.12: Comparison of Protein A and Protein G Filter Interaction with Various Polyclonal IgG's (66)

	Protein A		Protein G	
Species	Capacity (mg)	Ratio	Capacity (mg)	Ratio
Human IgG (pure)	43.99	3.55	46.66	4.42
Human IgG (serum)	39.03	3.15	40.05	3.79
Bovine IgG (pure)	25.19	2.03	44.44	4.20

The Ratio is the eluted IgG/bound ligand.

Pungor and coworkers (67) developed a method of continuous affinity-recycle extraction, shown in Figure 5.34. The sample is fed to the adsorption stage where it contacts the affinity adsorbent. The desired product selectively adsorbs while contaminants are diluted with the addition of wash buffer. The affinity adsorbent with the product then passes to a desorbing stage where the addition of the eluting solvent causes the product to desorb. The regenerated affinity adsorbent is recycled to the adsorption stage while the product is removed continuously through a membrane filter. The filter unit may be used either internally or externally from the two stages. Continuous operation is achieved by keeping the adsorption and desorption vessels well agitated and operating under steady-state conditions. The recovery yield of β-galactosidase from unpretreated homogenized cells was 70% with a 35-fold purification and 4-times concentration increase using this purification technique.

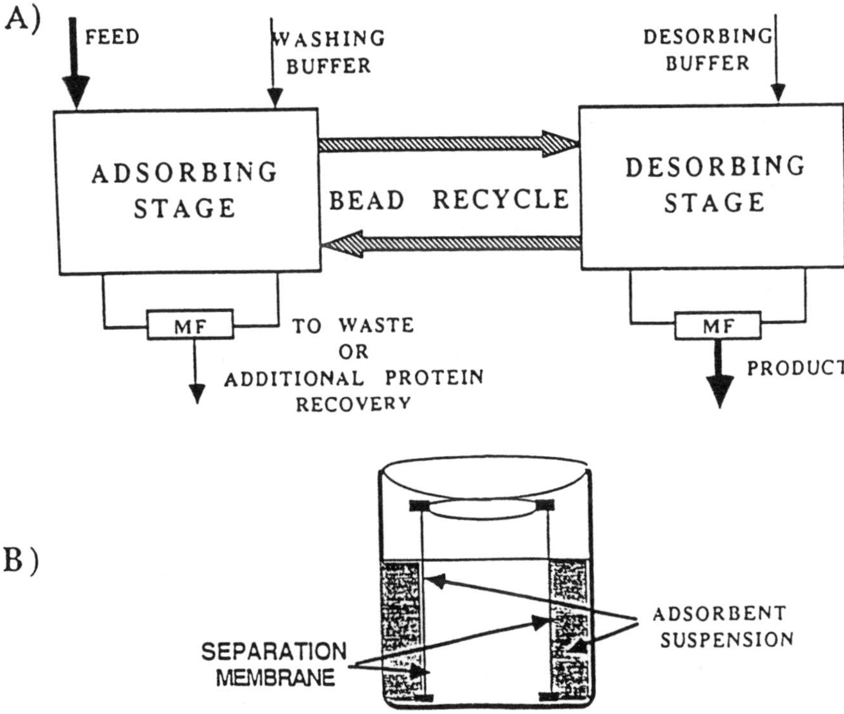

Figure 5.34. Schematic of continuous affinity-recycle extraction system (Reference 67).

A different approach to combining filtration and affinity purification was developed by Nigam and Wang (68). They modeled the use of small affinity adsorbent particles imbedded in hydrogel beads to treat whole broths. The hydrogel matrix allowed the very small particles to be used with relatively easy handling and reduced fouling of the adsorbent particles. The use of the small adsorbent particles also reduced the internal mass transfer resistance which allowed increased adsorption, as shown in Figure 5.35, for different numbers of immobilized particles imbedded in the hydrogel bead.

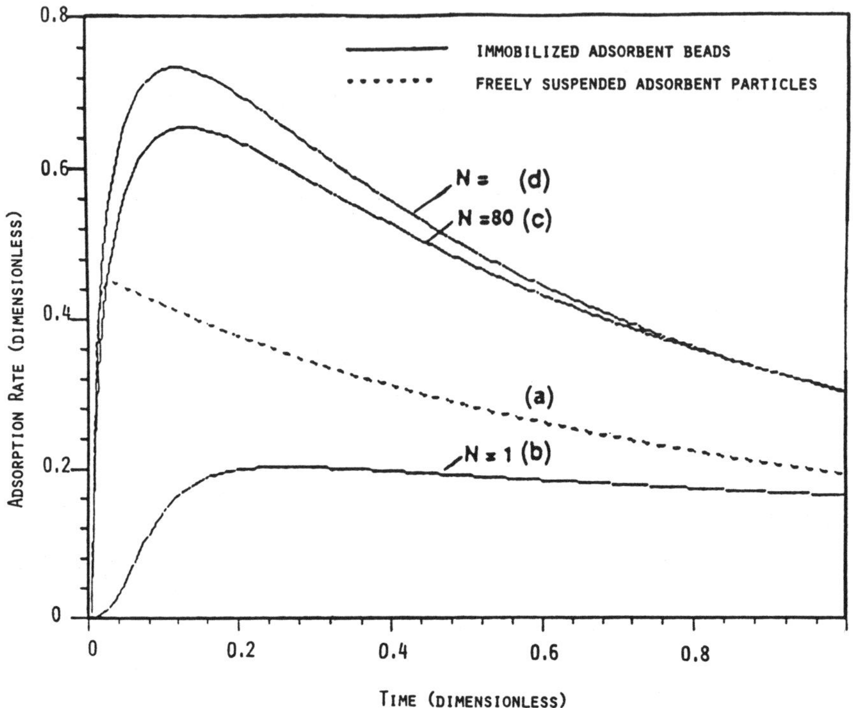

Figure 5.35. Adsorption rate as a function of time for four simulated cases (Reference 68).

5.6.4 Ligand Costs

When specific ligands are used to purify an enzyme, the cost of the affinity chromatography adsorbent can be quite expensive. Whether or not it is too expensive depends on the end use of the purified material. Hamman and Calton (69) have pointed out that although the cost of the monoclonal immunosorbent used to purify urokinase is \$600/liter, this only amounts to \$4.90/g of urokinase purified. Since urokinase drug treatment charges at hospital pharmacies are \$91,000/g, affinity chromatography is obviously not too expensive in this case.

5.7 AFFINITY CHROMATOGRAPHY APPLICATIONS

Affinity chromatography was originally devised for protein isolation and purification. Its success in this role is demonstrated by the proteins purified using the ligands shown in Table 5.13 (2). However, the use of this technique is certainly not limited to protein purification. Carbohydrates and glycoporteins have been purified by the use of immobilized lectins (Table 5.14) (3). Hormone receptors have been used for hormone purification and monoclonal antibodies have been used to purify antigens. Cell organelles and even whole cells have been purified by affinity chromatography. Some illustrative examples from different affinity purification applications will now be presented.

5.7.1 Protein Purification

Wickerhauser and coworkers (70) used heparin-Sepharose to perform a large-scale purification of antithrombin III from human plasma. The purification sequence is shown in Figure 5.36. While heparin-Sepharose is very effective in extracting antithrombin III from other plasma proteins, it will also bind

Table 5.13: Proteins Purified Using Affinity Chromatography (2)

Proteins Purified	Ligand
Succinic thiokinase	Coenzyme A
Phosphofructokinase	Adenosine monophosphate
Transcobalamin I and II	Coenzyme B_{12}
Methylmalonyl - CoA mutase	Coenzyme B_{12}
Luciferase	Flavin
Dihydrofolate reductase	Folate
Tyrosine aminotransferase	Pyridoxal phosphate
Aspartate aminotransferase	Pyridoxal phosphate
Glutamate oxaloacetate transaminase	Pyridoxal phosphate
Aldolase	Nucleotide phosphates
Citrate synthetase	Nucleotide phosphates
Ribonuclease reductase	Nucleotide phosphates
Sialyltransferase	Nucleotide phosphates
Galactosyltransferase	Nucleotide phosphates
Glycogen synthetase	Nucleotide phosphates

Table 5.14: Lectin-Agarose Affinity Chromatography Purifications of Glycoproteins (2)

Lectin Ligand	Glycoprotein Purified
Concanavalin A	α - Antitrypsin
	Herpes-specific membrane glycoproteins
	Horseradish peroxidase
	Human alkaline phosphatase
	Interferon
	Immunoglobulin A
	Receptors for insulin
	Receptors for epidermal growth factor
	Porcine enteropeptidase
	Rat brain glycopeptides
	Thyrotropin
Ricinus communis	α - Fetoprotein
Wheat germ agglutin	Erythropoietin
	Glycophorin A
	Receptors for insulin
	Receptors for somatomedin C
	Retinal glycoproteins

lipoproteins. Therefore, it was essential to remove the lipoproteins by precipitation with 20% polyethylene glycol prior to the affinity purification step. The pasteurization step was included to eliminate any possibility of infectivity remaining due to the hepatitis antigen. Table 5.15 shows the volumes of plasma treated along with the amount of product recovered with a 2M NaCl elution.

Several β-lactamases, enzymes that play an important part in antibiotic resistance, have been purified by Cartwright and Waley (71) using boronic acid-agarose columns with hydrophilic or hydrophobic spacer arms. The results are presented in Table 5.16.

For those β-lactamases with a relatively high affinity for m-aminophenylboronic acid ($K_i < 200\ \mu M$), gels having a hydrophilic spacer arm provide adequate binding. For those β-lactamases with low affinity for m-aminophenylboronic acid ($K_i > 200\ \mu M$), an additional binding force is required. This is provided by using a hydrophobic spacer arm. The additive nature of the hydrophobic interaction and the ligand specific interaction was illustrated for the β-lactamases that had K_i's in the millimolar region.

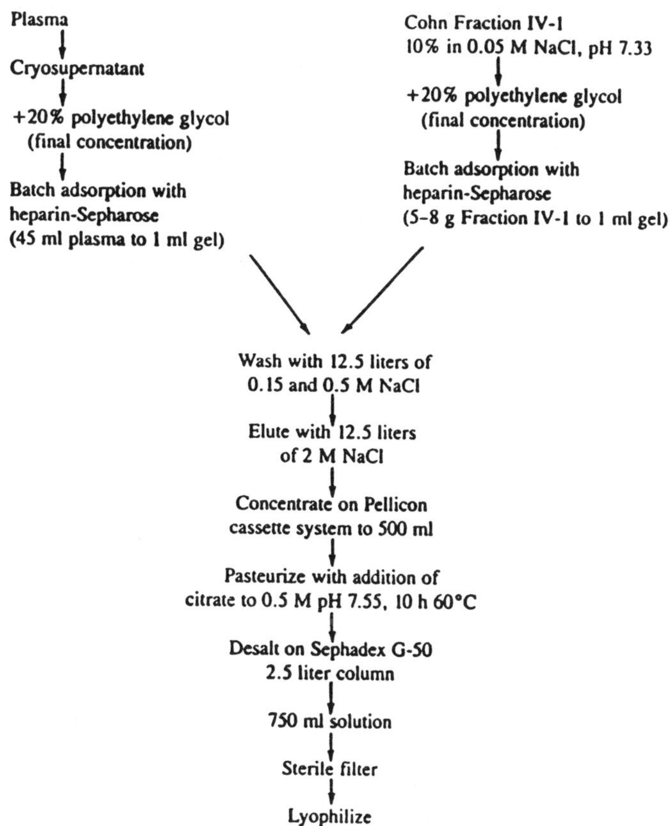

Figure 5.36. Purification of antithrombin III from human plasma or Cohn Fraction IV-1 (Reference 70).

Table 5.15: Recovery of Antithrombin III from Human Plasma (70)

Material	Volume (liters)	Antithrombin III activity		
		PE/ml	Total PE	% Recovery
Plasma	113	0.949	107,237	100.0
Cryosupernatant	111	0.939	104,229	97.2
20% PEG supernatant	115	0.655	75,325	70.2
DEAE unadsorbed + wash	215	0.321	69,015	64.4
HS unadsorbed	215	0.055	11,825	11.0
0.15 M NaCl wash	12.5	0.102	1,275	1.2
0.5 M NaCl wash	12.5	1.150	14,375	13.4
2.0 M NaCl eluate	12.0	4.133	49,596	46.2
Pasteurized product	0.74	46.8	34,632	32.3

Table 5.16: Purification of β-Lactamases on Phenylboronic Acid Affinity Columns (71)

β - Lactamase	Purification (fold)		K_i (μM) m-aminophenylboronic acid	Borate
Pseudomonas aeruginose	220	(a)	5	630
Entereobacter cloacae	12	(a)	60	5000
Pseudomonas maltophilia	2000	(a)	200	370
Bacillus cereus, I	15	(b)	2000	1000
Bacillus cereus, II	---	(c)	Not inhibited	
Bacillus cereus, III	60	(b)	1650	4400
Klebsiella aerogenes	40	(c)	770	600

(a) Reporter substrate was cephalosporin C.

(b) Reporter substrate was nitrocefin.

(c) Reporter substrate was penicillin G.

Fisher and Newsholme (72) developed a procedure which directly combined the elution of adenosine kinase from a gel filtration column with the adsorption of the enzyme onto a 5'-AMP-Sepharose 4B affinity chromatography column. The buffer used throughout the purification procedure consisted of 4mM $Na_2H_2PO_4$, 2mM EDTA and 5% (v/v) glycerol at pH 7.0 and 4°C. Adenosine kinase was eluted from the affinity column with 0.6 mM adenosine in the buffer. The specific activity of adenosine kinase in the peak activity fraction from the affinity column was about 2 μmol/min per mg of protein, a purification of 870-fold.

Affinity chromatography on immobilized calmodulin has been shown (73) to be a very convenient method for the rapid purification of bull seminalplasmin to over 99% purity. Many aspects of the complex formation between seminalplasmin and calmodulin have been shown to be identical with that of melittin, mastoparans and synthetic model peptides. This similarity is shown in Figure 5.37. The 1:1 Ca^{+2}-dependent complex is fully resistant to urea and displays an affinity constant of approximately 10^9 M^{-1}. The elution of seminalplasmin from the immobilized calmodulin requires the presence of both EDTA and 4M urea in the elution buffer.

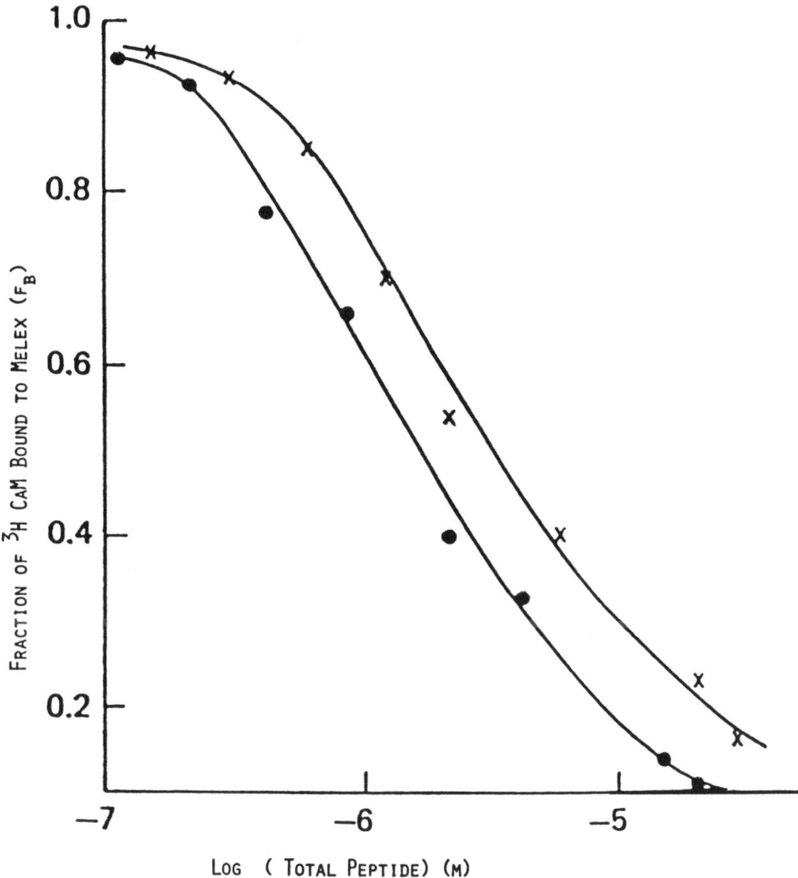

Figure 5.37. Displacement of Melex bound [^{3}H] calmodulin by bull seminalplasmin (•) as compared with soluble melittin (x) (Reference 73).

Malate dehydrogenase and 3-hydroxybutyrate dehydrogenase have been purified from *Rhodopseudomonas spheroides* extract using two sequential dye ligand affinity chromatography columns (74). The cell-free extract was applied to a 1.5 liter column of agarose-bound Procion Red H-3B. The 3-hydroxybutyrate dehydrogenase was eluted with the 1M KCl buffer wash and the malate dehydrogenase was subsequently eluted by including 2mM NADH in the buffer. The dialyzed fractions containing malate dehydrogenase activity were then applied to a 1 liter column of immobilized Procion Blue MX-4GD. The malate dehydrogenase was eluted with a linear gradient of KCl (0 to 700mM, 8 liters) to yield 1 g of homogeneous enzyme at a 70% overall yield.

The Procion Red H-3B column wash fractions which contained the 3-hydroxybutyrate dehydrogenase were applied to a 1 liter column of immobilized Procion Blue MX-4GD and eluted with 2mM NADH to yield 300 mg of homogeneous enzyme with an 80% overall yield. It should be noted that the non-affinity chromatography scheme for isolating this enzyme involves nine separate steps and has an overall yield of only 9%.

5.7.2 Interferon Purification

Knight and Fahey (75) have used Blue Sepharose Cl-4B in an affinity purification of homogeneous human fibroblast interferon in yields ranging from 20 to 40%. The interferon was produced by diploid fibroblast cells (FS-4) cultured in a serum-free medium to give crude interferon of high specific activity. The interferon adsorbs on the affinity material in the presence of 1 or 2M NaCl, whereas the contaminating proteins do not. Pure interferon is eluted from the Blue Sepharose using 50% ethylene glycol. It is collected into a buffered solution to make a final concentration of 37% ethylene glycol because interferon is much more stable in 37% ethylene glycol. A second column is used for concentration purposes, as shown in Table 5.17. The final product has a specific activity of 5 x 10^8 units/mg.

Table 5.17: Purification of Human Fibroblast Interferon (75)

Fraction	Volume (ml)	Units, Interferon	Protein (mg)	Specific activity	Recovered (%)
Crude	10,000	85 X 10^6	500	1.7 X 10^5	100
First Blue Sepharose column, 1M NaCl, 50% ethylene glycol	160	68 X 10^6	0.6	1 X 10^8	80
Second Blue Sepharose column, 2 M NaCl, 40% ethylene glycol	10	30 X 10^6	0.06	5 X 10^8	35

Staehelin and coworkers (76) at Hoffman-LaRoche have produced a range of monoclonal antibodies for the affinity chromatography purification of leukocyte interferon. The process for isolating recombinant human leukocyte interferon from cultures of *E. coli* is shown in Figure 5.38 and a summary of the purification is shown in Table 5.18. A 17 ml monoclonal antibody column was found to be able to process the extract from 1 kg bacterial. The affinity purification provided over 1,000-fold purification in a single step and yielded a homogeneous final product.

5.7.3 Receptor-Binding Affinity Chromatography

A receptor is a molecule that recognizes a specific chemical entity, binds to it and initiates a series of biochemical events resulting in a characteristic physiological response (77). Because hormone receptors are present in mammalian cells only in exceedingly small quantities (less than 10 ppm on a dry weight basis) and because they are relatively unstable, traditional methods for isolating and purifying membrane-bound receptors have been of only limited value. Only after the introduction of affinity chromatography did such isolation become experimentally feasible. Even with affinity chromatography, only a few cell receptors, like those shown in Table 5.19 (2), have been successfully purified.

The interaction between hormone, neurotransmitter or drug and the respective receptor is selective and specific. This selectivity causes the interaction between the ligand and its receptor to be one of high affinity, among the highest known among biological interactions. This means that harsh

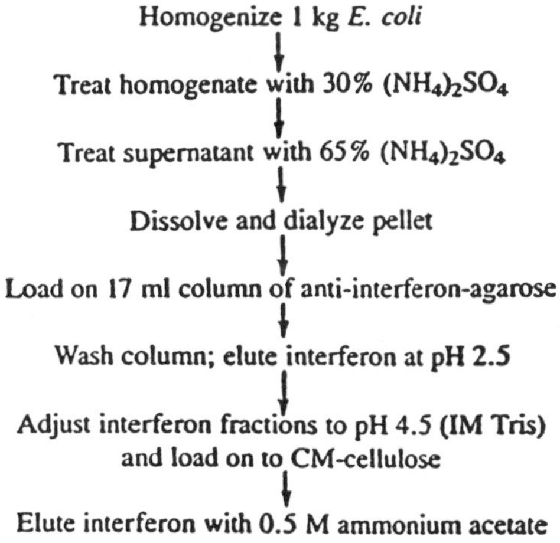

Homogenize 1 kg *E. coli*

↓

Treat homogenate with 30% $(NH_4)_2SO_4$

↓

Treat supernatant with 65% $(NH_4)_2SO_4$

↓

Dissolve and dialyze pellet

↓

Load on 17 ml column of anti-interferon-agarose

↓

Wash column; elute interferon at pH 2.5

↓

Adjust interferon fractions to pH 4.5 (IM Tris)
and load on to CM-cellulose

↓

Elute interferon with 0.5 M ammonium acetate

Figure 5.38. Process for isolating recombinant human leukocyte interferon from *E. coli* cultures (Reference 76).

Table 5.18: Purification of Leukocyte Interferon (76)

Step	Volume (ml)	Total protein (mg)	Total activity (units)	Specific activity	Purification factor	Recovered (%)
Ammonium sulfate fraction	700	37,100	7.4×10^9	2.0×10^5	1.0	100
Antibody column pool	27	30	7.0×10^9	2.3×10^8	1,150	95
CM 52	30	20	6.0×10^9	3.0×10^8	1,500	81

conditions are necessary to elute receptors from columns containing a selective affinity ligand. Monoclonal antibodies are now being used to allow elution to be carried out under milder conditions (78). Table 5.20 (38) shows examples of the use of specific ligands and monoclonal antibodies for the isolation of receptors and other binding proteins.

Table 5.19: Membrane-Bound Receptors Purified by Affinity Chromatography

Receptor	Source	Ligand
Acetylcholine	Electric eel	Cobratoxin
Epidermal growth factor	Placenta	Plant lectins
Epinephrine (adrenaline)	Erythrocytes	Alprenolol
Growth hormone	Liver	Growth hormone
Human chorionic gonadotrophin	Testes	Human chorionic gonadotrophin
Immunoglobulin E	Basophilic cells	Immunoglobulin E
Insulin	Liver, fat cells	Concanavalin A
Prolactin	Mammary gland	Growth hormone
Thyrotropin-stimulating hormone	Thyroid gland	Thyrotropin-stimulating hormone

Table 5.20: Receptor and Binding Proteins Purified by Affinity Chromatography (38)

Receptor	Ligand	Eluent
β- Andrenergic	Alprenolol, acebutolol	Isoproterenol, alprenolol
Estradiol	Estradiol derivative oligo (dT)	Estradiol, KCl
Fc	IgG	Urea
Insulin	Insulin	Guanidine, KSCN
Intrinsic factor	Intrinsic factor - B_{12}	EDTA
Transcobalamin II	Transcobalamin II	EDTA
	Binding Protein	
C3b component	Factor H	NaCl
Elongation factors Tu, Ts	GDP	GDP
Hemopexin	Heme, hematin	Gly-HCl, urea
Migration inhibitory factor	Glycolipids	KSCN
m RNA cap	m^7GDP	KCl

5.7.4 Immuno Affinity Chromatography

Affinity chromatography has been used for the purification of antibodies by employing antigens bound to carriers. Immobilized antibodies have also been used for the purification of proteins, including enzymes. Under ideal conditions, these columns allow separation of specific antibodies from crude mixtures in a one-step procedure. Both conventional and monoclonal antibodies have been used. A major difficulty in the use of immobilized antibodies is the affinity of an antibody for antigens, thereby making the recovery of active enzymes difficult owing to the drastic conditions required for their elution. It is not unusual to find eluents that consist of chaotrophs (5M KSCN), low pH (2.2) or high pH (11.5) buffers, urea (3.5-8M) or guanidine (6M). Despite such harsh conditions, it is of interest that antigenicity is retained in most cases. When its antigenicity is not retained, immobilized protein A (79) or antihapten antibody (80) may be used to bind antibody-antigen or hapten antibody-antigen complexes. It is a purified complex, not the antigen, that is eluted from the support.

The use of monoclonal antibodies (81) decreases the possibility of contamination with antibodies directed against proteins other than that desired. Since the number of monoclonal antibodies against a specific protein are many, it is advantageous to select low affinity antibodies for immobilization. The lower affinity will allow the use of milder elution conditions and reduces the possibility of irreversible denaturation. Some of the proteins purified by conventional and monoclonal antibodies are listed in Table 5.21.

5.7.5 Cell Separation by Affinity Chromatography

Immunoadsorbent techniques are capable of efficient and specific retention of cells and allow the recovery of unbound cells in a viable, unaltered state. In these methods, antibodies specific for a lymphocyte surface component are bonded to solid phase matrices to provide a selective ligand for isolating given cell subpopulations. Thus, B cells can be conveniently isolated from a mixed population using antiimmunoglobulin, T cells by using antibodies to Thy 1 antigens and null cells are taken as those that remain after removal of both B and T cells. Commercial monoclonal reagents can be used to separate T cells on the basis of Lyt

Table 5.21: Proteins Purified on Immuno Affinity Columns (38)

Protein	Eluent
Adenosine deaminase	Urea
α- Fetoprotein	Urea
Angiotensin I converting enzyme	$Mg\ Cl_2$
Choriogonadotropin	$Mg\ Cl_2$
Estrogen receptor	Na SCN
Factor VIII	NH_4SCN
Glucocorticoid receptor	Acetic acid
Insulin receptor	$Mg\ Cl_2$
Ribonuclease H	$Mg\ Cl_2$
Serine acetyltransferase	O- Acetylserine
Somatostatin	Acetic acid
Thyrotropin	Guanidine
Trypsin inhibitor	Propionic acid
Urokinase	Glycine - H Cl

antigens, for example, into killer, helper or suppressor cells (82).

The most important parameter determining the success of immunosorbent chromatography of cells is the level of antibody substitution on the matrix. For adsorption of unwanted cells, the level of the antibody ligand must be high enough to allow adequate retention of the bound cells. For positive selection, the level of the antibody ligand must be low enough so that desorption of the cells is not made difficult.

Hubbard and coworkers (83) used the globulin fraction of rabbit anti-mouse IgM serum for the separation of Ig^+ and Ig^- cells. The cell suspension (1-5 x 10^7 cells/mL; 10^8 cells maximum) was added to the column, followed by 2 bed volumes of the buffer. This fraction contained the unbound cells. After washing the column with 10 to 20 additional bed volumes of buffer, the bound cells were specifically eluted with 1 mg/mL of purified mouse IgM in the buffer. Release of the bound cells occurs through competition between the cell surface immunoglobulin and the free immunoglobulin for the immobilized antibody ligand. About 80 to 90% of the Ig^- cells and no Ig^+ cells were recovered in the unbound fraction. Likewise, 80 to 90% of the eluted bound cells possessed surface immunoglobulin, but were only about 35 to 40% of the Ig^+ cells applied to the column.

The interaction between *Staphylococcus aureus* (SpA) and cell surface IgG has been used in this technique to isolate lymphocytes and other cells (84). Table 5.22 shows the decrease in the percentage of Ig-bearing mouse B lymphocytes in the nonadherent cell population after affinity chromatography with SpA-Sepharose 6MB. The antibody bearing subpopulation is eluted from the column with IgG or by mechanical treatment.

Table 5.22: Isolation of Ig-Bearing Mouse Lymphocytes by Affinity Chromatography on Sp A-Sepharose 6MB (84)

Cell Fraction	Ig - bearing cells (%)
Before chromatography	48.8 ± 1.7
After Chromatography	
Nonadherent to column	10.9 ± 1.2
Adherent to column	90.3 ± 1.8

Separation of well-defined cell populations by affinity chromatography can only be achieved if antibody ligands against specific surface antigens are available.

The subpopulation enriched in cells bearing the specific antigen has interacted with antibody and SpA during the affinity chromatography operation. This is an unavoidable consequence of methods using antibody as a tool for cell separation. The antibody present on the surface of the eluted cells may affect their normal function. For this reason, affinity chromatography of cells is the method of choice for depletion of a cell population rather than for enrichment of a population in cells bearing a specific antigen.

5.8 REFERENCES

1. Cuatrecasas, P., Wilcheck, M., Anfinsen, C.B., Proc Natl Acad Sci, 61:636 (1968)

2. Parikh, I., Cuatrecasas, P., Chem Eng News, 63(34):17 (1985)

3. Lowe, C.R., An Introduction to Affinity Chromatography, Elsevier Biomedical Press, New York (1979)

4. Scouten, W.H., Affinity Chromatography: Bioselective Adsorption on Inert Matrices, John Wiley & Sons, New York (1981)

5. Larson, P.O., Glad, M., Hansson, L., et al., In: Advances in Chromatography, (Giddings, J.C., ed.) Marcel Dekker, New York, Vol 21 (1983)

6. Colowick, S.P., Kaplan, N.O., In: Methods in Enzymology (Jakoby, W.B., ed.) Academic Press, Inc., Orlando, Fl, Vol 104, Part C (1984)

7. Chaiken, I., Wilcheck, M., Parikh, I., eds., Affinity Chromatography and Biological Recognition, Academic Press, New York (1984)

8. Nishikawa, A.H., Chem Tech, 5:564 (1975)

9. Graves, D.J., Wu, Y.T., In: Methods in Enzymology (Jakoby, W.B., Wilchek, M., eds.) Academic Press, Inc., Orlando, FL, Vol 84, p 150 (1974)

10. Nishikawa, A.H., Bailon, P., Ramel, A.H., Adv Expl Med Biol, 42:33 (1974)

11. Lowe, C.R., J Biotechnol, 1:3 (1984)

12. Bottomley, R.C., Storer, A.C., Trayer, I.P., Biochem J, 159:667 (1976)

13. van Deemter, J.J., Zuiderweg, F.J., Klinkenberg, A., Chem Eng Sci, 5:271 (1956)

14. Yang, C.M., Tsao, G.T., Adv Biochem Eng, 25:1 (1982)

15. Arnold, F.H., Blanch, H.W., Wilke, C.R., The Chem Eng J, 30:B9 (1985)

16. Chung, I., Ph.D. Thesis, University of Tennessee, 1976

17. Katoh, S., Kambayashi, T., Deguchi, R., et al., Biotechn and Bioeng, 20:267 (1978)

18. Hashimoto, K., Miura, K., J Chem Eng Japan, 9:388 (1976)

19. Crank, J., Nicolson, J., J Proc Cambridge Philos Soc, 43:50 (1947)

20. Arnold, F.H., Chalmers, J.J., Saunders, M.S., et al., In: Purification of Fermentation Products (LeRoith, D., Shiloach, J., Leahy, T.J., eds.) American Chemical Society, Washington, DC, p 113 (1985)

21. Graham, E.E., Fook, C.F., AIChE J, 28:245 (1982)

22. Cuatrecasas, P., Anfinsen, C.B., In: Methods in Enzymology (Jakoby, W.B., ed.) Academic Press, New York, Vol 22, p 345 (1971)

23. May, S.W., Landgraff, L.M., In: Recent Developments in Separation Science (Li, N.N., Dranoff, J.S., Schultz, J.S., et al., eds.) CRC Press, Boca Raton, FL, Vol 5, p 227 (1979)

24. Voser, W., J Chem Tech Biotechnol, 32:109 (1982)

25. Porath, J., In: Methods in Enzymology (Jakoby, W.B., Wilchek, M., eds.) Academic Press, New York, Vol 34, p 13 (1974)

26. Weliky, N., Weetall, H.H., Immunochem, 2:293 (1965)

27. Axen, R., Porath, J., Ernback, S., Nature (London), 214:1302 (1967)

28. Inman, J.K., Dintzis, H.M., Biochemistry, 8:4074 (1969)

29. Ultrogel is available from LKB, Inc., Rockville, MD

30. Ohlson, S., Hansson, L., Larsson, P.O., et al., FEBS Lett, 93:5 (1978)

31. Haller, W., Nature (London), 206:693 (1965)

32. Weetall, H.H., Hersh, L.S., Biochim Biophys Acta, 185:464 (1969)

33. Weetall, H.H., Havewala, N.B., Biotech Bioeng Symp, 3:241 (1972)

34. Weetall, H.H., Filbert, A.M., In: Methods in Enzymology (Jakoby, W.B., Wilchek, M., eds.) Academic Press, New York, Vol 34, p 59 (1974)

35. Cuatrecasas, P., J Biol Chem, 245:3059 (1970)

36. Shaltiel, S., In: Methods in Enzymology (Jakoby, W.B., Wilchek, M., eds.) Academic Press, New York, Vol 34, p 126 (1974)

37. Er-el, Z., Zaidenzaig, Y., Shaltiel, S., Biochem. Biophys Res Commun, 49:383 (1972)

38. Wilchek, M., Miron, T., Kohn, J., In: Methods in Enzymology (Jakoby, W.B., ed.) Academic Press, New York, Vol 104, p 3 (1984)

39. Mosbach, K., Adv Enzymol Relat Areas Mol Biol, 46:205 (1978)

40. Lis, H., Lotan, R., Sharon, N., Ann N Y Acad Sci, 234:232 (1974)

41. Guilford, H., Chem Soc Rev, 2:249 (1973)

42. Dean, P.D.G., Watson, D.H., J Chromatog, 165:301 (1979)

43. Lowe, C.R., Small, D.A.P., Atkinson, A., Int J Biochem, 13:33 (1981)

44. Gionazza, E., Arnaud, P., Biochem J, 201:129 (1982)

45. Lowe, C.R., Pearson, J.C., In: Methods in Enzymology (Jakoby, W.B., ed.) Academic Press, New York, Vol 104, p 97 (1984)

46. Bruton, C.J., Atkinson, A., Nucleic Acids Res, 7:1579 (1979)

47. Lowe, C.R., Hans, M., Spibey, N., et al., Anal Biochem, 104:23 (1980)

48. Clonis, Y.D., Lowe, C.R., Biochim Biophys Acts, 659:86 (1981)

49. Clonis, Y.D., Lowe, C.R., Biochem J, 191:247 (1980)

50. Hill, E.A., Hirtenstein, M.D., In: Advances in Biotechnological Processes, Alan R. Liss, Inc., New York, p 31 (1983)

51. Eveleigh, J.W., Levy, D.E., J Solid-phase Biochem, 2:47 (1977)

52. Inman, J.K., Dintzis, H.M., Biochemistry, 8:4074 (1969)

53. Absolom, D.R., Sep Purif Methods, 10:239 (1981)

54. Byfield, P.G.H., Copping, S., Bartlett, W.A., Biochem Soc Trans, 10:104 (1982)

55. Singh, P., Lewis, S.D., Shafer, J.A., Arch Biochem Biophys, 193:284 (1979)

56. Singh, P., Lewis, S.D., Shafer, J.A., Arch Biochem Biophys, 203:776 (1980)

57. Suzuki, S., Karube, I., In: Recent Developments in Separation Science (Li, N.N., ed.) CRC Press, Boca Raton, FL, Vol 6, p 117 (1981)

58. Eveleigh, J.W., presented at the Fourth International Symposium on Affinity Chromatography and Related Techniques, Veldhoven, Holland (1981)

59. Sundberg, L., Porath, J., Aspberg, K., Biochim Biophys Acta, 221:394 (1970)

60. Wong, B.L., Charm, S.E., J Macromol Sci Chem, A10:53 (1976)

61. Chase, H.A., In: Discovery and Isolation of Microbial Products (Verrall, M.S., ed.) Ellis Horwood Ltd., Chichester, UK, p 129 (1985)

62. Baker, B.R., Siebenieck, H.U., J Med Chem, 14:799 (1971)

63. Begovich, J.M., Rep ORNL-5915, NTIS, U.S. Dept. Commerce, Springfield, VA (1982)

64. Mattiasson, B., Ramstorp, M., Ann N Y Acad Sci, 13:307 (1983)

65. Luong, J.H.T., Nguyen, A.L., Male, K.B., Bio/Technology, 5:564 (1987)

66. Mandaro, R.M., Roy, S., Hou, K.C., Bio/Technology, 5:928 (1987)

67. Pungor, E., Jr., Afeyan, N.B., Gordon, N.F., et al., Bio/Technology, 5:604 (1987)

68. Nigam, S.G., Wang, H.Y., In: Separation, Recovery and Purification in Biotechnology, American Chemical Society, Washington, DC, p 153 (1986)

69. Hammon, J.P., Calton, G.J., In: Purification of Fermentation Products (LeRoith, D., Shiloach, J., Leahy, T.J., eds.) American Chemical Society, Washington, DC, p 105 (1985)

70. Wickerhauser, M., Williams, C., Mercer, J., Vox Sang, 36:281 (1979)

71. Cartwright, S.J., Waley, S.G., Biochem J, 221:505 (1984)

72. Fisher, M.N., Newsholme, E.A., Biochem J, 221:521 (1984)

73. Comte, M., Malnoe, A., Cox, J.A., Biochem J, 240:567 (1986)

74. Scawen, M.D., Darbyshire, J., Harvey, M.J., et al., Biochem J, 203:699 (1982)

75. Knight, E., Fahey, D., J Biol Chem, 256:3609 (1981)

76. Staehelin, T., Hobbs, D.S., Kung, H., J Biol Chem, 256:9750 (1981)

77. Hollenberg, M.D., Cuatrecasas, P., Methods Cancer Res, 12:317 (1976)

78. Stein, B.S., Sussman, H.H., J Biol Chem, 258:2668 (1983)

79. Schneider, C., Newman, R.A., Sutherland, D.R., et al., J Biol Chem, 257:10766 (1982)

80. Kanellopoulos, J., Rossi, G., Metzger, N., J Biol Chem, 254:7691 (1979)

81. Goding, J.W., J Immunol Methods, 39:285 (1980)

82. Matossian-Rogers, A., Rogers, P., Herzenberg, L.A., Cell Immunol, 69:91 (1982)

83. Hubbard, R.A., Schulter, S.F., Marchalonis, J.J., In: Methods in Enzymology (Jakoby, W.B., ed.) Academic Press, New York, Vol 104, p 139 (1984)

84. Ghetie, V., Sjoguist, J., In: Methods in Enzymology (Jakoby, W.B., ed.) Academic Press, New York, Vol 104, p 132 (1984)

Index

1 day